Multivalued Analysis and Nonlinear Programming Problems with Perturbations

Nonconvex Optimization and Its Applications

Volume 66

The titles published in this series are listed at the end of this volume.

Multivalued Analysis and Nonlinear Programming Problems with Perturbations

by

Bernd Luderer

Chair of Business Mathematics,
Faculty of Mathematics, Chemnitz University of Technology, Germany

Leonid Minchenko

Chair of Informatics,
Byelorussian State University of Informatics & Radioelectronics, Byelorussia

and

Tatyana Satsura

Chair of Informatics,
Byelorussian State University of Informatics & Radioelectronics, Byelorussia

KLUWER ACADEMIC PUBLISHERS
DORDRECHT / BOSTON / LONDON

A C.I.P. Catalogue record for this book is available from the Library of Congress.

ISBN 978-1-4419-5236-3

Published by Kluwer Academic Publishers,
P.O. Box 17, 3300 AA Dordrecht, The Netherlands.

Sold and distributed in North, Central and South America
by Kluwer Academic Publishers,
101 Philip Drive, Norwell, MA 02061, U.S.A.

In all other countries, sold and distributed
by Kluwer Academic Publishers,
P.O. Box 322, 3300 AH Dordrecht, The Netherlands.

Printed on acid-free paper

Contents

Preface

This book is concerned with topological and differential properties of multivalued mappings and marginal functions. Beside this applications to the sensitivity analysis of optimization problems, in particular nonlinear programming problems with perturbations, are studied. The elaborated methods are primarily obtained by theories and concepts of two former Soviet Union researchers, Demyanov and Rubinov. Consequently, a significant part of the presented results have never been published in English before. Based on the use of directional derivatives as a key tool in studying nonsmooth functions and multifunctions, these results can be considered as a further development of quasidifferential calculus created by Demyanov and Rubinov.

In contrast to other research in this field, especially the recent publication by Bonnans and Shapiro, this book analyses properties of marginal functions associated with optimization problems under quite general constraints defined by means of multivalued mappings. A unified approach to directional differentiability of functions and multifunctions forms the base of the volume.

Different notions of the derivative of a multivalued mapping play an important role in the investigation of a large number of problems. Various kinds of derivatives of multivalued mappings were used in studying economic problems (Rubinov, Aubin), constrained minimax problems (Demyanov), control problems described by differential inclusions (Clarke, Pshenichny, Aubin, Kurzhanski, Frankowska, Mordukhovich and others) and differential games (Krasovski, Subbotin). Moreover, a great field of research is in sensitivity and stability analysis of nonlinear programming problems under perturbations (Aubin, Rockafellar, Ioffe, Shapiro, Bonnans, Auslender, Cominetti and others). Here, as in other works, the weakened and non-unique nature of such an important concept as derivative and differentiability in general, is an expression of one of the most essential features of multivalued analysis in comparison with classical analysis.

In the present book the authors develop a theory of directional differential calculus of multivalued mappings that includes the calculus of higher-order directional derivatives. This approach differs from the one brought in by Bonnans and Shapiro. Its origin goes back to a publication by Demyanov in 1974, where he proposes the concept of tangent (feasible) directions playing the role of the directional derivative of a multivalued mapping. In general, multivalued mappings are not directionally differentiable in the sense described in the book. Therefore, the authors pay a lot of attention to the specification of classes of multivalued mappings being directionally differentiable. These results have not yet been published in English.

In studying the properties of multivalued mappings the fundamental idea of this volume is to use the close relationship between these mappings and their so-called marginal or optimal value functions (extreme value functions defined on the values of multivalued mappings). In particular, differential (as well as topological) properties of multivalued mappings can completely be derived from associated properties of marginal functions and vice versa. Therefore, under quite general assumptions concerning multivalued mappings, such properties as upper and lower semicontinuity, (Lipschitz) continuity or directional differentiability are equivalent to their counterpart in marginal functions theory, in particular, to properties of the distance function being the simplest marginal function.

In marginal functions theory a wide range of results is accessible. The most precise approximations of nonsmooth functions, for example, are accomplished by directional derivatives (provided they exist). First results dealing with calculation of directional derivatives for marginal functions were obtained in linear programming (Mills). Further requirements to study this subject emerged from minimax theory and perturbation analysis of mathematical programming problems. Beginning with the seventies of the last century, research in differential properties of marginal functions started to achieve even more attention (see e. g. Golshtein, Rockafellar, Rubinov, Hiriart-Urruty, Shapiro, Ioffe, Auslender and many others). Examples here are the study of second-order directional derivatives of marginal functions initiated by Demyanov and the second-order derivatives of a more general type introduced by Ben-Tal and Zowe. The study of their existence and constructive computation was topic of recent papers by Auslender, Cominetti, Ioffe, Shapiro, Bonnans and others. Notable in this context is that marginal functions are not necessarily differentiable. Thus, in many papers the trend appeared to specify certain classes of problems (e. g. convex or regular) with at least directionally differentiable marginal functions.

The description and stability analysis of extremum problems with the help of multivalued mappings is helpful in different manner. The description allows not only to obtain very natural assumptions and general results but often also to get considerable advantages in deriving these results. The same is true for the stability analysis of extremum problems with respect to perturbations, where the central problem is the study of differential properties of marginal functions. Moreover, generalized derivatives of marginal functions and their estimates (approximations) can be quite useful for sensitivity analysis of perturbations in extremum problems.

Divided into 5 chapters, the book starts with basic concepts and problems of convex and nonsmooth analysis, which are required as background knowledge for the further treatment.

The second chapter describes topological and differential properties of multivalued mappings. In the first part topological notions, such as uniform boundedness, upper and lower semicontinuity and continuity, Lipschitz and pseudo-Lipschitz properties of mappings and their marginal functions are introduced. The mutual connection between them is established. In the second part different concepts and properties of differentiability and approximation techniques for multivalued mappings are considered. The description of derivatives of mappings in terms of the distance function is obtained. The important lemma about the removal of constraints is proved.

Chapter 3 is devoted to subdifferentials of marginal functions. Estimates of the Clarke subdifferential of an arbitrary marginal function are obtained under quite general assumptions. Moreover, for locally convex optimization problems an exact formula for the calculation of the subdifferential of their marginal functions is obtained as a consequence of the general method proposed above. For the important class of quasidifferentiable functions in the sense of Demyanov and Rubinov basic knowledge (quasidifferential calculus) is given and optimum conditions are described.

In the next chapter a number of general theorems concerning the existence of the directional derivative of the maximum function are proved. These results enable to specify broad classes of directionally differentiable mappings for which we succeed in describing their directional derivatives as well as the directional derivatives of their marginal functions in a constructive way. So-called strongly differentiable multivalued mappings introduced earlier by Tyurin, Banks and Jacobs are studied in detail. The obtained results related to such mappings reveal their connection with other classes of differentiable mappings. Furthermore, a theorem about the directional differentiability of their marginal functions is established.

Chapter 5 is completely devoted to the study of nonlinear mathematical programming problems. A survey of regularity conditions is given at the beginning of the chapter. Certain regularity conditions (linear independence of gradients, Mangasarian-Fromowitz constraint qualification, (R)-regularity condition) are treated and generalized. Special attention will be paid to the study of interdependence between different regularity conditions. A second part of the fifth chapter is concerned with nonlinear programming problems involving perturbations and contains results on the Lipschitz or Hölder behaviour. Differentiability properties of their optimal solutions, upper and lower estimates of Dini derivatives as well as estimates for the Clarke subdifferential of marginal functions are given. Various conditions for the existence of first- and second-order directional derivatives of optimal value functions are described. Quasidifferentiable problems in nonlinear programming are intensively studied and, again, estimates for the potential directional derivative are given.

It remains to say that the book contains results of research work carried out by the authors in recent years in collaboration with colleagues and students of the mathematical departments of the Byelorussian State University of Informatics and Radioelectronics and the Chemnitz University of Technology. Known results from literature are also taken into account (see Bibliographical Comments). We want to note, however, and this was indeed not our object that not all questions connected with this area of research have been covered.

The book is intended for students and experts of mathematics specializing in optimization. It is based on lectures held at Byelorussian State University of Informatics and Radioelectronics, Byelorussia, and Chemnitz University of Technology, Germany. A number of examples have been included in the text for a better understanding of the material.

Lastly, we want to thank German Academic Exchange Service (DAAD) and INTAS program for the grants awarded to the authors that allowed us to lay the groundwork for this book. We would also like to thank the editorial and production staff of Kluwer Academic Publishers, who did a remarkable professional job and enabled the present work to see the light of day.

The authors would be very grateful to readers who draw their attention to errors or obscurities in the book or suggest any improvements.

Chemnitz / Minsk
February 2002

Bernd Luderer
Leonid Minchenko
Tatyana Satsura

Chapter 1

BASIC NOTATION

$\overset{\triangle}{=}$ –   equal by definition

$\langle x^*, x \rangle$ –   scalar product of two vectors x^* and x

$|x|$ –   Euclidean norm of the vector x

B –   open unit ball with centre at 0

cl C –   closure of the set C

int C –   interior of the set C

co C –   convex hull of the set C

$\overline{\text{co}}\, C$ –   convex closure of the set C

ri C –   relative interior of the set C

$0^+C = \{\bar{x} \in X \mid x + \lambda\bar{x} \in C \text{ for all } x \in C,\ \lambda > 0\}$

      –   recession cone of set C

$K^+ = \{x^* \in X \mid \langle x^*, x \rangle \geq 0 \text{ for all } x \in K\}$

      –   cone conjugate to the cone K

$CS(Y)$ –   family of all non-empty compact subsets of the space Y

$CCS(Y)$ –   family of all non-empty convex compact subsets of the space Y

$\rho(x, C)$ –   distance function between the point x and the set C

$S_C(x^*) = \sup\{\langle x^*, x \rangle \mid x \in C\}$ –   support function of the set C

$x_k \overset{C}{\to} x$ –   means $x_k \to x$ and $x_k \in C$ for all $k = 1, 2, \ldots$

$x' \overset{f}{\to} x$ –   means $x' \to x$ and $f(x') \to f(x)$

dom f –   effective domain of the function f

epi f –   epigraph of the function f

$\operatorname{gr} F = \{(x, y) \mid x \in X, \ y \in F(x)\}$

       –   graph of the multivalued mapping F

$d_F(x, y)$ –   distance function between the point y and the image $F(x)$ of the point x

$T_E^L(x)$ –   lower tangent cone to the set E at the point x

$T_E^U(x)$ –   upper tangent cone to the set E at the point x

$T_E^C(x)$ –   Clarke tangent cone to the set E at the point x

$\gamma_E(x)$ –   cone of feasible directions of the set E at the point x

$D^+ f(x; \bar{x})$ –   upper Dini derivative of the function f

$D_+ f(x; \bar{x})$ –   lower Dini derivative of the function f

$T_E^L(x)$ –   lower tangent cone to the set E at the point x

$T_E^U(x)$ –   upper tangent cone to the set E at the point x

$T_E^C(x)$ –   Clarke tangent cone to the set E at the point x

$\gamma_E(x)$ –   cone of feasible directions of the set E at the point x

$D^+ f(x; \bar{x})$ –   upper Dini derivative of the function f

$D_+ f(x; \bar{x})$ –   lower Dini derivative of the function f

$D^\oplus f(x; \bar{x})$ –   upper directional derivative of the function f in the sense of Hadamard

$D_\oplus f(x; \bar{x})$ –   lower directional derivative of the function f in the sense of Hadamard

$N_E(x) = -[T_E^C(x)]^+$

       –   Clarke normal cone to the set E at the point x

$\partial^0 f(x) = \{x^* \in X \mid (x^*, -1) \in N_{\operatorname{epi} f}(x, f(x))\}$

       –   Clarke subdifferential of the function f at the point x

$f^\uparrow(x; \bar{x}) = \lim_{\delta \downarrow 0} \ \limsup_{x' \overset{f}{\to} x, \ \varepsilon \downarrow 0} \ \inf_{|\hat{x} - \bar{x}| < \delta} \ \varepsilon^{-1}[f(x' + \varepsilon \hat{x}) - f(x')]$

       –   directional subderivative of the function f

$f^0(x; \bar{x}) = \limsup_{x' \to x, \ \varepsilon \downarrow 0} \ \varepsilon^{-1}[f(x' + \varepsilon \bar{x}) - f(x')]$

       –   generalized directional derivative of the function f

$\partial^\infty f(x) = \{x^* \in X \mid (x^*, 0) \in N_{\operatorname{epi} f}(x, f(x))\}$

       –   asymptotic Clarke subdifferential of the function f at the point x

$D_S F(z; \bar{x}) = \{\bar{y} \in Y \mid (\bar{x}, \bar{y}) \in T^S_{\mathrm{gr}\, F}(z)\}, \quad S = U, L, C,$

    – upper, lower and Clarke derivative, resp., of the multivalued mapping F at the point z in the direction $\bar{x}$

$\hat{D}_L F(z; \bar{x}) = \liminf_{\varepsilon \downarrow 0} \varepsilon^{-1}(F(x + \varepsilon \bar{x}) - y)$

    – lower direct derivative of the mapping F at the point $z \in \mathrm{gr}\, F$ in the direction $\bar{x}$

$\hat{D}_U F(z; \bar{x}) = \limsup_{\varepsilon \downarrow 0} \varepsilon^{-1}(F(x + \varepsilon \bar{x}) - y)$

    – upper direct derivative of the mapping F at the point $z \in \mathrm{gr}\, F$ in the direction $\bar{x}$

$(\bar{P}_x)$ – denotes the mathematical programming problem $\inf\{\, f(x, y) \mid y \in F(x)\}$

(P_x) – denotes the problem of nonlinear programming $\inf\{\, f(x, y) \mid h_i(x, y) \leq 0, \ i = 1, \ldots, r, \ h_i(x, y) = 0, \ i = r+1, \ldots, p \}$

$I = \{1, \ldots, r\}$ – set of indices belonging to inequality constraints

$I_0 = \{r+1, \ldots, p\}$ – set of indices belonging to equality constraints

$I(x, y) = \{i \in I \mid h_i(x, y) = 0\}$

    – set of indices belonging to inequality constraints which are active

$\varphi(x) = \inf\{f(x, y) \mid y \in F(x)\}$

    – optimal value function associated with the problems $(\bar{P}_x)$ and (P_x)

$\omega(x) = \{y \in F(x) \mid \varphi(x) = f(x, y)\}$

    – set of optimal solutions associated with the problems $(\bar{P}_x)$ and (P_x)

$L(x, y, \lambda)$ – Lagrangian of the problem (P_x) at the point $y \in \omega(x)$

$\Lambda(x, y)$ – set of Lagrange multipliers of the problem (P_x) at the point $y \in \omega(x)$

$\omega_\varepsilon(x_0 + t\bar{x}) = \{y \in F(x_0 + t\bar{x}) \mid f(x_0 + t\bar{x}, y) \leq \varphi(x_0 + t\bar{x}) + \varepsilon\}$

    – set of ε-optimal (suboptimal) solutions

$$\Gamma_F(z) = \{\bar{z} \in X \times Y \mid \langle \nabla h_i(z), \bar{z} \rangle \leq 0,\ i \in I(z),\ \langle \nabla h_i(z), \bar{z} \rangle = 0,\ i \in I_0\}$$

      – linearized tangent cone to the set gr F at the point $z = (x, y) \in \mathrm{gr}\, F$

$$D(z) = \{\bar{y} \in \hat{D}F(z; 0) \mid f'_y(z; \bar{y}) \leq 0\}$$

      – cone of critical directions of the problem ($\bar{P}_x$) at the point $z = (x, y) \in \mathrm{gr}\, F$

$$\Gamma_F(z; \bar{x}) = \{\bar{y} \in R^m \mid (\bar{x}, \bar{y}) \in \Gamma_F(z)\}$$

$$I^2(z_0, \bar{z}_1) = \{i \in I(z_0) \mid \langle \nabla h_i(z_0), \bar{z}_1 \rangle = 0\}$$

$$\Gamma_F^2(z_0, \bar{z}_1) = \Big\{ \bar{z}_2 = (\bar{x}_2, \bar{y}_2) \mid \langle \nabla h_i(z_0), \bar{z}_2 \rangle + \tfrac{1}{2}\langle \bar{z}_1, \nabla^2 h_i(z_0)\bar{z}_1 \rangle = 0,\ i \in I_0,$$
$$\langle \nabla h_i(z_0), \bar{z}_2 \rangle + \tfrac{1}{2}\langle \bar{z}_1, \nabla^2 h_i(z_0)\bar{z}_1 \rangle \leq 0,\ i \in I^2(z_0, \bar{z}_1) \Big\}$$

$$\Gamma_F^2(z_0, \bar{z}_1; \bar{x}_2) = \{\bar{y}_2 \in Y \mid (\bar{x}_2, \bar{y}_2) \in \Gamma_F^2(z_0, \bar{z}_1)\}$$

Chapter 2

BASIC CONCEPTS AND PROBLEMS OF MULTIVALUED ANALYSIS

In this chapter we describe the main concepts and problems of convex and nonsmooth analysis, which are required as background knowledge for the further treatment. Some of the reviewed results are classical and well-known, other are more specific. For simplicity, we suppose the underlying space X to be finite-dimensional, i.e., we consider $X = R^n$. By $\langle x^*, x \rangle$ we denote the scalar product of two vectors x^* and x from X, while $|x|$ is the Euclidean norm of the vector x. Finally, B denotes the open unit ball with centre at 0, i.e. $B = \{x \in X | \ |x| < 1\}$.

1. Basic Concepts of Convex Analysis

A review of the following and other main results from convex analysis one can find in [149], [154], [172] as well as in any other standard textbook on the subject.

1.1 Convex Sets

A set $C \subset X$ is called *convex* if together with any two of its points $x_1, x_2 \in C$ it contains the segment between them, i. e. $\lambda x_1 + (1-\lambda)x_2 \in C$ for $\lambda \in [0, 1]$.

The empty set $\emptyset$ is supposed to be convex by definition. It is not hard to verify that for a convex set C its closure $\mathrm{cl}\, C$, its interior $\mathrm{int}\, C$ and the set

$$\lambda C = \{\lambda x \mid x \in C\}, \quad \lambda \in R,$$

are also convex.

If the sets C_1 and C_2 are convex, then their intersection $C_1 \cap C_2$ and their algebraic sum

$$C_1 + C_2 = \{x_1 + x_2 \mid x_1 \in C_1, \ x_2 \in C_2\}$$

5

are also convex.

Let $M \subset X$. The intersection of all convex (or convex and closed) sets in X containing the set M is called the *convex hull* (*convex closure*) of the set M and denoted by $\operatorname{co} M$ ($\overline{\operatorname{co}} M$, resp.)

It is obvious that the sets $\operatorname{co} M$ and $\overline{\operatorname{co}} M$ are convex. In addition, $\operatorname{co} M \subset \overline{\operatorname{co}} M$ and, hence, $\operatorname{cl} \operatorname{co} M \subset \overline{\operatorname{co}} M$. Vice versa, $\overline{\operatorname{co}} M \subset \operatorname{cl} \operatorname{co} M$. Thus, $\overline{\operatorname{co}} M = \operatorname{cl} \operatorname{co} M$.

A linear combination $\lambda_1 x_1 + \lambda_2 x_2 + \ldots + \lambda_m x_m$ is called the *convex combination* of the points $x_1, x_2, \ldots, x_m$ if $\lambda_1 + \lambda_2 + \ldots + \lambda_m = 1$, $\lambda_i \geq 0$, $i = 1, \ldots, m$.

THEOREM 2.1 *(**Carathéodory**) The convex hull* $\operatorname{co} M$ *coincides with the set of all convex combinations of points from M. In addition, in the space $X = R^n$ any point in $\operatorname{co} M$ can be represented as the convex combination of no more than $n + 1$ points of M.*

It follows directly from Theorem 2.1 that the set M is convex if and only if $\operatorname{co} M = M$.

In the case under study, when the space X is finite-dimensional, a convex set C has either a non-empty interior or it is contained in a set (linear manifold) being the translation of a subspace of smaller dimension. The intersection of all subspaces containing $C - x_0$, where x_0 is an arbitrary point in C, is called the *generating subspace* of the set C and denoted by $\operatorname{Lin} C$. The set of points $x \in C$ such that $x + \operatorname{Lin} C \cap \varepsilon B \subset C$ for some $\varepsilon > 0$ is called the *relative interior* of the convex set C and is denoted by $\operatorname{ri} C$.

Note that if C is a non-empty convex set, then $\emptyset \neq \operatorname{ri} C \subset C$.

THEOREM 2.2 *The following statements are equivalent:*
1. $\operatorname{ri} \operatorname{cl} C = \operatorname{ri} C$, $\quad \operatorname{cl} \operatorname{ri} C = \operatorname{cl} C$;
2. *if* $\operatorname{ri} C_1 \cap \operatorname{ri} C_2 \neq \emptyset$, *then* $\operatorname{ri} C_1 \cap \operatorname{ri} C_2 = \operatorname{ri}(C_1 \cap C_2)$.

The following theorem is of great importance in convex analysis and applications.

THEOREM 2.3 *(**Separation theorem**) Let C_1 and C_2 be non-empty closed convex sets and one of them be bounded. If $C_1 \cap C_2 = \emptyset$, then there exist a vector x^* and a number $\varepsilon > 0$ such that $\langle x^*, x_1 \rangle \leq \langle x^*, x_2 \rangle - \varepsilon$ for all $x_1 \in C_1$, $x_2 \in C_2$.*

A set $K \subset X$ is called a *cone*, if the inclusion $x \in K$ implies $\lambda x \in K$ for all $\lambda > 0$. It is not hard to verify that a cone K is convex if and only if from $x_1, x_2 \in K$ the inclusion $x_1 + x_2 \in K$ follows.

Note that for every set $M \subset X$ the set $\text{cone}\, M = \bigcup_{\lambda \geq 0} \lambda M$ can be defined, which is obviously a cone. In the case, when the set M is convex, the cone $\text{cone}\, M$ is convex too and coincides with the intersection of all convex cones containing M and the origin of co-ordinates.

Every convex cone K can be associated with the *conjugate cone*

$$K^+ = \{x^* \in X \mid \langle x^*, x \rangle \geq 0 \text{ for all } x \in K\}.$$

It is easy to see that K^+ is a convex cone containing the origin of co-ordinates. It is also known about K^+ that:

1. the cone K^+ is closed;
2. $(K^+)^+ = \text{cl}\, K$;
3. for $x \in \text{int}\, K$ the inequality $\langle x^*, x \rangle > 0$ holds for all $x^* \in K^+ \backslash \{0\}$.

Furthermore, if the cones K_1 and K_2 are convex, then $K_1 + K_2$ is also convex and $(K_1 + K_2)^+ = K_1^+ \cap K_2^+$. For closed convex cones K_1 and K_2 the equality

$$(K_1 \cap K_2)^+ = \text{cl}\,(K_1^+ + K_2^+)$$

is valid, too.

Let C be a non-empty convex set in X. The set

$$0^+ C \overset{\triangle}{=} \{\bar{x} \in X \mid x + \lambda \bar{x} \in C \text{ for all } x \in C, \ \lambda > 0\}$$

is called the *recession cone* of the set C.

Since $0 \in 0^+ C$, then $0^+ C$ is non-empty. It can be also proved that the recession cone $0^+ C$ is a convex set and

$$0^+ C = \{\bar{x} \in X \mid C + \bar{x} \subset C\}.$$

THEOREM 2.4 *Let C be a non-empty closed convex set. Then $0^+ C$ is closed and*

1. $0^+ C = \{\bar{x} \in X \mid x_0 + \lambda \bar{x} \in C \ \forall \ \lambda > 0\}$, *where x_0 is an arbitrary point in C;*
2. $0^+ C$ *coincides with the set of the limits of all sequences $\{\lambda_k x_k\}$, where $x_k \in C$ and $\lambda_k \downarrow 0$.*

From this theorem we get that a non-empty closed convex set C is bounded if and only if $0^+ C = \{0\}$. Another consequence of Theorem 2.4 is the validity of the relation $0^+ (\bigcap_{i \in I} C_i) = \bigcap_{i \in I} 0^+ C_i$ for an arbitrary family of closed convex sets C_i with non-empty intersection.

Recession cones are used to get some criteria for the closedness of convex sets. In particular, the following theorem is valid.

THEOREM 2.5 *Let C_1 and C_2 be non-empty closed convex sets in X. If $0^+C_1 \cap (-0^+C_2) = \{0\}$, then $C_1 + C_2$ is a closed set and $0^+(C_1 + C_2) = 0^+C_1 + 0^+C_2$.*

An important role among convex sets play so-called polyhedral sets.

A set C is called *polyhedral* if its points x satisfy a system of linear inequalities

$$\langle x_i^*, x \rangle \leq \alpha_i, \quad i = 1, \ldots, m, \tag{2.1}$$

where $\alpha_i \in R$, $x_i^* \in X$, $i = 1, \ldots, m$, are fixed.

A particular case of a polyhedral set is a *convex polyhedron*, i.e. a bounded set described by system (2.1). A convex polyhedron is the convex hull of a finite number of points.

A cone K is called *polyhedral*, if there exists a finite m-tuple of vectors $x_1, \ldots, x_m$ such that

$$K = \left\{ x \in X \mid x = \sum_{i=1}^{m} \lambda_i x_i, \quad \lambda_i \geq 0, \ i = 1, \ldots, m \right\}.$$

It is well-known that a polyhedral cone can always be specified by a finite system of linear homogeneous inequalities

$$\langle x_i^*, x \rangle \leq 0, \quad i = 1, \ldots, m. \tag{2.2}$$

Vice versa, the set of solutions of system (2.2) is always a polyhedral cone.

We would like to mention the following properties of polyhedral cones. A polyhedral cone is always closed; the sum and the intersection of polyhedral cones are also polyhedral cones. In addition, for polyhedral cones K_1 and K_2 we have $(K_1 \cap K_2)^+ = K_1^+ + K_2^+$.

THEOREM 2.6 *If a cone K is defined by the system of linear inequalities (2.2), then*

$$K^+ = \{ x^* \in X \mid x^* = -\sum_{i=1}^{m} \lambda_i x_i^*, \ \lambda_i \geq 0, \ i = 1, \ldots, m \}.$$

THEOREM 2.7 *A polyhedral set C is the sum of a convex polyhedron C_0 and a polyhedral cone K, i.e.*

$$C = C_0 + K. \tag{2.3}$$

Note also that if the polyhedral set C is given by the system of inequalities (2.1), then

$$0^+C = \{ x \mid \langle x_i^*, x \rangle \leq 0, \quad i = 1, \ldots, m \}$$

and it is easy to prove that 0^+C coincides with the cone K from the representation (2.3).

1.2 Convex Functions

Let us consider a function $f : X \to R \cup \{\pm\infty\}$. To this function f there correspond the sets

$$\mathrm{dom}\, f = \{x \in X \mid f(x) < +\infty\},$$
$$\mathrm{epi}\, f = \{(x, \lambda) \in X \times R \mid f(x) \leq \lambda\}$$

called the *effective domain* and the *epigraph* of the function f, respectively.

It is not hard to see that

$$f(x) = \inf_\lambda \{\lambda \mid (x, \lambda) \in \mathrm{epi}\, f\},$$

i.e., the epigraph completely defines the corresponding function f.

A function f is called *convex* if the set $\mathrm{epi}\, f$ is convex in the space $X \times R$. If $\mathrm{dom}\, f \neq \emptyset$ and $f(x) > -\infty$ for all x, then f is called a *proper* function. Necessary and sufficient for a proper function f to be convex is the validity of Jensen's inequality

$$f(\lambda x_1 + (1 - \lambda)x_2) \leq \lambda f(x_1) + (1 - \lambda)f(x_2) \tag{2.4}$$

for all $x_1, x_2 \in X$, $\lambda \in [0, 1]$.

We suppose further that the following natural rules of operations with symbols $+\infty$ and $-\infty$ are valid:

$$+\infty + \alpha = +\infty, \quad -\infty + \alpha = -\infty,$$
$$\alpha \cdot (+\infty) = +\infty, \quad \alpha \cdot (-\infty) = -\infty \quad \text{if} \ \ 0 < \alpha \leq +\infty,$$
$$\alpha \cdot (+\infty) = -\infty, \quad \alpha \cdot (-\infty) = +\infty \quad \text{if} \ \ -\infty \leq \alpha < 0,$$
$$+\infty + \infty = +\infty, \quad -\infty - \infty = -\infty,$$

completing them as in [96] by the rules $0 \cdot (+\infty) = 0 \cdot (-\infty) = 0$, $+\infty - \infty = -\infty + \infty = +\infty$, $\inf \emptyset = +\infty$, $\sup \emptyset = -\infty$.

With the help of these rules it can be shown that the convexity of a function (not only of a proper one) is equivalent to the validity of inequality (2.4).

Examples of convex functions are:

the *affine* function

$$f(x) = \langle x^*, x \rangle + \alpha, \quad \text{where } x^* \in X, \ \alpha \in R;$$

the *indicator function* of a convex set $C \subset X$

$$\delta(x \mid C) = \begin{cases} 0 & \text{if } x \in C, \\ +\infty & \text{if } x \notin C; \end{cases}$$

the *support function* of a convex set $C \subset X$

$$S_C(x^*) = \sup\{\langle x^*, x \rangle \mid x \in C\}.$$

Let us consider some operations on convex functions. It is easy to see that the sum $f + g$ of proper convex functions f and g is also a convex function.

A function f is called *closed* if the set epi f is closed in $X \times R$. The closedness of a function f is equivalent to the lower semicontinuity of f, which in turn is equivalent to the closedness of the level sets $\{x \mid f(x) \leq \alpha\}$ in X for all $\alpha \in R$. With any function f we can associate the function cl f defined as epi(cl f) $=$ cl(epi f) and called the *closure* of the function f. Since the closure of a convex set is convex, then the closure of a convex function is always a convex function.

The function $f(x) = \sup\{f_i(x) \mid i \in I\}$ is called the *supremum* of the family of functions f_i, $i \in I$. Obviously epi $f = \bigcap_{i \in I}$ epi f_i, thus, the supremum of a family of convex functions is also a convex function. In the same way, the supremum of a family of closed functions is closed.

Note that from what was said above it immediately follows that the support function $S_C(x^*)$ of a convex set C is convex and closed.

THEOREM 2.8 *Let f_i, $i \in I$, be proper convex functions and let $f(x) = \sup\{f_i(x) \mid i \in I\}$. Then cl $f = \sup\{$ cl $f_i \mid i \in I\}$ provided that there exists a point $x_0 \in \bigcap_{i \in I} \operatorname{ri} \operatorname{dom} f_i$ such that $|f(x_0)| < \infty$.*

Let f and g be convex functions. The function $f \bigtriangledown g$ defined as

$$(f \bigtriangledown g)(x) = \inf_{\lambda}\{\lambda \mid (x, \lambda) \in \operatorname{epi} f + \operatorname{epi} g\}$$

is called the *infimal convolution* of the functions f and g. From the definition it follows that $f \bigtriangledown g$ is a convex function. In the case of proper convex functions f and g the relationship $(f \bigtriangledown g)(x) = \inf_{y}\{f(x-y)+g(y)\}$ is valid.

We will frequently make use of the *distance function* between a point x and a set C, which is defined by

$$\rho(x, C) = \rho_C(x) \overset{\triangle}{=} \inf\{|x - y| \mid y \in C\},$$

and by $+\infty$ if $C = \emptyset$.

EXAMPLE 2.9 *Let C be a convex set. Trying to find the infimal convolution of the functions $f(x) = |x|$ and $g(x) = \delta(x \mid C)$, we get*

$$(f \triangledown g)(x) = \inf_y \{|x - y| + \delta(y \mid C)\} = \inf\{|x - y| \mid y \in C\} = \rho(x, C).$$

From this relation it follows that the distance function to a convex set is convex.

With any function $f : X \to R \cup \{\pm\infty\}$ we can associate the *conjugate function* $f^*(x^*) = \sup_x \{\langle x^*, x \rangle - f(x)\}$. The function f^* is always convex and closed.

THEOREM 2.10 *(Fenchel-Moreau) Let f be a proper convex closed function. Then $f(x) = f^{**}(x)$, where $f^{**}(x) = \sup_x \{\langle x^*, x \rangle - f^*(x^*)\}$.*

EXAMPLE 2.11 *Let C be a convex and closed set. Its indicator function $\delta(\cdot \mid C)$ and support function $S_C(\cdot)$ are conjugate to each other, i. e.*

$$\delta^*(\cdot \mid C) = S_C(\cdot), \quad S_C^*(\cdot) = \delta(\cdot \mid C).$$

According to this fact, one often uses the notation $\delta^*(\cdot \mid C)$ for the support function.

EXAMPLE 2.12 *Let C be a non-empty, convex and closed set. It is not hard to verify that*

$$[\rho_C(\cdot)]^* = \begin{cases} S_C(x^*) & \text{if } |x^*| \leq 1, \\ +\infty & \text{if } |x^*| > 1. \end{cases}$$

From the last relationship and Theorem 2.10 it follows that

$$\rho_C(x) = \sup\{\langle x^*, x \rangle - S_C(x^*) \mid |x^*| \leq 1\}. \tag{2.5}$$

Let A and C be non-empty and compact sets in X. The value

$$\rho_H(A, C) = \max\{\sup_{x \in A} \rho_C(x), \sup_{x \in C} \rho_A(x)\} \tag{2.6}$$

is called the *Hausdorff metric* between the sets A and C. If A and C are convex, then from (2.6) in view of (2.5) and the definition of the support function it follows that

$$\rho_H(A, C) = \sup_{|x^*| \leq 1} |S_A(x^*) - S_C(x^*)|. \tag{2.7}$$

THEOREM 2.13 *Let f and g be proper convex functions. Then*

$$(f \triangledown g)^* = f^* + g^*, \quad (\operatorname{cl} f + \operatorname{cl} g)^* = \operatorname{cl}(f \triangledown g)^*.$$

In addition, if $\operatorname{ri} \operatorname{dom} f \cap \operatorname{ri} \operatorname{dom} g \neq \emptyset$*, then in the second equality the closure operation can be omitted.*

A function f is called *positively homogeneous* if $f(\lambda x) = \lambda f(x)$ for all $\lambda > 0$, $x \in X$. It is obvious that a function f is convex and positively homogeneous if and only if its epigraph $\operatorname{epi} f$ is a convex cone. If this function is closed, then obviously $f(0) = 0$.

Note that the support function $S_C(x^*)$ is always positively homogeneous, convex and closed. Vice versa, every convex, closed and positively homogeneous function f appears to be the support function of a convex set, namely, of the effective domain $\operatorname{dom} f^*$. Indeed, for a proper function f this follows from the Fenchel-Moreau Theorem. If f fails to be proper, then this statement can be proved directly.

Now we state some properties of support functions. Let A and C be non-empty, convex and closed sets in X. The following statement follows from the Separation Theorem and the properties of operations on convex functions.

THEOREM 2.14 *1. The inclusion $x \in C$ is equivalent to the condition $\langle x^*, x \rangle \leq S_C(x^*)$ for all $x^* \in X$.*
2. $A \subset C$ is equivalent to $S_A(\cdot) \leq S_C(\cdot)$.
3. $S_{A+C}(\cdot) = S_A(\cdot) + S_C(\cdot)$.
4. Let the sets A and C not coincide with X and $A \cap C \neq \emptyset$. Then

$$S_{A \cap C}(\cdot) = \operatorname{cl}[S_A(\cdot) \triangledown S_C(\cdot)].$$

(If $\operatorname{ri} A \cap \operatorname{ri} C \neq \emptyset$, then the closure operation can be omitted.)
5. Let C_i, $i \in I$, be certain sets in X, where I is an arbitrary index set. Denoting $C = \operatorname{cl} \operatorname{co} \bigcup_{i \in I} C_i$, we have $S_C(\cdot) = \sup_{i \in I} S_{C_i}(\cdot)$.

1.3   Topological and Differential Properties of Convex Functions

It is well-known that a convex function f is continuous at all points of $\operatorname{ri} \operatorname{dom} f$. Moreover, the following assertion is true.

THEOREM 2.15 *Let f be a proper convex function. Then the following statements are equivalent:*
1. f is bounded above in a neighbourhood of a point x;

2. f is continuous at x;

3. f is Lipschitz continuous in a neighbourhood of the point x, i. e. there exists a constant $l > 0$ such that $|f(x_1) - f(x_2)| \leq l|x_1 - x_2|$ for all x_1, x_2 from some neighbourhood of x.

The value

$$f'(x; \bar{x}) = \lim_{\varepsilon \downarrow 0} \varepsilon^{-1}[f(x + \varepsilon \bar{x}) - f(x)]$$

is called the *directional derivative* of the function f at the point x in the direction $\bar{x} \in X$ if this limit (finite or infinite) exists. A function possessing a directional derivative at every point of $x \in \text{dom} f$ in all directions is called *directionally differentiable*.

For a proper convex function f at every point $x \in \text{dom} f$ the directional derivative $f'(x; \bar{x})$ exists in any direction $\bar{x} \in X$. This derivative appears to be a convex positively homogeneous function of the variable $\bar{x}$.

For a convex function f the set

$$\partial f(x_0) \overset{\triangle}{=} \{x^* \in X \mid f(x) - f(x_0) \geq \langle x^*, x - x_0 \rangle \text{ for all } x \in X\}$$

is called the *subdifferential* of the convex function f at the point $x_0 \in \text{dom} f$ (consisting of *subgradients*). The subdifferential $\partial f(x)$ is a convex and closed set in X; its support function is $\text{cl} f'(x; \bar{x})$, where the closure is taken with respect to $\bar{x}$. Thus, $\partial f(x) = \partial_{\bar{x}} f'(x; 0)$, where the notation $\partial_{\bar{x}} f'(x; 0)$ means the subdifferential with respect to the second variable. If f is convex and continuous at the point x, then $\partial f(x)$ is a non-empty and compact set and

$$f'(x; \bar{x}) = \max\{\langle x^*, \bar{x} \rangle \mid x^* \in \partial f(x)\}. \tag{2.8}$$

In particular, if f is convex and differentiable at x, then its subdifferential $\partial f(x)$ contains only one element, the gradient $\nabla f(x)$.

THEOREM 2.16 *(**Moreau-Rockafellar***) Let f_1 and f_2 be proper convex functions and let the function f_1 be continuous at some point $x_0 \in \text{dom} f_1 \cap \text{dom} f_2$. Then*

$$\partial(f_1 + f_2)(x) \subset \partial f_1(x) + \partial f_2(x).$$

REMARK. For f_1 and f_2 being proper convex functions, the inclusion $\partial(f_1 + f_2)(x) \supset \partial f_1(x) + \partial f_2(x)$ is always true. Thus, under the assumptions of the Moreau-Rockafellar theorem equality holds.

THEOREM 2.17 *Let* $\{f_k\}$ *be a sequence of finite convex functions converging to the convex finite function* f *at every point of the open convex set* C. *Then for any sequence* $\{x_k\}$, $x_k \in C$, $x_k \to x \in C$ *and any sequence* $\{\bar{x}_k\}$, $\bar{x}_k \to \bar{x} \in X$ *the following relation is true:*

$$\limsup_{k\to\infty} f_k'(x_k; \bar{x}_k) \leq f'(x; \bar{x}).$$

A function $f : X \to R \cup \{\pm\infty\}$ is called *concave* if the function $-f$ is convex. A function $f : X \times Y \to R \cup \{\pm\infty\}$ is called *convex-concave* if the function $x \mapsto f(x, y)$ is convex for every y and the function $y \mapsto f(x, y)$ is concave for every x. In the sequel we need the following statement.

THEOREM 2.18 *(**Minimax theorem**) Let* $X_0 \subset X$ *and* $Y_0 \subset Y$ *be convex closed sets and at least one of them be compact. Then*

$$\inf_{x\in X_0} \sup_{y\in Y_0} f(x, y) = \sup_{y\in Y_0} \inf_{x\in X_0} f(x, y).$$

Note that the inequality

$$\inf_{x\in X_0} \sup_{y\in Y_0} f(x, y) \geq \sup_{y\in Y_0} \inf_{x\in X_0} f(x, y)$$

holds for arbitrary sets X_0 and Y_0 and for any function $f(x, y)$.

2. Elements of Nonsmooth Analysis and Optimality Conditions

2.1 Tangent Cones

Let $X = R^n$, $E \subset X$. We define the *lower tangent cone* to the set E at the point $x \in E$ as the set $T_E^L(x)$ of elements $\bar{x} \in X$ such that for every $\bar{x}$ there exists a n-dimensional vector function $o(\varepsilon)$ with $x + \varepsilon\bar{x} + o(\varepsilon) \in E$ for every $\varepsilon \geq 0$ and $o(\varepsilon)/\varepsilon \to 0$ for $\varepsilon \downarrow 0$.

Under the *upper tangent cone* to the set E at the point $x \in E$ we understand the set $T_E^U(x)$ of elements $\bar{x} \in X$ such that for every $\bar{x}$ there exist sequences $\varepsilon_k \downarrow 0$ and $\hat{x}_k \to \bar{x}$ with $x + \varepsilon_k\hat{x}_k \in E$, $k = 1, 2, \ldots$ The upper tangent cone $T_E^U(x)$ is often called the *contingent cone*.

Finally, we define the *Clarke tangent cone* to E at the point $x \in E$ as the set $T_E^C(x)$ of elements $\bar{x} \in X$ such that for any sequences $\varepsilon_k \downarrow 0$ and $x_k \xrightarrow{E} x$ one can find a sequence $\hat{x}_k \to \bar{x}$ with $x_k + \varepsilon_k\hat{x}_k \in E$, $k = 1, 2, \ldots$ (Here and later on the notation $x_k \xrightarrow{E} x$ means $x_k \to x$ and $x_k \in E$ for all $k = 1, 2, \ldots$)

Note that one can state an equivalent definition of the cone $T_E^L(x)$ similar to the definitions of the cones $T_E^U(x)$ and $T_E^C(x)$. Namely, $\bar{x} \in$

$T_E^L(x)$ if and only if for any sequence $\varepsilon_k \downarrow 0$ there exists a sequence $\hat{x}_k \to \bar{x}$ such that $x + \varepsilon_k \hat{x}_k \in E$, $k = 1, 2, \ldots$

Note also that the tangent cones $T_E^L(x)$, $T_E^U(x)$ and $T_E^C(x)$ are always non-empty and closed. In addition, $T_E^C(x)$ is convex. It is not hard to prove that

$$T_E^C(x) \subset T_E^L(x) \subset T_E^U(x).$$

If the set E is convex, then

$$T_E^C(x) = T_E^L(x) = T_E^U(x) = \operatorname{cl} \operatorname{cone}(E - x).$$

Together with the tangent cones defined above we consider the cone $\gamma_E(x)$ called the *cone of feasible directions* of the set E at the point x. It is the set of vectors $\bar{x} \in X$ such that for every $\bar{x}$ there exists a number $\varepsilon_0 > 0$ with $x + \varepsilon \bar{x} \in E$, $\varepsilon \in [0, \varepsilon_0]$.

2.2 Directional Derivatives

Let $X = R^n$, $f : X \to R$. Together with the usual directional derivative $f'(x; \bar{x})$ of the function f at the point x in the direction $\bar{x} \in X$ we consider the *upper* and *lower derivatives* at the point x in the direction $\bar{x}$

$$D_+ f(x; \bar{x}) = \liminf_{\varepsilon \downarrow 0} \varepsilon^{-1}[f(x + \varepsilon \bar{x}) - f(x)]$$

and

$$D^+ f(x; \bar{x}) = \limsup_{\varepsilon \downarrow 0} \varepsilon^{-1}[f(x + \varepsilon \bar{x}) - f(x)].$$

The derivatives $D_+ f(x; \bar{x})$ and $D^+ f(x; \bar{x})$ are also called the *upper* and *lower Dini derivatives*. Obviously, these derivatives always exist although they are not necessarily finite.

Together with the Dini derivations we consider the *upper* and *lower directional derivatives in the sense of Hadamard* of the function f defined as

$$D_\oplus f(x; \bar{x}) = \liminf_{\varepsilon \downarrow 0, \hat{x} \to \bar{x}} \varepsilon^{-1}[f(x + \varepsilon \hat{x}) - f(x)],$$

$$D^\oplus f(x; \bar{x}) = \limsup_{\varepsilon \downarrow 0, \hat{x} \to \bar{x}} \varepsilon^{-1}[f(x + \varepsilon \hat{x}) - f(x)],$$

respectively.

It is obvious that if a function f is locally Lipschitz continuous, then the Hadamard derivatives coincide with the corresponding Dini derivatives.

There exists a close connection between directional derivatives and tangent cones. Thus, for example,

$$T_{\operatorname{epi} f}^U(x, f(x)) = \operatorname{epi} D_\oplus f(x; \cdot).$$

The following statement can be easily proved.

LEMMA 2.19 *Let the function $f : X \to R$ attain its minimum (maximum) on the set $M \subset X$ at the point $x_0 \in M$. Then for every feasible direction $\bar{x} \in \gamma_M(x_0)$, we have*

$$D_+ f(x_0; \bar{x}) \geq 0 \quad (D^+ f(x_0; \bar{x}) \leq 0).$$

2.3   Clarke Subdifferentials

Let $E \subset X$. We consider the Clarke tangent cone $T_E^C(x)$ to the set E at the point x and define the *Clarke normal cone* to the set E at x as the set $N_E(x) = -[T_E^C(x)]^+$. From this definition it follows that the Clarke normal cone is non-empty, convex and closed in X.

Let $f : X \to R \cup \{\pm\infty\}$ be a lower semicontinuous function having a finite value at x. The set

$$\partial^0 f(x) = \{x^* \in X \mid (x^*, -1) \in N_{\text{epi } f}(x, f(x))\}$$

is called the *Clarke subdifferential (Clarke generalized gradient)* of the function f at the point x.

The Clarke subdifferential $\partial^0 f(x)$ is convex and closed in X; its support function is the directional subderivative

$$f^\uparrow(x; \bar{x}) = \lim_{\delta \downarrow 0} \ \limsup_{x' \xrightarrow{f} x, \ \varepsilon \downarrow 0} \ \inf_{|\hat{x} - \bar{x}| < \delta} \ \varepsilon^{-1}[f(x' + \varepsilon \hat{x}) - f(x')].$$

(Here and later on the notation $x' \xrightarrow{f} x$ means that $x' \to x$ and $f(x') \to f(x)$.)

In the special case if f is locally Lipschitz continuous in a neighbourhood of every point x, the subderivative $f^\uparrow(x; \bar{x})$ reduces to the generalized directional derivative

$$f^0(x; \bar{x}) = \limsup_{x' \to x, \ \varepsilon \downarrow 0} \varepsilon^{-1}[f(x' + \varepsilon \bar{x}) - f(x')]$$

and the Clarke subdifferential is a non-empty and compact set in X with the support function $f^0(x; \bar{x})$. Note that the relation

$$\text{epi } f^\uparrow(x; \cdot) = T_{\text{epi} f}^C(x, f(x))$$

holds.

For a locally Lipschitz continuous function f defined on $X = R^n$ the Clarke subdifferential has a quite simple form. It is known that in this case the function f is differentiable almost everywhere in X (in the sense of the Lebesgue measure). Let us denote by X_f the set of all points, at which f is differentiable, and by $\bigtriangledown f(x)$ we denote its gradient. Then

$$\partial^0 f(x) = \operatorname{co} \{ \lim \bigtriangledown f(x_i) \mid x_i \to x, \ x_i \in X_f \}.$$

If we deal with a continuous convex function f, then the Clarke subdifferential $\partial^0 f(x)$ coincides with the subdifferential $\partial f(x)$ in the sense of convex analysis. In the case of a continuously differentiable function f the Clarke subdifferential coincides with the gradient of the function f. Thus, the Clarke subdifferential provides a generalization of the subdifferential concept in the sense of convex analysis for nonconvex functions as well as a generalization of the strict derivative to the nondifferentiable case.

Together with $\partial^0 f(x)$ we consider the *asymptotic Clarke subdifferential*

$$\partial^\infty f(x) = \{ x^* \in X \mid (x^*, 0) \in N_{\mathrm{epi}\, f}(x, f(x)) \}.$$

The asymptotic Clarke subdifferential $\partial^\infty f(x)$ is a convex and closed cone in X. If X is a finite-dimensional space and $|f(x)| < \infty$, then $\partial^0 f(x) \cup (\partial^\infty f(x) \backslash \{0\}) \neq \emptyset$.

THEOREM 2.20 *The following statements are equivalent:*
1. *$\partial^0 f(x)$ is non-empty and bounded;*
2. *f is Lipschitz continuous in a neighbourhood of x;*
3. *$\partial^\infty f(x) = \{0\}$;*
4. *$f^\uparrow(x; \bar{x}) < +\infty$ for all $\bar{x} \in X$.*

Let us recall several simple properties of the Clarke subdifferential.

THEOREM 2.21 *The following statements hold:*
1. *for a locally Lipschitz continuous function f and any $\alpha \in R$, we have*
$$\partial^0 (\alpha f)(x) = \alpha \partial^0 f(x);$$
2. *for locally Lipschitz continuous functions f_1 and f_2, we have*
$$\partial^0 (f_1 + f_2)(x) \subset \partial^0 f_1(x) + \partial^0 f_2(x),$$

moreover, if one of the functions f_1, f_2 is strictly differentiable, then instead of the inclusion equality holds;
3. *if f is a locally Lipschitz continuous function and attains its minimum at the point x, then $0 \in \partial^0 f(x)$;*
4. *if $x \in E \subset X$, then $N_E(x) = \operatorname{cl} \operatorname{cone} \partial^0 \rho_E(x)$.*

THEOREM 2.22 *If the function f attains a local minimum or maximum at the point x_0, then*

$$0 \in \partial^0 f(x).$$

A locally Lipschitz continuous function f is called *regular* at a point x if for all $\bar{x} \in X$ there exists $f'(x; \bar{x})$ coinciding with $f^0(x; \bar{x})$.

THEOREM 2.23 *Let $f(x) = \max\{f_i(x) \,|\, i = 1, \ldots, m\}$, where the functions f_i are locally Lipschitz continuous. Then the function f is also locally Lipschitz continuous and*

$$\partial^0 f(x) \subset \mathrm{co}\{\partial^0 f_i(x) \mid i \in I(x)\},$$

where $I(x) = \{i = 1, \ldots, m \mid f_i(x) = f(x)\}$. In addition, if all the functions f_i, $i \in I(x)$, are regular at x, then f is regular at x either and in the relationship mentioned above equality holds.

3.   Quasidifferentiable Functions and Problems

An important class of functions in nonsmooth optimization consists of so-called quasidifferentiable functions introduced by Demyanov and Rubinov in 1979 at the Conference on Dynamic Control in Sverdlovsk, see [58]. They are peculiar in being directionally differentiable, however, the well-known formula of the directional derivative being the support function of some kind of subdifferential as it is true e. g. in the convex case (see relation (2.8)) or in the case of Lipschitz continuous functions (see Clarke's calculus) has to be understood in a more general sense now. Namely, the directional derivative is now the difference of two maximum expressions, and the subdifferential changes into a pair of sets.

3.1   Elements of Quasidifferential Calculus

DEFINITION 2.24 *A function $f \colon R^n \to R$ is called quasidifferentiable at the point x if it is directionally differentiable at this point and there exist two compact convex subsets $\underline{\partial} f(x)$ and $\overline{\partial} f(x)$ of R^n such that*

$$f'(x; \bar{x}) = \lim_{t \downarrow 0} \frac{1}{t}[f(x + t\bar{x}) - f(x)] = \max_{z \in \underline{\partial} f(x)} \langle z, \bar{x} \rangle + \min_{z \in \overline{\partial} f(x)} \langle z, \bar{x} \rangle.$$

The pair of sets $Df(x) = [\underline{\partial} f(x), \overline{\partial} f(x)]$ is said to be the quasidifferential of the function at the point x, consisting of the subdifferential and the superdifferential.

REMARK. The quasidifferential of a function is not uniquely defined, since for an arbitrary convex compact set A, $[\underline{\partial} f(x) - A, \overline{\partial} f(x) + A]$

is also a quasidifferential. Therefore, it makes sense to speak about equivalence classes of quasidifferentials.

DEFINITION 2.25 *If an equivalence class of quasidifferentials contains an element of the type $[\underline{\partial}f(x),\{0\}]$ (resp. $[\{0\},\overline{\partial}f(x)]$), then f is referred to as subdifferentiable (resp. superdifferentiable) in the sense of Demyanov and Rubinov.*

REMARK. The class of quasidifferentiable functions contains convex, concave and differentiable functions, but also convex-concave, maximum and other functions, among them functions being the difference of two convex functions.

Properties of quasidifferentiable functions are described in the following statement.

LEMMA 2.26 *Let the functions f_i, $i=1,\ldots,k$, be quasidifferentiable at the point x and have the quasidifferentials $Df_i(x)=[\underline{\partial}f_i(x),\overline{\partial}f_i(x)]$, let $\lambda \in R$. Then the following functions are also quasidifferentiable at x:*

(i) $f = f_1 + f_2$, where

$$Df(x) = [\underline{\partial}f_1(x) + \underline{\partial}f_2(x), \overline{\partial}f_1(x) + \overline{\partial}f_2(x)];$$

(ii) $f = \lambda f_1$, where

$$Df(x) = \begin{cases} [\lambda\underline{\partial}f_1(x), \lambda\overline{\partial}f_1(x)] & \text{if} \quad \lambda \geq 0, \\ [\lambda\overline{\partial}f_1(x), \lambda\underline{\partial}f_1(x)] & \text{if} \quad \lambda < 0; \end{cases}$$

(iii) $f = \max\limits_{i=1,\ldots,k} f_i$, where

$$Df(x) = \left[\bigcup_{i\in I(x)} \left(\underline{\partial}f_i(x) - \sum_{j\in I(x), j\neq i} \overline{\partial}f_j(x) \right), \sum_{i\in I(x)} \overline{\partial}f_i(x) \right]$$

with $I(x) = \{i \mid f_i(x) = f(x)\}$.

For the proof see [63], Section 2.

An important generalization of statement (iii) developed in [101] concerns the continual maximum of functions, i.e. the maximum of infinitely many functions

$$\varphi_1(x) = \max_{y\in Y} f(x,y), \tag{2.9}$$

where $f : R^n \times R^m \to R$. This extension plays an important role in decomposition theory when studying optimal value functions.

DEFINITION 2.27 *The function* $f : X \rightarrow R$, $X \subset R^n$, *is said to be uniformly directionally differentiable at the point* x *if, for any* $\hat{x} \in R^n$ *and* $\varepsilon > 0$, *there exist numbers* $\delta > 0$ *and* $\alpha_0 > 0$ *such that the inequality*

$$|f(x + \alpha\hat{x}) - f(x) - \alpha f'(x; \bar{x})| < \alpha\varepsilon$$

holds for every $\hat{x} \in B(\bar{x}, \delta)$, $\alpha \in [0, \alpha_0]$.

LEMMA 2.28 *Let* $Y \subset R^m$ *be a compact set, and let* $f(x, y)$ *be a function on* $R^n \times Y$ *continuous in* y *for every* x *from a neighbourhood of the point* x^*. *Moreover, it is assumed that for any* $\bar{x} \in R^n$ *and* $\varepsilon > 0$ *one can find numbers* $\delta > 0$ *and* $\alpha_0 > 0$ *such that the inequality*

$$f(x^* + \alpha x, y) - f(x^*, y) - \alpha f'_y(x^*; \bar{x})| < \alpha\varepsilon \qquad (2.10)$$

is fulfilled for all $y \in Y$ *whenever* $||x - \bar{x}|| < \delta$, $0 < \alpha < \alpha_0$. *Then the function* φ_1 *defined by (2.9) is uniformly directionally differentiable at* x^* *with respect to an arbitrary direction* $\bar{x}$ *and*

$$\varphi'_1(x^*; \bar{x}) = \max_{y \in \omega_1(x^*)} f'_y(x^*; \bar{x}), \qquad (2.11)$$

where $\omega_1(x^*) = \{y \in Y \mid f(x^*, y) = \varphi_1(x^*)\}$.

Here $f_y(x)$ denotes the function $f(x, y)$ understood as a function of x, and $f'_y(x^*; \bar{x})$ is its directional derivative with respect to x for fixed y (at the point x^* in the direction $\bar{x}$). The proof can be found in [83].

COROLLARY 2.29 *Let the function* $f(x, y)$ *be additive with respect to* x *and* y, *i. e.* $f(x, y) = f_1(x) + f_2(y)$. *Furthermore, let* Y *be compact,* f_1 *uniformly directionally differentiable at* x^* *and* f_2 *continuous. Then the assumptions of Lemma 2.28 are fulfilled.*

REMARK. The uniform directional differentiability of a function f at a point x can be ensured by assuming that f is directionally differentiable at x and locally Lipschitz continuous in a neighbourhood of x. This statement can be applied to f_1 in Corollary 2.29. If the function $f(x, y)$ is differentiable in x and y, then condition (2.10) can be guaranteed if one supposes that $f_y(x)$ is locally Lipschitz continuous near x^*.

REMARK. Lemma 2.28 provides a sufficient condition for equality in (2.11) as well as for existence of the directional derivative of φ_1 at x^*. Generally speaking, $\varphi'_1(x^*; \cdot)$ may or may not exist and, even in case of existence, only the inequality

$$\varphi'_1(x^*; \bar{x}) \geq \max_{y \in \omega_1(x^*)} f'_y(x^*; \bar{x}) \qquad (2.12)$$

holds. The following example demonstrates this phenomenon.

EXAMPLE 2.30 $f(x,y) = x - 2|y - x|$, $Y = [-2, 2]$. *We have*

$$\varphi_1(x) = \begin{cases} 3x + 4 & \text{if} & x < -2, \\ x & \text{if} & -2 \leq x \leq 2, \\ 4 - x & \text{if} & x > 2 \end{cases}$$

and, for $x^* \in [-2, 2]$, $\omega_1(x^*) = \{x^*\}$. *Thus, for* $x^* \in [-2, 2]$, *we have* $\varphi_1(x) = x$ *and, consequently,* $\varphi_1'(x^*; \bar{x}) = \bar{x}$. *On the other hand,*

$$\max_{y \in \omega_1(x^*)} f_y'(x^*; \bar{x}) = f_{x^*}'(x^*; \bar{x}) = \begin{cases} -\bar{x} & \text{if} & \bar{x} \geq 0 \\ 3\bar{x} & \text{if} & \bar{x} < 0 \end{cases}$$

so that in relation (2.12) strict inequality holds. This is due to the violation of condition (2.10).

Now we are going to formulate conditions under which the function φ_1 is quasidifferentiable. Moreover, a representation of the quasidifferential of φ_1 will be given. This statement will be used later on. By $D_x f(x, y)$ we denote the quasidifferential of f with respect to x.

THEOREM 2.31 *Let, in addition to the assumptions of Lemma 2.28, the function* $f(x, y)$ *be quasidifferentiable with respect to* x *at the point* x^* *for any* $y \in Y$, *where* $D_x f(x^*, y) = [\underline{\partial}_x f(x^*, y), \overline{\partial}_x f(x^*, y)]$. *Moreover, let there exist convex compact sets* $B(x^*)$ *and* $A_y(x^*)$ *such that* $B(x^*) = \overline{\partial}_x f(x^*, y) + A_y(x^*)$ $\forall$ $y \in \omega_1(x^*)$. *Then the function* φ_1 *is quasidifferentiable at the point* x^* *and*

$$D\varphi_1(x^*) = [\operatorname{co} \bigcup_{y \in \omega_1(x^*)} (\underline{\partial}_x f(x^*, y) - A_y(x^*), B(x^*)].$$

The proof is given in [101]. Some generalizations using so-called quasidifferential kernels have been described in [177], [178].

Since in the general case equation (2.11) need not necessarily hold, (2.12) may be fulfilled as a strict inequality. In this case one can only claim that $D\varphi_1(x^*)$ has a representation of the form

$$\overline{\partial}\varphi_1(x^*) = B(x^*), \quad \underline{\partial}\varphi_1(x^*) \supset \operatorname{co} \bigcup_{y \in \omega_1(x^*)} (\underline{\partial}_x f(x^*, y) - A_y(x^*)).$$

Applied to Example 2.30 this yields the following. Since $\omega_1(x^*) = \{x^*\}$ and $D_x f(x^*, x^*) = [1, [-2, 2]]$, we can choose $B(x^*) = [-2, 2]$ and thus $A_y(x^*) = 0$. Then we get $\overline{\partial}\varphi_1(x^*) = [-2, 2]$ and $\underline{\partial}\varphi_1(x^*) \supset \{1\}$. In fact, as we know, $\varphi_1(x) = x$ for $x \in [-2, 2]$. Therefore $D\varphi_1(x^*) = [1, 0] \cong [[-1, 3], [-2, 2]]$.

Later on we repeatedly need an auxiliary proposition concerning homogeneous programs.

LEMMA 2.32 *In the problem*

$$f(z) \to \inf_{z,\chi}, \quad G_i(z) + a_i + \chi b_i \leq 0, \ i = 1,\dots,k, \qquad (2.13)$$

let $F, G_i : R^n \to R$ *be positively homogeneous functions of order one and* $a_i, b_i, \chi \in R$, $i = 1,\dots,k$. *Then the Lagrange dual problem to (2.13) is*

$$\sum_{i=1}^{k} u_i a_i \to \sup, \quad u \in K, \qquad (2.14)$$

where $K = \{u \in R^k \,|\, u_i \geq 0, \ u_i b_i = 0, \ i = 1,\dots,k, \ F(z) + \sum_{i=1}^{k} u_i G_i(z) \geq 0 \,\forall\, z\}$.

Proof. The Lagrange dual problem to (2.13) can be written as

$$\sup_{u \geq 0} \inf_{z,\chi} \left[F(z) + \sum_{i=1}^{k} u_i (G_i(z) + a_i + \chi b_i) \right].$$

In view of the homogeneity of F and G_i, the inner problem (minimization over z and χ for fixed u) can be solved explicitly and yields $-\infty$ if there exists a $\hat{z}$ with $F(\hat{z}) + \sum_{i=1}^{k} u_i G_i(\hat{z}) < 0$ or an index i_0 such that $u_{i_0} b_{i_0} \neq 0$. Otherwise, the minimum is equal to $\sum_{i=1}^{k} u_i a_i$. From this we directly get problem (2.14). ∎

3.2   Necessary Optimality Conditions

In quasidifferential calculus necessary conditions for a local minimum are different from those for a maximum. Thus, to a certain extent, it is possible to distinguish between local maxima and minima.

LEMMA 2.33 *Let the function* $f \colon R^n \to R$ *be quasidifferentiable at* x. *Then, if* x *is a local minimizer, the inclusion* $-\overline{\partial} f(x) \subset \underline{\partial} f(x)$ *holds. Analogously, if* x *is a local maximizer of* f, *then* $-\underline{\partial} f(x) \subset \overline{\partial} f(x)$.

For the proof, see [56].

DEFINITION 2.34 *If the first or second inclusion of Lemma 2.33 is fulfilled, then* x *is said to be an inf-stationary or an sup-stationary point of* f.

Note that inf-stationarity of x is equivalent to the relation $f'(x; \bar{x}) \geq 0$.

Let us consider the constrained problem

$$f(x) \to \inf, \qquad g(x) \leq 0 \qquad (2.15)$$

with $f : R^n \to R$, $g : R^n \to R^k$. Necessary conditions for (2.15) are derived e. g. in [63], but here we need another type of conditions involving certain "Lagrange multipliers". The corresponding conditions are first given for f and g being subdifferentiable functions, then for the general quasidifferentiable case. As usual, the *Lagrangian* associated with (2.15) is of the form

$$L(x, u) = L(x, u_0, u_1, \ldots, u_k) = u_0 f(x) + \sum_{i=1}^{k} u_i g_i(x).$$

THEOREM 2.35 *Let x^* be a local minimizer of problem (2.15), and assume f and g to be subdifferentiable at x^*. Then there exist scalars $u_i^* \geq 0$, $i = 1, \ldots, k$, not all zero, such that $u_i^* g_i(x^*) = 0$, $i = 1, \ldots, k$, and*

$$0 \in \underline{\partial}_x L(x^*, u_0^*, \ldots, u_k^*) = u_0^* \underline{\partial} f(x^*) + \sum_{i=1}^{k} u_i^* \underline{\partial} g_i(x^*).$$

If, in addition, there exists a vector $\hat{x}$ with

$$(GSC) \qquad g_i'(x^*; \hat{x} - x^*) < 0 \quad \forall i \in I(x^*)$$

(generalized Slater condition), then actually $u_0^* \neq 0$.

Proof. It can be found e. g. in [107]. There are used standard arguments of convex analysis, above all the Separation Theorem. ∎

While Theorem 2.35 is in accordance with the well-known Lagrange principle, in the general case, when we deal with quasidifferentiable functions, only a weakened form of the Lagrange principle holds (cf. [107]). This is used in the following statement.

THEOREM 2.36 *Let x^* be a local minimizer of problem (2.15), and assume f and g to be quasidifferentiable at the point x^* with the quasidifferentials $Df(x^*) = [\underline{\partial} f(x^*), \overline{\partial} f(x^*)]$, $Dg_i(x^*) = [\underline{\partial} g_i(x^*), \overline{\partial} g_i(x^*)]$. Then, for any elements $v \in \overline{\partial} f(x^*)$, $w_i \in \overline{\partial} g_i(x^*)$, $i \in I(x^*)$, there exist Lagrange multipliers $u_i^*(v, w_I) \geq 0$, $i = 0, 1, \ldots, k$, not all zero, such that $u_i^*(v, w_I) g_i(x^*) = 0$, $i = 1, \ldots, k$, and*

$$0 \in u_0^*(v, w_I)(\underline{\partial} f(x^*) + v) + \sum_{i \in I(x^*)} u_i^*(v, w_I)(\underline{\partial} g_i(x^*) + w_i), \qquad (2.16)$$

where (v, w_I) *denotes the vector composed of the elements* v *and* w_i, $i \in I(x^*)$. *If, in addition, the regularity condition*

$$(RC) \qquad \exists \, \hat{x}: \quad \max_{z \in \underline{\partial} g_i(x^*)} \langle z, \hat{x} \rangle + \max_{z \in \overline{\partial} g_i(x^*)} \langle z, \hat{x} \rangle < 0 \quad \forall \, i \in I(x^*)$$

is satisfied, then actually $u_0^*(v, w_I) \neq 0 \; \forall v \in \overline{\partial} f(x^*)$, $w_i \in \overline{\partial} g_i(x^*)$, $i \in I(x^*)$.

Proof. Fix $v \in \overline{\partial} f(x^*)$, $w_i \in \overline{\partial} g_i(x^*)$, $i \in I(x^*)$ and let f_v (analogously g_{iw_i}) be a function associated with x^*, defined via the relation

$$f_v(x) = f(x^*) + \max \{ \langle z, x - x^* \rangle \mid z \in \underline{\partial} f(x^*) + v \} \qquad (2.17)$$

and having the properties $f_v(x^*) = f(x^*)$, $D f_v(x^*) = [\underline{\partial} f(x^*) + v, \{0\}]$, $f'(x^*; \bar{x}) = \min \{ f_v'(x^*; \bar{x}) \mid v \in \overline{\partial} f(x^*) \} \; \forall \bar{x} \in R^n$. It is easy to see that, at the point x^*, there cannot exist a direction $\tilde{x}$ satisfying simultaneously the conditions $f_v'(x^*; \tilde{x}) < 0$ and $g_{iw_i}'(x^*; \tilde{x}) < 0$, $i \in I(x^*)$. In fact, if we could indicate such a direction, then, by what was said above, $f'(x^*; \tilde{x}) < 0$, $g_i'(x^*; \tilde{x}) < 0 \; \forall i \in I(x^*)$. This, however, contradicts the assumption that x^* provides a local minimum in (2.15). Thus, considering the subdifferentiable problem

$$f_v(x) \to \inf, \quad g_{iw_i(x)} \leq 0, \; i \in I(x^*), \qquad (2.18)$$

and using again the Separation Theorem (cf. [105]), we conclude the existence of multipliers $u_i^*(v, w_I)$ satisfying (2.16). Finally, taking the fixed elements $w_i \in \overline{\partial} g_i(x^*)$, $i \in I(x^*)$, condition (RC) guarantees at x^* the validity of the generalized Slater condition (GSC) for every function g_{iw_i}, which in turn ensures $u_0^*(v, w_I) \neq 0$ (see Theorem 2.35). ∎

DEFINITION 2.37 *Vectors* $u^*(v, w_I) = (u_0^*(v, w_I), \dots, u_k^*(v, w_I)$ *satisfying the conditions of Theorem 2.36 are called Lagrange multipliers, regular if* $u_0^*(v, w_I) \neq 0$, *and the set of all regular multipliers is denoted by* $K(x^*, v, w_I)$.

As mentioned above, the validity of (RC) ensures the existence of regular Lagrange multipliers. Unfortunately, (RC) depends on the concrete form of the quasidifferentials $D g_i(x^*)$.

EXAMPLE 2.38 *Let* $g(x) = \min\{-x, -2x\}$. *Then* $Dg(0) = [\{0\}, [-2, -1]] \cong [[-1, 1], [-3, 0]]$ *provide two equivalent quasidifferential representations. The first meets condition (RC), while the second fails.*

Fortunately, it can be shown (see [102]) that regular multipliers do exist for any representation of the quasidifferential $D g_i(x^*)$, $i \in I(x^*)$, if (RC) holds for at least one representation.

LEMMA 2.39 *In problem (2.15) , suppose that f and g are quasidifferentiable at the point x^* being a local minimizer to (2.15), and let (RC) hold at this point. Then for any $v \in \overline{\partial} f(x^*)$, $w_i \in \overline{\partial} g_i(x^*)$, $i \in I(x^*)$, the set $K(x^*, v, w_I)$ is non-empty, compact and convex.*

Proof. The existence of regular Lagrange multipliers is stated in Theorem 2.36. Due to the convexity of the subdifferentials and the nonnegativity of Lagrange multipliers, the convexity can be shown using standard arguments of convex analysis. While closedness of $K(x^*, v, w_I)$ is evident, boundedness remains to be proved. We fix $v \in \overline{\partial} f(x^*)$, $w_i \in \overline{\partial} g_i(x^*)$, $i \in I(x^*)$, and consider the pair of problems

$$f'_v(x^*; \bar{x}) \to \inf_{\bar{x}}, \quad g'_{iw_i}(x^*; \bar{x}) \leq -1, \quad i \in I(x^*) \tag{2.19}$$

and

$$\sum_{i=1}^{k} u_i \to \sup_u, \quad u = (u_1, \ldots, u_k) \in K(x^*, v, w_I), \tag{2.20}$$

which are Lagrange dual to each other as can easily be seen from Lemma 2.32. Condition (RC) implies the existence of feasible elements in (2.19). With regard to the fact that x^* is a local minimizer to (2.15) one can conclude the boundedness of the optimal value of (2.19). Taking advantage of the weak duality theorem and the nonnegativity of the Lagrange multipliers u_i, we get the boundedness of the set $K(x^*, v, w_I)$. For more detail, see [105]. ∎

Theorem 2.36 has been generalized by weakening the regularity condition (RC) and deriving a new generalized Kuhn-Tucker condition (see Kuntz and Scholtes [93]). Another weakened variant can be found in Ward [175].

In Yin and Xu [179] Clarke's generalized gradient of the constraints plays an important role, and the Lagrange multipliers involved are independent of particular elements of the superdifferentials of the objective function and the constraints.

THEOREM 2.40 **(Yin, Xu)** *Let x^* be a local minimizer of problem (2.15), and assume f and g to be quasidifferentiable at the point x^* with the quasidifferentials $Df(x^*) = [\underline{\partial} f(x^*), \overline{\partial} f(x^*)]$ and $Dg_i(x^*) = [\underline{\partial} g_i(x^*), \overline{\partial} g_i(x^*)]$, $i = 1, \ldots, k$. If $0 \notin \partial^0 g'(x^*; 0)$, then for all elements $v \in \overline{\partial} f(x^*)$, $w_i \in \overline{\partial} g_i(x^*)$, $i \in I(x^*)$, there exist nonnegative numbers u_i^*, $i \in I(x^*)$, depending on v and w_i, $i \in I(x^*)$, such that*

$$0 \in \underline{\partial} f(x^*) + v + \left[\left(\sum_{i \in I(x^*)} \right) \partial^0 g'(x^*; 0) \right] \cap \left[\sum_{i \in I(x^*)} u_i^*(\underline{\partial} g_i(x^*) + w_i) \right],$$

where $\partial^0 g'(x^*; 0)$ is Clarke's generalized gradient of the directional derivative $g'(x^*; \cdot)$ at the point 0.

In Xia [177], [178] other interesting results concerning necessary minimum conditions in Fritz-John form are stated, where the Lagrange multipliers depend on direction. Namely, it is shown that, under the same assumptions as in Theorem 2.36, for any local minimizer x^* to problem (2.15) and each direction $\bar{x} \in R^n$ there exists a vector $\lambda^* = (\lambda_0^*, \lambda_1^*, \ldots, \lambda_k^*) \neq 0$ depending on direction $\bar{x}$ such that

$$\delta^*(\bar{x} \mid -\overline{\partial}_x L(x^*, \lambda^*)) \leq \delta^*(\bar{x} \mid \underline{\partial}_x L(x^*, \lambda^*))$$

and $\lambda_i^* g_i(x^*) = 0$, $i = 1, \ldots, k$, $\lambda_i^* \geq 0$, $i = 0, 1, \ldots, k$, where δ^* denotes the support function.

Chapter 3

TOPOLOGICAL AND DIFFERENTIAL PROPERTIES OF MULTIVALUED MAPPINGS

In this chapter topological notions such as uniformly boundedness, upper and lower semicontinuity and continuity, Lipschitz and pseudo-Lipschitz continuity of mappings and their marginal functions are dealt with. The interrelations between them are established. In the second part of this chapter different concepts and properties of differentiability and approximation of multivalued maps are considered. A description of derivatives of mappings in terms of the distance function is given.

1. Topological Properties of Multivalued Mappings and Marginal Functions

1.1 Multivalued Mappings

Let $X = R^n$, $Y = R^m$. A mapping $F : X \to 2^Y$ is called *multivalued mapping (multifunction)* if every point $x \in X$ is associated with a set $F(x) \subset Y$.

To a multivalued mapping F one can assign the sets

$$\operatorname{gr} F = \{(x, y) \mid x \in X, \ y \in F(x)\},$$
$$\operatorname{dom} F = \{x \in X \mid F(x) \neq \emptyset\},$$

called the *graph* and the *effective set* of the multivalued mapping F, respectively. Clearly, $y \in F(x) \Leftrightarrow (x, y) \in \operatorname{gr} F$, i. e. the multivalued mapping F is completely described by its graph.

A set $F(x)$ is called the *image of the point* x and the set

$$F(M) = \bigcup_{x \in M} F(x)$$

is referred to as the *image of the set* $M \subset X$ under the mapping F.

If the images $F(x)$ of all points $x \in X$ are closed (compact, convex), then the multivalued mapping F is said to be *closed-valued (compact-valued, convex-valued)*.

A multivalued mapping F is said to be *convex* if its graph gr F is a convex set in $X \times Y$.

Let $F : X \to 2^Y$. The *upper (resp. lower) topological limit* of the mapping F at a point $x_0 \in X$ are the sets

$$\limsup_{x \to x_0} F(x) \overset{\triangle}{=} \{y \in Y \mid \exists\, x_k \to x_0,\ \exists\, \{y_k\}: y_k \in F(x_k),$$
$$k = 1, 2, \ldots, y_k \to y\},$$

$$\liminf_{x \to x_0} F(x) \overset{\triangle}{=} \{y \in Y \mid \forall\, x_k \to x_0,\ \exists\, \{y_k\}: y_k \in F(x_k),$$
$$k = 1, 2, \ldots, y_k \to y\}.$$

One can give equivalent definitions of topological limits with the help of the distance function:

$$\limsup_{x \to x_0} F(x) = \{y \in Y \mid \liminf_{x \to x_0} \rho(y, F(x)) = 0\},$$
$$\liminf_{x \to x_0} F(x) = \{y \in Y \mid \limsup_{x \to x_0} \rho(y, F(x)) = 0\}.$$

The upper (lower) topological limit is always closed, where obviously the inclusion

$$\liminf_{x \to x_0} F(x) \subset \limsup_{x \to x_0} F(x).$$

is always true.

A multivalued mapping F is said to be *upper semicontinuous* (u.s.c.) or *closed* at the point $x_0 \in X$ if

$$\limsup_{x \to x_0} F(x) \subset F(x_0).$$

If F is u.s.c. at every point $x \in X$, then it is called a *closed mapping*. As can be easily proved, the closedness of a mapping F is equivalent to the closedness of gr F in $X \times Y$.

A multivalued mapping F is said to be *lower semicontinuous* (l.s.c.) at the point $x_0 \in X$ if

$$\liminf_{x \to x_0} F(x) \supset F(x_0).$$

It is said to be *continuous* at $x_0 \in X$ if it is u.s.c. and l.s.c. at this point. In this case

$$\liminf_{x \to x_0} F(x) = \limsup_{x \to x_0} F(x) = F(x_0).$$

Furthermore, the mapping F is called *upper semicontinuous in the sense of Hausdorff* (H.u.s.c.) at a point $x_0 \in X$ if for all $\varepsilon > 0$ there exists a $\delta > 0$ such that

$$F(x) \subset F(x_0) + \varepsilon B \quad \text{for all} \quad x \in x_0 + \delta B.$$

Similarly, a multivalued mapping F is called *lower semicontinuous in the sense of Hausdorff* (H.l.s.c.) at a point $x_0 \in X$ if for all $\varepsilon > 0$ there exists a $\delta > 0$ such that

$$F(x_0) \subset F(x) + \varepsilon B \quad \text{for all} \quad x \in x_0 + \delta B.$$

Finally, a multivalued mapping F is said to be *continuous in the sense of Hausdorff* at the point $x_0 \in X$ if it is both H.u.s.c. and H.l.s.c. at this point.

Let us denote by $CS(Y)$ and $CCS(Y)$, respectively, the family of all non-empty compact subsets and the family of all non-empty convex compact subsets of the space Y.

Let $F : X \to CS(Y)$. It is not hard to see that the mapping F is continuous in the sense of Hausdorff at a point $x_0 \in X$ if and only if

$$\lim_{x \to x_0} \rho_H(F(x), F(x_0)) = 0.$$

A multivalued mapping F is said to be *uniformly bounded* at the point $x_0 \in X$ if there exist a compact set $Y_0 \subset Y$ and a neighbourhood X_0 of x_0 such that $F(X_0) \subset Y_0$.

LEMMA 3.1 *Let the multivalued mapping $F : X \to CS(Y)$ be uniformly bounded at the point x_0. Then the mapping F is continuous in the sense of Hausdorff (H.u.s.c., H.l.s.c.) at the point x_0 if and only if it is continuous (u.s.c., l.s.c.) at this point.*

Proof. 1) Let F be u.s.c. at the point x_0. We want to show that F is H.u.s.c. at x_0. Suppose the opposite. Then there exists $\varepsilon_0 > 0$ such that for every $k = 1, 2, \ldots$ there are points x_k, y_k satisfying the conditions

$$|x_k - x_0| < \frac{1}{k}, \quad y_k \in F(x_k), \quad y_k \notin F(x_0) + \varepsilon_0 B. \qquad (3.1)$$

Due to the uniform boundedness of F at x_0 we can assume that the sequence $\{y_k\}$ is bounded and, therefore, has a convergent subsequence. Without loss of generality we assume $y_k \to y_0$. Then from the upper semicontinuity of F it follows that $y_0 \in F(x_0)$. On the other hand, (3.1) shows that $y_0 \in Y \backslash \{F(x_0) + \varepsilon_0 B\}$, and hence $y_0 \notin F(x_0)$. The obtained contradiction shows that F must be H.u.s.c. The opposite statement is quite obvious.

2) Let F be l.s.c. at the point x_0. Let us suppose that the mapping F is not H.l.s.c. at x_0. This means that there exist $\varepsilon_0 > 0$, $x_k \to x_0$ and $y_0 \in F(x_0)$ such that $y_0 \notin F(x_k) + \varepsilon_0 B$, $k = 1, 2, \ldots$, i.e. $|y_k - y_0| \geq \varepsilon_0$ for all $y_k \in F(x_k)$, $k = 1, 2, \ldots$ The latter means that there does not exist a sequence $\{y_k\}$ such that $y_k \in F(x_k)$, $k = 1, 2, \ldots$, $y_k \to y_0$. Hence, F is not l.s.c. at the point x_0. The contradiction proves that F is H.l.s.c. at x_0.

3) The equivalence of continuity of F at the point x_0 and its continuity in the sense of Hausdorff follows directly from 1) and 2). ∎

Note that the requirement of uniform boundedness in the assumptions of the lemma is essential. Furthermore, uniform boundedness itself is a consequence of upper semicontinuity of the mapping F in the sense of Hausdorff.

LEMMA 3.2 *Let the multivalued mapping $F : X \to CS(Y)$ be H.u.s.c. at every point of X, and let X_0 be a compact set in X. Then $F(X_0)$ is a compact set in Y.*

Proof. The validity of this lemma follows from the existence of a finite covering of X_0 with neighbourhoods $V(x_i)$, $i = 1, \ldots, p$, such that

$$F(V(x_i)) \subset F(x_i) + B, \quad i = 1, \ldots, p.$$

Consequently, the set $F(X_0) \subset \bigcup_{i=1,\ldots,p} \{F(x_i) + B\}$ is bounded. On the other hand, $F(X_0)$ is a closed set. In fact, let $y_k \in F(X_0)$ and $y_k \to y_0$. Then there exist points $x_k \in X_0$ such that $y_k \in F(x_k)$, $k = 1, 2, \ldots$ Without loss of generality we can assume that $x_k \to x_0 \in X_0$. Then in view of the upper semicontinuity it follows that $y_0 \in F(x_0) \subset F(X_0)$. ∎

EXAMPLE 3.3 *The mapping $F(x) = g(x)$ is u.s.c. (l.s.c.) if and only if the function g is continuous.*

EXAMPLE 3.4 *The constant mapping $F(x) = U$, where U is a closed set, is u.s.c. and l.s.c., hence, continuous.*

EXAMPLE 3.5 *The mapping $F(x) = \{y \mid h(x, y) = 0\}$ is u.s.c. if the function h is continuous.*

EXAMPLE 3.6 *The mapping $F(x) = \{y \mid h(x, y) \leq 0\}$ is u.s.c. if h is u.s.c.*

EXAMPLE 3.7 *The mapping $F(x) = co\{f_1(x), \ldots, f_r(x)\}$ with continuous functions f_i, $i = 1, \ldots, p$, is continuous.*

EXAMPLE 3.8 *If $f : X \times U \to Y$ is continuous and U is compact, then the multivalued mapping $F(x) = f(x, U)$ is H.u.s.c.*

EXAMPLE 3.9 *If the mapping F is u.s.c. and uniformly bounded at a point x_0, then the mapping $x \mapsto coF(x)$ is u.s.c. at x_0.*

EXAMPLE 3.10 *If the mapping F is l.s.c. at x_0, then the mapping $x \mapsto coF(x)$ is l.s.c. at x.*

DEFINITION 3.11 *The sets $F^-(M) = \{x \mid F(x) \subset M\}$ and $F^{-1}(M) = \{x \mid F(x) \cap M \neq \emptyset\}$ are called strong and weak inverse images of the set M.*

LEMMA 3.12 *(Characterization of upper semicontinuous mappings) The following statements are equivalent:*
   1. F is H.u.s.c.;
   2. the set $F^-(G)$ is open for every open set G in Y;
   3. the set $F^{-1}(M)$ is closed for every closed set M in Y.

Proof. 1. $\Leftrightarrow$ 2. Let G be open and $x_0 \in F^-(G)$. Then from the upper semicontinuity of the mapping F it follows that there exists a neighbourhood $V(x_0)$ such that $F(V(x_0)) \subset G$, i.e. $V(x_0) \subset F^-(G)$. Vice versa, for every neighbourhood $U = F(x_0) + \varepsilon B$ of the set $F(x_0)$ there exists an open set G such that $F(x_0) \subset G \subset U$. According to condition 2, $F^-(G)$ is open and for all $x \in V(x_0) = F^-(G)$ the condition $F(x) \subset F(x_0) + \varepsilon B$ is valid.

2. $\Leftrightarrow$ 3. This follows immediately from the fact that for any $M \subset Y$ the relation $X \backslash F^{-1}(M) = F^-(X \backslash M)$ holds. $\blacksquare$

It is also possible to characterize upper semicontinuous mappings in a local way, which can be obtained from Lemma 3.12.

COROLLARY 3.13 *The following statements are equivalent:*
   1. F is H.u.s.c. at a point x_0;
   2. for every open set G such that $F(x_0) \subset G$, its strong inverse image $F^-(G)$ contains some neighbourhood of the point x_0.

An analogous characterization is valid for lower semicontinuous mappings.

LEMMA 3.14 *(Characterization of lower semicontinuous mappings) The following statements are equivalent:*
   1. F is l.s.c.;
   2. the set $F^-(M)$ is closed for every closed M in Y;
   3. the set $F^{-1}(G)$ is open for every open G in Y.

Proof. It is similar to the proof of Lemma 3.12. $\blacksquare$

An important class of multivalued mappings continuous in the sense of Hausdorff are Lipschitz continuous mappings. Let the multivalued mapping F have non-empty values on the set $D \subset X$. The mapping F is said to be *Lipschitz continuous* on D if there exists a constant $l > 0$ such that, for any $x_1, x_2 \in D$,

$$F(x_1) \subset F(x_2) + l|x_1 - x_2|B. \tag{3.2}$$

Note that for a mapping $F : X \to CS(Y)$ condition (3.2) can be written in an equivalent form using the Hausdorff metric. Thus, a mapping F is Lipschitz continuous on D whenever, for all $x_1, x_2 \in D$,

$$\rho_H(F(x_1), F(x_2)) \le l|x_1 - x_2|.$$

A multivalued mapping F is said to be *locally Lipschitz continuous*, if for every point $x \in X$ there exists a neighbourhood, where F is Lipschitz continuous.

EXAMPLE 3.15 *Let the function* $g : R^n \times R^p \to R^p$ *be locally Lipschitz continuous and let the set* $U \subset R^p$ *be non-empty, bounded and closed. Then the mapping* $F(x) = g(x, U) = \{g(x, u) \mid u \in U\}$ *is locally Lipschitz continuous.*

EXAMPLE 3.16 *Let the functions* $g_i : R^n \to R^m$, $i = 1, \ldots, r$, *be locally Lipschitz continuous. Then the mapping* $F(x) = \mathrm{co} \{g_1(x), \ldots, g_r(x)\}$ *is locally Lipschitz continuous.*

EXAMPLE 3.17 *Let the multivalued mapping* $F : X \to 2^Y$ *be non-empty, convex and uniformly bounded at the point* x_0. *Then* F *is Lipschitz continuous in a neighbourhood of* x_0.

1.2   Marginal Functions

Let $F : X \to 2^Y$ be a multivalued mapping and $f : X \times Y \to R$ be a function. We consider the functions $\varphi : X \to R \cup \{\pm\infty\}$ and $\Phi : X \to R \cup \{\pm\infty\}$ defined as follows (suppose that $\inf \emptyset = +\infty$, $\sup \emptyset = -\infty$):

$$\varphi(x) = \inf\{f(x, y) \mid y \in F(x)\},$$
$$\Phi(x) = \sup\{f(x, y) \mid y \in F(x)\}.$$

The functions φ and Φ are called *marginal* (or *optimal value*) *functions* of the multivalued mapping F. Together with marginal functions we consider the *marginal* (or *optimal solution*) *mappings*

$$\omega(x) = \{y \in F(x) \mid f(x, y) = \varphi(x)\},$$
$$\Omega(x) = \{y \in F(x) \mid f(x, y) = \Phi(x)\}.$$

Simple examples of marginal functions are the distance and the support functions of a mapping F, where $p \in Y$:

$$d_F(x,y) \overset{\triangle}{=} \inf\{|y - v| \mid v \in F(x)\},$$

$$S_F(x,p) = S_{F(x)}(p) \overset{\triangle}{=} \sup\{\langle p,y \rangle \mid y \in F(x)\}.$$

LEMMA 3.18 1. Let the mapping F be l.s.c. at the point x_0 and
  a) the function f be u.s.c.; then the function φ is u.s.c. at x_0;
  b) the function f be l.s.c.; then the function Φ is l.s.c. at x_0.
  2. Let the mapping F be u.s.c. and uniformly bounded at the point x_0 and
  a) the function f be l.s.c.; then the function φ is l.s.c. at x_0;
  b) the function f be u.s.c.; then the function Φ is u.s.c. at x_0.
  3. If the mapping F is continuous and uniformly bounded at the point x_0 and the function f is continuous, then the functions φ and Φ are continuous at x_0.
  4. If the mapping F is Lipschitz continuous on the set $D \subset X$ (with Lipschitz constant l_1) and the function f is Lipschitz continuous on $D \times F(D)$ (with constant l_2), then the functions φ and Φ are Lipschitz continuous on D (with the constant $l = (l_1 + 1)l_2$).

Proof. 1. Let $x_k \to x_0$. Suppose $\varphi(x_0) > -\infty$. Then for every $\varepsilon > 0$ there exists an element $y_\varepsilon \in F(x_0)$ such that $f(x_0, y_\varepsilon) - \varphi(x_0) \leq \varepsilon$. Because of the lower semicontinuity of F there exists a sequence $\{y_k\}$ such that $y_k \to y_\varepsilon$ and $y_k \in F(x_k)$, $k = 1, 2, \ldots$ Consequently, $f(x_k, y_k) \geq \varphi(x_k)$ and, hence,

$$\limsup_{k \to \infty} \varphi(x_k) \leq \limsup_{k \to \infty} f(x_k, y_k) \leq f(x_0, y_\varepsilon) \leq \varphi(x_0) + \varepsilon.$$

Thus, $\limsup \varphi(x_k) \leq \varphi(x_0)$ and φ is u.s.c. at x_0.

If $\varphi(x_0) = -\infty$, then for every $\mu > 0$ there exists a point $y_\mu \in F(x_0)$ such that $f(x_0, y_\mu) \leq -\mu$. Then, like in the first case, there exists a sequence $\{y_k\}$ such that $y_k \to y_\mu$, $y_k \in F(x_k)$, $k = 1, 2, \ldots$, and, hence, $f(x_k, y_k) \geq \varphi(x_k)$. This implies

$$\limsup_{k \to \infty} \varphi(x_k) \leq \limsup_{k \to \infty} f(x_k, y_k) \leq f(x_0, y_\mu) \leq -\mu,$$

i. e. $\limsup \varphi(x_k) \leq -\infty = \varphi(x_0)$ and φ is u.s.c. at x_0. The second statement concerning Φ can be proved similarly.
  2. If $\varphi(x_0) = -\infty$ then the assertion is obvious. Let $\varphi(x_0) > -\infty$. Two cases are possible: $F(x_0) = \emptyset$ or $F(x_0) \neq \emptyset$. At first, suppose $F(x_0) = \emptyset$. Then $\varphi(x_0) = +\infty$ by definition. By Corollary 3.13 there

exists a neighbourhood $V(x_0)$ such that $F(x_0) = \emptyset$ for $x \in V(x_0)$. In other words, $\varphi(x) = +\infty$ for any $x \in V(x_0)$. Hence φ is l.s.c. at x_0.

Assume now $F(x_0) \neq \emptyset$. Let us take an arbitrary sequence $x_k \to x_0$. If $F(x_k) = \emptyset$, $k = 1, 2, \ldots$, then statement 2 is valid. Suppose now $F(x_k) \neq \emptyset$. If $\varphi(x_k) = -\infty$ for an infinite number of points, then there would exist a sequence $y_k \in F(x_k)$, $k = 1, 2, \ldots$, such that $f(x_k, y_k) \le -k$ for an infinite number of points. Due to the uniform boundedness of F, this sequence is bounded and without loss of generality we can assume that $y_k \to y_0$, where $y_0 \in F(x_0)$ by the upper semicontinuity of F at x_0. Taking into account the lower semicontinuity of f, from $f(x_k, y_k) \le -k$ for $k \to \infty$ we obtain

$$-\infty \ge \liminf_{k \to \infty} f(x_k, y_k) \ge f(x_0, y_k).$$

Therefore $\varphi(x_0) = -\infty$ and φ is l.s.c. at x_0.

It remains to consider the case $\varphi(x_k) > -\infty$ for every $k = 1, 2, \ldots$ Let us take an arbitrary $\varepsilon > 0$ and choose a sequence $y_k \in F(x_k)$ such that $\varphi(x_k) \ge f(x_k, y_k) - \varepsilon$ and $\liminf \varphi(x_k) = \lim \varphi(x_k)$. Without loss of generality we can assume that $y_k \to y_0$, $y_0 \in F(x_0)$. Then passing to the limit in the last inequality and taking account of the lower semicontinuity of f, we obtain

$$\lim_{k \to \infty} \varphi(x_k) - \varepsilon \ge \liminf_{k \to \infty} f(x_k, y_k) \ge f(x_0, y_0) \ge \varphi(x_0).$$

Since $\varepsilon > 0$ and the sequence $x_k \to x_0$ were chosen arbitrarily, this implies that φ is l.s.c. at x_0. The statement concerning Φ can be proved similarly.

3. This statement follows directly from 1. and 2.

4. For any (x, y), $(\bar{x}, \bar{y}) \in D \times F(D)$, the inequality

$$|f(\bar{x}, \bar{y}) - f(x, y)| \le l_2|\bar{x} - x| + l_2|\bar{y} - y| \qquad (3.3)$$

holds. In view of the Lipschitz continuity of F, for any point $\bar{y} \in F(\bar{x})$ we can find a point $y(\bar{y}) \in F(x)$ such that

$$|\bar{y} - y(\bar{y})| \le l_1|\bar{x} - x|. \qquad (3.4)$$

From (3.3) and (3.4) for $y = y(\bar{y})$ we obtain

$$f(\bar{x}, \bar{y}) \le f(x, y(\bar{y})) + l_2|\bar{x} - x| + l_2|\bar{y} - y(\bar{y})|$$
$$\le f(x, y(\bar{y})) + l_2|\bar{x} - x| + l_1 l_2|\bar{x} - x| \le \Phi(x) + l_2(l_1 + 1)|\bar{x} - x|$$

for any $\bar{y} \in F(\bar{x})$. Therefore $\Phi(\bar{x}) \le \Phi(x) + l_2(l_1 + 1)|\bar{x} - x|$. Quite similar the inverse inequality $\Phi(x) \le \Phi(\bar{x}) + l_2(l_1 + 1)|\bar{x} - x|$ can be derived, which implies

$$|\Phi(\bar{x}) - \Phi(x)| \le l_2(l_1 + 1)|\bar{x} - x|$$

for all $\bar{x}, x \in D$, i.e. Φ is Lipschitz continuous in D. The statement concerning φ can be proved in an analogous way. ∎

From Lemma 3.18 it follows that topological properties of marginal functions are defined by corresponding properties of multivalued mappings. The inverse effect is valid too, i.e., marginal functions can be used for a complete topological description of multivalued mappings.

LEMMA 3.19 *Let the multivalued mapping F be closed-valued. Then*

1. if the function d_F is l.s.c. at $\{x_0\} \times Y$, then the mapping F is u.s.c. at x_0;

2. if the function d_F is u.s.c. at $\{x_0\} \times Y$, then the mapping F is l.s.c. at x_0;

3. if the function d_F is continuous at $\{x_0\} \times Y$, then the mapping F is continuous at x_0;

4. if the function d_F is Lipschitz continuous on the set $D \times Y$, then the mapping F is Lipschitz continuous on D.

Proof. 1. Let $x_k \to x_0$, $y_k \in F(x_k)$, $k = 1, 2, \ldots$, and $y_k \to y_0$. Then the passage to the limit in inequality $d_F(x_k, y_k) \leq 0$ yields

$$d_F(x_0, y_0) \leq \liminf_{k \to \infty} d_F(x_k, y_k) \leq 0,$$

i.e. $y_0 \in F(x_0)$.

2. Let us suppose the opposite, i.e., let F fail to be l.s.c. at x_0. This means that there exist a sequence $x_k \to x_0$, a point $y \in F(x_0)$ and a number $\varepsilon > 0$ such that $d_F(x_k, y) \geq \varepsilon$ for all $k = 1, 2, \ldots$ In this case, by the upper semicontinuity we obtain $d_F(x_0, y) \geq \varepsilon$, i.e. $y \notin F(x_0)$. Thus, F is l.s.c. at the point x_0.

3. This statement follows immediately from the first two.

4. By the Lipschitz continuity of d_F there exists a number $l > 0$ such that

$$|d_F(x_1, y) - d_F(x_2, y)| \leq l|x_1 - x_2|$$

for any $x_1, x_2 \in D$, $y \in Y$.

Let $\delta > 0$ be arbitrarily given. Then we can find a point $y_2(x_2) \in F(x_2)$ such that

$$\sup_{y \in F(x_2)} d_F(x_1, y) \leq d_F(x_1, y_2(x_2)) + \delta$$
$$= d_F(x_1, y_2(x_2)) - d_F(x_2, y_2(x_2)) + \delta \leq l|x_1 - x_2| + \delta.$$

In a similar way we can find an element $y_1(x_1) \in F(x_1)$ such that

$$\sup_{y \in F(x_1)} d_F(x_2, y) \leq d_F(x_2, y_1(x_1)) + \delta$$
$$= d_F(x_2, y_1(x_1)) - d_F(x_1, y_1(x_1)) + \delta \leq l|x_1 - x_2| + \delta.$$

From these two inequalities one gets $\rho_H(F(x_1), F(x_2)) \leq l|x_1 - x_2| + \delta$. Because of the arbitrary choice of $\delta > 0$ the lemma is proved. ∎

REMARK. If in statement 4 of Lemma 3.19 the mapping F satisfies the condition $F(D) \subset Y_0$, where Y_0 is a compact set in Y, then in this statement Y can be replaced by Y_0.

Uniting Lemmas 3.18 and 3.19, the following proposition can be formulated.

LEMMA 3.20 *Let the multivalued mapping* $F : X \to CS(Y)$ *be uniformly bounded at the point* x_0. *Then*

1. F *is l.s.c. at* x_0 *if and only if* d_F *is u.s.c. at* $\{x_0\} \times Y$;
2. F *is u.s.c. at* x_0 *if and only if* d_F *is l.s.c. at* $\{x_0\} \times Y$;
3. F *is continuous at* x_0 *if and only if* d_F *is continuous at* $\{x_0\} \times Y$;
4. F *is Lipschitz continuous on* $D \subset X$ *if and only if* d_F *is Lipschitz continuous on* $D \times Y$.

REMARK. In assertion 4 of Lemma 3.20 the function d_F satisfies a Lipschitz condition in the form

$$|d_F(x, y) - d_F(\bar{x}, \bar{y})| \leq l|x - \bar{x}| + |y - \bar{y}|$$

for $x, \bar{x} \in D$; $y, \bar{y} \in Y$, where l is the Lipschitz constant of the mapping F.

A similar statement can be established for the support function of a convex-valued mapping.

LEMMA 3.21 *Let the multivalued mapping* $F : X \to CCS(Y)$ *be uniformly bounded at the point* x_0. *Then*

1. F *is l.s.c. at* x_0 *if and only if* $S_F(\cdot, p)$ *is l.s.c. at* x_0;
2. F *is u.s.c. at* x_0 *if and only if* $S_F(\cdot, p)$ *is u.s.c. at* x_0;
3. F *is continuous at* x_0 *if and only if* $S_F(\cdot, p)$ *is continuous at* x_0;
4. F *is Lipschitz continuous on* $D \subset X$ *if and only if* $S_F(\cdot, p)$ *is Lipschitz continuous on* D *for any* $p \in Y$.

The following lemma concerns upper semicontinuity of marginal mappings.

LEMMA 3.22 *Let the multivalued mapping* F *be continuous and uniformly bounded at the point* x_0, *and let the function* f *be continuous. Then the mappings* $\omega(\cdot)$ *and* $\Omega(\cdot)$ *are u.s.c. at* x_0.

Proof. By virtue of Lemma 3.18 the function φ is continuous at the point x_0. Let $x_k \to x_0$, $y_k \to y_0$ and $y_k \in \omega(x_k)$, $k = 1, 2, \ldots$ Then $f(x_k, y_k) = \varphi(x_k)$, $y_k \in F(x_k)$, $k = 1, 2, \ldots$ If we let $k \to \infty$, by the continuity of F we obtain $f(x_0, y_0) = \varphi(x_0)$, $y_0 \in F(x_0)$, i. e. $y_0 \in \omega(x_0)$. The statement concerning Ω can be proved in a similar way. ∎

LEMMA 3.23 *Let the multivalued mapping F be u.s.c. at the point x_0, let φ be u.s.c. at x_0 (Φ be l.s.c. at x_0, resp.) and f be l.s.c. (u.s.c., resp.). Then the mappings $\omega(\cdot)$ and $\Omega(\cdot)$ are u.s.c. at x_0.*

Proof. Let $x_k \to x_0$, $y_k \to y_0$ and $y_k \in \omega(x_k)$, $k = 1, 2, \ldots$ Then passing to the limit in equality $f(x_k, y_k) = \varphi(x_k)$, we obtain $f(x_0, y_0) \le \varphi(x_0)$. Because of $y_0 \in F(x_0)$ and the upper semicontinuity of F, this inequality implies $y_0 \in \omega(x_0)$. The upper semicontinuity of Ω can be proved analogously. ∎

1.3 Pseudolipschitz and Pseudohölder Continuity of Multivalued Mappings

In many problems it suffices to require that the multivalued mapping under study has a property which is less restrictive than Lipschitz continuity.

DEFINITION 3.24 *A multivalued mapping $F : X \to 2^Y$ is called pseudolipschitz continuous at a point $z_0 = (x_0, y_0) \in$ gr F with respect to $M \subset X$ if there exist neighbourhoods $V(x_0)$ and $V(y_0)$ of the points x_0 and y_0 as well as a constant $l > 0$ such that*

$$F(x_1) \cap V(y_0) \subset F(x_2) + l|x_1 - x_2|B \qquad (3.5)$$

for any $x_1, x_2 \in V(x_0) \cap M$. The mapping F is called pseudolipschitz continuous at the point z_0 if $M = X$.

Let $V(x_0) = x_0 + \delta_0 B$, $V(y_0) = y_0 + \delta B$ be neighbourhoods of the points x_0 and y_0. We want to study, under which conditions the relation $F(x) \cap V(y_0) \ne \emptyset$ is valid. Suppose that for some $\bar{x} \in V(x_0)$, we have $F(\bar{x}) \cap V(y_0) = \emptyset$. This means that $\rho(y_0, F(\bar{x})) \ge \delta$. In view of (3.5), $y_0 \in F(x_0) \cap V(y_0)$ and, hence, $y_0 \in F(\bar{x}) + l|\bar{x} - x_0|B$. This implies

$$\delta \le \rho(y_0, F(\bar{x})) \le l|\bar{x} - x_0| < l\delta_0.$$

In this way, the following result has been proved.

LEMMA 3.25 *If $\delta \ge l\delta_0$, then $F(x) \cap V(y_0) \ne \emptyset$ for any $x \in V(x_0)$.*

The following lemma is a supplement to Lemma 3.20.

LEMMA 3.26 *Let F be a closed-valued mapping. Then the following statements are equivalent:*

   1. F is pseudolipschitz continuous at the point $z_0 = (x_0, y_0) \in \operatorname{gr} F$ with respect to M;

   2. the function d_F is Lipschitz continuous on $(V(x_0) \cap M) \times V(y_0)$.

Proof. 1. $\Rightarrow$ 2. Without loss of generality we can assume $V(x_0) = x_0 + \delta_0 B$, $V_\delta(y_0) = y_0 + \delta B$ and $\delta \geq 2l\delta_0$. Then for any $x, \bar{x} \in V(x_0) \cap M$ we have $F(\bar{x}) \cap V_{\delta/2}(y_0) \neq \emptyset$ and $F(\bar{x}) \cap V_\delta(y_0) \subset F(x) + l|\bar{x} - x|B$. Therefore for every $x, \bar{x} \in V(x_0) \cap M$ and any $\bar{v} \in F(\bar{x}) \cap V_{\delta/2}(y_0)$ there exists a point $v(\bar{v}) \in F(x)$ such that $|v(\bar{v}) - \bar{v}| \leq l|\bar{x} - x|$. Furthermore, for any $y, \bar{y} \in V_{\delta/2}(y_0)$ and for arbitrary $\bar{v} \in F(\bar{x}) \cap V_{\delta/2}(y_0)$, we obtain

$$\begin{aligned} |\bar{y} - \bar{v}| &\geq |y - v(\bar{v})| - |\bar{y} - y| - |\bar{v} - v(\bar{v})| \\ &\geq |y - v(\bar{v})| - l|\bar{x} - x| - |\bar{y} - y| \qquad (3.6) \\ &\geq d_F(x, y) - l|\bar{x} - x| - |\bar{y} - y|. \end{aligned}$$

Since $\delta > |\bar{y} - \bar{v}| \geq \rho(\bar{y}, F(\bar{x}))$, we get

$$\inf\{|\bar{y} - v| \mid v \in F(\bar{x})\} = \inf\{|\bar{y} - \bar{v}| \mid \bar{v} \in F(\bar{x}) \cap V_\delta(y_0)\}.$$

Then from (3.6) we derive

$$d_F(\bar{x}, \bar{y}) \geq d_F(x, y) - l|\bar{x} - x| - |\bar{y} - y|$$

for any $x, \bar{x} \in V(x_0) \cap M$ and $y, \bar{y} \in V_{\delta/2}(y_0)$. Replacing x by $\bar{x}$ and y by $\bar{y}$, we obtain

$$d_F(x, y) \geq d_F(\bar{x}, \bar{y}) - l|\bar{x} - x| - |\bar{y} - y|.$$

Thus, for any $x, \bar{x} \in V(x_0) \cap M$ and $y, \bar{y} \in V_{\delta/2}(y_0)$

$$|d_F(\bar{x}, \bar{y}) - d_F(x, y)| \leq l|\bar{x} - x| + |\bar{y} - y|. \qquad (3.7)$$

The latter means that d_F is Lipschitz continuous on $(V(x_0) \cap M) \times V_{\delta/2}(y_0)$.

   2. $\Rightarrow$ 1. Let d_F be Lipschitz continuous on $(V(x_0) \cap M) \times V_{\delta/2}(y_0)$, i.e. let condition (3.7) be valid. Then for any $y = \bar{y} \in F(\bar{x}) \cap V_{\delta/2}(y_0)$ the inequality $\rho(y, F(x)) \leq l|\bar{x} - x|$ results. For any $x, \bar{x} \in V(x_0)$, this implies

$$F(\bar{x}) \cap V_{\delta/2}(y_0) \subset F(x) + l|\bar{x} - x|B. \qquad \blacksquare$$

REMARK. From the proof of Lemma 3.26 it follows that if (3.5) is valid for some neighbourhoods $V(x_0)$ and $V_\delta(y_0)$, then d_F is Lipschitz continuous on $V(x_0) \times V_{\delta/2}(y_0)$.

   A generalization of the notion of pseudolipschitz continuity was considered in [38] and [171]. In the following, let ν be a positive constant.

DEFINITION 3.27 *A multivalued mapping F is called pseudohölder continuous of order ν at the point $z_0 = (x_0, y_0) \in \mathrm{gr}\, F$, if there exist neighbourhoods $V(x_0)$ and $V(y_0)$ of the points x_0 and y_0 as well as a constant $l > 0$ such that, for any $x_1, x_2 \in V(x_0)$,*

$$F(x_1) \cap V(y_0) \subset F(x_2) + l|x_1 - x_2|^\nu B.$$

Similarly to the statements proved above it can be shown for a pseudohölder continuous mapping F that if $\delta \geq l\delta_0{}^\nu$, then $F(x) \cap V(y_0) \neq \emptyset$ for all $x \in V(x_0)$, where $V(x_0) = x_0 + \delta_0 B$ and $V(y_0) = y_0 + \delta B$. An analogue to Lemma 3.26 for pseudohölder continuous mappings can be proved as well.

LEMMA 3.28 *Let F be a closed-valued mapping. The following statements are equivalent:*

1. the mapping F is pseudohölder continuous of order ν at the point $z_0 = (x_0, y_0) \in \mathrm{gr}\, F$;

2. for any $x, \bar{x} \in V(x_0)$ and $y, \bar{y} \in V_{\delta/2}(y_0)$, one has

$$|d_F(x, y) - d_F(\bar{x}, \bar{y})| \leq l|x - \bar{x}|^\nu + |y - \bar{y}|. \tag{3.8}$$

Proof. The argument is very close to that of Lemma 3.26. ∎

1.4 Properties of Convex Mappings

Let us remind that a multivalued mapping $F : X \to 2^Y$ is convex, if its graph $\mathrm{gr}\, F$ is a convex set.

LEMMA 3.29 *The following statements are equivalent:*

1. F is a convex mapping;

2. the inclusion

$$F(\lambda_1 x_1 + \lambda_2 x_2) \supset \lambda_1 F(x_1) + \lambda_2 F(x_2) \tag{3.9}$$

holds for all $x_1, x_2 \in X$ and $\lambda_1 \geq 0$, $\lambda_2 \geq 0$, $\lambda_1 + \lambda_2 = 1$.

Proof. Let $(x_1, y_1) \in \mathrm{gr}\, F$ and $(x_2, y_2) \in \mathrm{gr}\, F$. Then due to the convexity $(\lambda_1 x_1 + \lambda_2 x_2, \lambda_1 y_1 + \lambda_2 y_2) \in \mathrm{gr}\, F$, i.e. $\lambda_1 y_1 + \lambda_2 y_2 \in F(\lambda_1 x_1 + \lambda_2 x_2)$. Since y_1 and y_2 are arbitrary points from $F(x_1)$ and $F(x_2)$ respectively, (3.9) is valid. Arguing inversely, from (3.9) we obtain the convexity of $\mathrm{gr}\, F$. ∎

LEMMA 3.30 *Let F be a convex mapping. Then its support function $S_F(x, p)$ is concave with respect to x for every $p \in Y$. Conversely, if F is closed-valued and $S_F(\cdot, p)$ is concave for every p, then F is convex.*

Proof. If F is convex, then by virtue of Lemma 3.29 the inclusion (3.9) is valid. By the properties of the support function (Theorem 1.9) it follows that

$$S_F(\lambda_1 x_1 + \lambda_2 x_2, p) \geq \lambda_1 S_F(x_1, p) + \lambda_2 S_F(x_2, p) \qquad (3.10)$$

for all $x_1, x_2 \in X$ and $\lambda_1 \geq 0$, $\lambda_2 \geq 0$, $\lambda_1 + \lambda_2 = 1$. The latter is equivalent to the concavity of $S_F(\cdot, p)$. Vice versa, from (3.10) and the closed-valuedness of F, by Theorem 2.14 we obtain the inclusion (3.9). ∎

EXAMPLE 3.31 *The mapping $F(x) = C$ is convex if C is a convex set.*

EXAMPLE 3.32 *The mapping $F(x) = \{y \in Y \mid h_i(x, y) \leq 0, \; i = 1, \ldots, r\}$ is convex if the functions $h_i : X \times Y \to R$, $i = 1, \ldots, r$, are convex.*

Now we consider the marginal function

$$\varphi(x) = \inf\{f(x, y) \mid y \in F(x)\}.$$

LEMMA 3.33 *Let $f : X \times Y \to R$ be a convex function and $F : X \to 2^Y$ be a convex multivalued mapping. Then the function φ is convex.*

Proof. Let $z_1 = (x_1, y_1)$, $z_2 = (x_2, y_2)$. By inclusion (3.9) we get

$$\begin{aligned}
\varphi(\lambda_1 x_1 + \lambda_2 x_2) &= \inf\{f(\lambda_1 x_1 + \lambda_2 x_2, y) \mid y \in F(\lambda_1 x_1 + \lambda_2 x_2)\} \\
&\leq \inf\{f(\lambda_1 z_1 + \lambda_2 z_2) \mid y_1 \in F(x_1), \; y_2 \in F(x_2)\} \\
&\leq \inf\{\lambda_1 f(z_1) + \lambda_2 f(z_2) \mid y_1 \in F(x_1), \; y_2 \in F(x_2)\} \\
&= \lambda_1 \varphi(x_1) + \lambda_2 \varphi(x_2), \qquad (3.11)
\end{aligned}$$

i.e., for all $x_1, x_2 \in X$ and $\lambda_1 \geq 0$, $\lambda_2 \geq 0$, $\lambda_1 + \lambda_2 = 1$ the inequality $\varphi(\lambda_1 x_1 + \lambda_2 x_2) \leq \lambda_1 \varphi(x_1) + \lambda_2 \varphi(x_2)$ is valid. Thus φ is convex. ∎

COROLLARY 3.34 *Let the assumptions of Lemma 3.33 be satisfied. Then $\operatorname{dom} \varphi = \operatorname{dom} F$ if $\operatorname{dom} f = X \times Y$.*

Proof. Let $x \in \operatorname{dom} \varphi$, i.e. $\varphi(x) < +\infty$. Suppose that $x \notin \operatorname{dom} F$. Then $F(x) = \emptyset$ and $\varphi(x) = \inf \emptyset = +\infty$. From this contradiction the inclusion $\operatorname{dom} \varphi \subset \operatorname{dom} F$ results. Inversely, let $x \in \operatorname{dom} F$. Then there exists a point $y_0 \in F(x)$. Consequently, $\varphi(x) \leq f(x, y_0) < +\infty$, i.e. $x \in \operatorname{dom} \varphi$ and $\operatorname{dom} F \subset \operatorname{dom} \varphi$. ∎

COROLLARY 3.35 *Let V be a convex set in R^r, the function $f : X \times Y \times V \to R$ be finite and convex and let $F : X \to 2^Y$ be a convex multivalued mapping. Then the function*

$$\varphi(x, v) = \inf\{f(x, y, v) \mid y \in F(x)\}$$

is convex, and $\operatorname{dom} \varphi = \operatorname{dom} F \times V$.

Proof. Denote $\tilde{x} = (x, v)$, $\tilde{y} = (y, v)$, $\tilde{F}(\tilde{x}) = F(x) \times \{v\}$. Applying Lemma 3.33 to f and $\tilde{F}$, we obtain the convexity of φ. Corollary 3.34 implies that $\operatorname{dom} \varphi = \operatorname{dom} F \times V$. ∎

LEMMA 3.36 *Let F be a closed-valued convex mapping and $x_0 \in \operatorname{ri} \operatorname{dom} F$. Then F is pseudolipschitz continuous with respect to $\operatorname{ri} \operatorname{dom} F$ at any point (x_0, y_0), where $y_0 \in F(x_0)$.*

Proof. By virtue of Corollary 3.35, the function d_F is convex for $z = (x, y)$. Furthermore, $\operatorname{dom} d_F = \{V(x_0) \cap \operatorname{dom} F\} \times Y$, where $V(x_0)$ is a neighbourhood of the point x_0. Then by Theorem 2.15 d_F is Lipschitz continuous on the set $\{V'(x_0) \cap \operatorname{ri} \operatorname{dom} F\} \times V(y_0)$, where $V'(x_0)$ and $V(y_0)$ are neighbourhoods of the points x_0 and y_0, respectively. Applying Lemma 3.26, we obtain the desired assertion. ∎

LEMMA 3.37 *Let the assumptions of Lemma 3.36 be satisfied and, in addition, the mapping F be uniformly bounded at x_0. Then the mapping F is Lipschitz continuous on $V'(x_0) \cap \operatorname{ri} \operatorname{dom} F$.*

Proof. We can argue in the same way as in Lemma 3.36, but instead of Theorem 2.15 and Lemma 3.26, we have to use Theorem 10.4 from [154] and Lemma 3.20.

1.5 Closed convex processes

A multivalued mapping $K(\cdot) : X \to 2^Y$ is called a *convex process* if its graph is a convex cone in $X \times Y$. Let us denote its graph $\operatorname{gr} K(\cdot)$ by K, i.e. K is a convex cone in $X \times Y$. If the cone K is convex and closed, then $K(\cdot)$ is called a *closed convex process*.

Let $K(\cdot)$ be a closed convex process and let $N \triangleq -K^+$. Then together with $K(\cdot)$ it makes sense to consider the adjoint process $N(\cdot)$ defined by

$$N(y^*) \triangleq \{x^* | (x^*, y^*) \in N\}.$$

LEMMA 3.38 $[\operatorname{dom} K(\cdot)]^+ = -N(0)$.

Proof. As can be easily seen, $\operatorname{dom} K(\cdot)$ is a convex cone in X, containing the point 0. From the definition of the adjoint cone it follows that

$$[\operatorname{dom} K(\cdot)]^+ = \{x^* \mid \langle x^*, x \rangle \geq 0 \, \forall x \in \operatorname{dom} K(\cdot)\}$$
$$= \{x^* | \langle x^*, x \rangle \geq 0 \, \forall x \text{ such that } \exists y : (x, y) \in K\}$$
$$= \{x^* | \langle (x^*, 0), z \rangle \geq 0 \, \forall z \in K\} = K^+ \cap \{(x^*, y^*) | y^* = 0\} = -N(0). \blacksquare$$

LEMMA 3.39 $0^+K(x) = K(0)$ *for all* $x \in \operatorname{dom} K(\cdot)$.

Proof. Let $\bar{y} \in 0^+K(x)$. Since $K(x)$ is a convex closed and non-empty set, then by virtue of Theorem 2.4, $\bar{y} = \lim \lambda_k y_k$, where $y_k \in K(x)$, $k = 1, 2, \ldots$ and $\lambda_k \downarrow 0$. Then $(x, y_k) \in K$ and, hence, $(\lambda_k x, \lambda_k y_k) \in K$. By the closedness of K we get $(0, \bar{y}) \in K$, i.e. $\bar{y} \in K(0)$. Thus, the inclusion $0^+K(x) \subset K(0)$ holds.

Let $\bar{y} \in K(x)$. This means that $(0, \bar{y}) \in K$. Then for any $y \in K(x)$ and any $\lambda > 0$ the relations $(x, y) \in K$, $(x, y) + \lambda(0, \bar{y}) = (x, y + \lambda \bar{y}) \in K$ are valid, which implies $y + \lambda \bar{y} \in K(x)$, i.e. $\bar{y} \in 0^+K(x)$. Thus $K(0) \subset 0^+K(x)$ and, hence, $0^+K(x) = K(0)$. ∎

LEMMA 3.40 *A multivalued mapping* $K(\cdot)$ *is compact-valued if and only if* $K(0) = \{0\}$.

Proof. If $K(0) = \{0\}$, then by Lemma 3.39 the sets $K(x)$ are bounded and, thus, $K(\cdot)$ is compact-valued. Inversely, from the compactness of $K(x)$ it follows that $0^+K(x) = \{0\}$. By Lemma 3.39 one obtains $K(0) = \{0\}$. ∎

LEMMA 3.41 *Let* $f : X \times Y \to R$ *be a convex positive homogeneous function with* $\operatorname{dom} f = X \times Y$, *and let* $K(\cdot)$ *be a closed convex process. Then the function* $\varphi(x) = \inf\{f(x, y) | y \in K(x)\}$ *is a convex positive homogeneous function and*

1. $\operatorname{dom} \varphi = \operatorname{dom} K(\cdot)$;
2. $\varphi^*(x^*) = \delta(x^* | \Lambda_0)$;
3. $\operatorname{cl} \varphi(x) = \delta^*(x^*, \Lambda_0)$, *where* $\Lambda_0 = \{x^* \in X | (x^*, 0) \in \partial f(0) + N\}$.

Proof. The convexity of φ follows from Lemma 3.33. The positive homogeneity can be checked directly, while $\operatorname{dom} \varphi = \operatorname{dom} K(\cdot)$ by Corollary 3.34. Let us prove condition 2. By definition

$$\varphi^*(x^*) = \sup_x\{\langle x^*, x \rangle - \varphi(x)\} = \sup_x\{\langle x^*, x \rangle - \inf_{y \in K(x)} f(z)\}$$
$$= \sup_{z \in K}\{\langle (x^*, 0), z \rangle - f(z)\}.$$

Now, using the definition of the subdifferential and the minimax theorem, we obtain

$$\varphi^*(x^*) = \sup_{z \in K} \inf_{\xi \in \partial f(0)} \{\langle (x^*, 0), z \rangle - \langle \xi, z \rangle\}$$

$$= \inf_{\xi \in \partial f(0)} \sup_{z \in K} \{\langle (x^*, 0), z \rangle - \langle \xi, z \rangle\} = \inf_{\xi \in \partial f(0)} \delta^*((x^*, 0) - \xi \,|\, K)$$

$$= \inf_{\xi \in \partial f(0)} \begin{cases} 0, & \text{if } (x^*, 0) \in \xi - K^+, \\ +\infty, & \text{if } (x^*, 0) \notin \xi - K^+ \end{cases} = \delta(x^* | \Lambda_0).$$

It remains to note that condition 3 follows immediately from the fact that $\operatorname{cl}\varphi(x) = \varphi^{**}(x) = \delta^*(x|\Lambda_0)$. ∎

EXERCISE 3.42 *Let $F : X \to 2^Y$ be u.s.c. and uniformly bounded at x_0 and $G : X \times Y \to 2^V$ be u.s.c. at the points $\{x_0\} \times F(x_0)$. Show that the mapping $H : x \mapsto \bigcup\limits_{y \in F(x)} G(x, y)$ is u.s.c. at x_0.*

EXERCISE 3.43 *Let $F : X \to 2^Y$ be u.s.c. and uniformly bounded at x_0. Show that $\operatorname{co} F : x \mapsto \operatorname{co} F(x)$ is u.s.c. at x_0.*

EXERCISE 3.44 *Let $F : X \to 2^Y$, $G : X \to 2^Y$ and $F(x_0) \cap G(x_0) \neq \emptyset$. Show that if the mappings F and G are u.s.c. at x_0 and one of them is uniformly bounded at x_0, then the mapping $F \cap G : x \mapsto F(x) \cap G(x)$ is u.s.c. at x_0.*

EXERCISE 3.45 *Let the function $f : X \to Y$ be locally Lipschitz continuous, and let the multivalued mapping $G : Y \to 2^V$ be closed and pseudolipschitz continuous at $(x_0, f(x_0))$. Show that the mapping $F : x \mapsto G(f(x))$ is pseudolipschitz continuous at $\{x_0\} \times G(f(x_0))$.*

2. Directional Differentiability of Multivalued Mappings

2.1 Tangent Cones and Derivatives of Multivalued Mappings

Let $X = R^n$, $Y = R^m$, $Z = X \times Y$. We consider the set $E \subset Z$. Let us define the *lower (resp. upper) tangent cone* to E at a point $z \in E$ as

$$T_E^L(z) \triangleq \liminf_{\varepsilon \downarrow 0} \varepsilon^{-1}(E - z) \quad \text{and} \quad T_E^U(z) \triangleq \limsup_{\varepsilon \downarrow 0} \varepsilon^{-1}(E - z).$$

In addition, we consider the *Clarke tangent cone*

$$T_E^C(z) \triangleq \liminf_{\varepsilon \downarrow 0, z' \xrightarrow{E} z} \varepsilon^{-1}(E - z')$$

to E at $z \in E$, where the notation $z' \xrightarrow{E} z$ means that $z' \to z$, $z' \in E$.

Since the upper and lower topological limits are closed sets, the cones $T_E^S(z)$ with $S = L, U, C$ are closed. Moreover, $T_E^C(z)$ is a convex cone. Note that the cone $T_E^U(z)$ is often called the *contingent cone*.

LEMMA 3.46 *Let $z \in E$. Then the following statements are equivalent:*
1. *$\bar{z} \in T_E^U(z)$;*
2. *$D_+\rho_E(z; \bar{z}) = 0$;*

3. there are sequences $\varepsilon_k \downarrow 0$ and $\bar{z}_k \to \bar{z}$ such that $z + \varepsilon_k \bar{z}_k \in E$, $k = 1, 2, \ldots$

Proof. 1. $\Leftrightarrow$ 2. In fact, due to the definition of the cone $T_E^U(z)$ the condition $\bar{z} \in T_E^U(z)$ is equivalent to

$$0 = \liminf_{\varepsilon \downarrow 0} \rho(\bar{z}, \varepsilon^{-1}(E - z)) = \liminf_{\varepsilon \downarrow 0} \varepsilon^{-1} \rho_E(z + \varepsilon \bar{z}) = D_+ \rho_E(z; \bar{z}).$$

1. $\Leftrightarrow$ 3. This equivalence follows from the definition of the upper topological limit. ∎

LEMMA 3.47 *Let $z \in E$. Then the following statements are equivalent:*
  1. $\bar{z} \in T_E^L(z)$;
  2. $D^+ \rho_E(z; \bar{z}) = 0$;
  3. there exists a function $o(\varepsilon)$ such that $z + \varepsilon \bar{z} + o(\varepsilon) \in E$ for $\varepsilon \geq 0$ and $o(\varepsilon)/\varepsilon \to 0$ if $\varepsilon \downarrow 0$;
  4. for any sequence $\varepsilon_k \downarrow 0$ there exists a sequence $\bar{z}_k \to \bar{z}$ such that $z + \varepsilon_k \bar{z}_k \in E$, $k = 1, 2, \ldots$

Proof. It is similar to the proof of Lemma 3.46.

LEMMA 3.48 *Let $z \in E$. Then the following statements are equivalent:*
  1. $\bar{z} \in T_E^C(z)$;
  2. $\rho_E^0(z; \bar{z}) = 0$;
  3. $\lim_{\bar{z} \overset{E}{\to} z} D_+ \rho_E(\bar{z}; \bar{z}) = 0$;
  4. for any sequences $\varepsilon_k \downarrow 0$, $z_k \overset{E}{\to} z$ one can find a sequence $\bar{z}_k \to \bar{z}$ such that $z_k + \varepsilon_k \bar{z}_k \in E$, $k = 1, 2, \ldots$

Proof. The equivalences 1. $\Leftrightarrow$ 2. and 1. $\Leftrightarrow$ 4. can be proved similarly to the proof of Lemma 3.46. The proof of 1. $\Leftrightarrow$ 3. can be found in [42]. ∎

Let $F: X \to 2^Y$, $z = (x, y) \in \mathrm{gr}\, F$. For any $\bar{x} \in X$ we define the sets

$$D_S F(z; \bar{x}) \overset{\triangle}{=} \{\bar{y} \in Y \mid (\bar{x}, \bar{y}) \in T_{\mathrm{gr}\, F}^S(z)\}, \quad S = L, U, C,$$

called *upper, lower and Clarke derivative*, resp., of the multivalued mapping F at the point z in the direction $\bar{x}$.

It is easy to prove that

$$\mathrm{gr}\, D_S F(z; \cdot) = T_{\mathrm{gr}\, F}^S(z), \quad S = L, U, C.$$

If the mapping F is convex, then the tangent cones $T_{\mathrm{gr}\, F}^S(z)$ with $S = L, U, C$, coincide (see Section 2). Therefore, all derivatives $D_S F(z; \bar{x})$, with $S = L, U, C$ coincide too.

In addition, from the closedness of the various tangent cones $T^S_{\mathrm{gr}\,F}(z)$, $S = L, U, C$, it follows that the sets $D_S F(z; \bar{x})$, $S = L, U, C$, are closed.

Apart from the derivatives considered above we introduce the *lower* and *upper direct derivatives* of the mapping F at the point $z \in \mathrm{gr}\,F$ in the direction $\bar{x}$:

$$\hat{D}_L F(z; \bar{x}) = \liminf_{\varepsilon \downarrow 0} \varepsilon^{-1}(F(x + \varepsilon \bar{x}) - y),$$
$$\hat{D}_U F(z; \bar{x}) = \limsup_{\varepsilon \downarrow 0} \varepsilon^{-1}(F(x + \varepsilon \bar{x}) - y). \tag{3.12}$$

It is not hard to see that if the mapping F is Lipschitz continuous in a neighbourhood of the point x, then the equalities

$$\hat{D}_L F(z; \bar{x}) = D_L F(z; \bar{x}), \quad \hat{D}_U F(z; \bar{x}) = D_U F(z; \bar{x})$$

hold for all $y \in F(x)$, $\bar{x} \in X$.

In particular, if the multivalued mapping F reduces to a singlevalued function $(F(x) = \{f(x)\})$, then

$$\hat{D}_L F(z; \bar{x}) = \lim_{\varepsilon \downarrow 0} \varepsilon^{-1}[f(x + \varepsilon \bar{x}) - f(x)] = f'(x; \bar{x}),$$

i. e., we get the ordinary directional derivative. At the same time

$$D_L F(z; \bar{x}) = \lim_{\varepsilon \downarrow 0, \hat{x} \to \bar{x}} \varepsilon^{-1}[f(x + \varepsilon \hat{x}) - f(x)] \stackrel{\triangle}{=} D_\oplus f(x; \bar{x}),$$

i. e., one gets the *directional derivative in the sense of Hadamard*. The derivatives $\hat{D}_U F(z; \bar{x})$ and $D_U F(z; \bar{x})$ lead to the contingent derivatives (ordinary and in the sense of Hadamard) of the function f in the direction $\bar{x}$.

Since there exist equivalent definitions of topological limits in terms of the distance function, one can give definitions of the tangent cones and the derivatives of multivalued mappings with the help of the distance function. In particular, we get

$$\hat{D}_L F(z; \bar{x}) = \{\bar{y} \in Y \mid (\bar{x}, \bar{y}) \in \hat{T}^L_{\mathrm{gr}\,F}(z)\},$$
$$\hat{D}_U F(z; \bar{x}) = \{\bar{y} \in Y \mid (\bar{x}, \bar{y}) \in \hat{T}^U_{\mathrm{gr}\,F}(z)\}, \tag{3.13}$$

where

$$\hat{T}^L_{\mathrm{gr}\,F}(z) \stackrel{\triangle}{=} \{\bar{z} \in Z \mid \limsup_{\varepsilon \downarrow 0} \varepsilon^{-1} d_F(z + \varepsilon \bar{z}) = 0\},$$
$$\hat{T}^U_{\mathrm{gr}\,F}(z) \stackrel{\triangle}{=} \{\bar{z} \in Z \mid \liminf_{\varepsilon \downarrow 0} \varepsilon^{-1} d_F(z + \varepsilon \bar{z}) = 0\}.$$

Analogously

$$T^L_{\mathrm{gr}\,F}(z) \triangleq \{\bar z \in Z \mid \limsup_{\varepsilon \downarrow 0} \varepsilon^{-1}\rho(z + \varepsilon\bar z, \mathrm{gr}\,F) = 0\},$$

$$T^U_{\mathrm{gr}\,F}(z) \triangleq \{\bar z \in Z \mid \liminf_{\varepsilon \downarrow 0} \varepsilon^{-1}\rho(z + \varepsilon\bar z, \mathrm{gr}\,F) = 0\}.$$

The derivative $\hat D_L F(z; \bar x)$ is also called the *set of tangential (feasible) directions*.

LEMMA 3.49 *Let $z_0 \in \mathrm{gr}\,F$. Then the following statements are equivalent:*

1. $\bar y \in \hat D_L F(z_0; \bar x)$;

2. $D^+ d_F(z_0; \bar z) = \limsup\limits_{\varepsilon \downarrow 0} \varepsilon^{-1}[d_F(z_0 + \varepsilon\bar z) - d_F(z_0)] = 0$;

3. there exists a function $o(\varepsilon)$ such that $y_0 + \varepsilon\bar y + o(\varepsilon) \in F(x_0 + \varepsilon\bar x)$ for $\varepsilon \geq 0$ and $o(\varepsilon)/\varepsilon \to 0$ as $\varepsilon \downarrow 0$;

4. for any sequence $\varepsilon_k \downarrow 0$ one can find a sequence $\bar y_k \to \bar y$ such that $y_0 + \varepsilon_k\bar y_k \in F(x_0 + \varepsilon_k\bar x)$, $k = 1, 2, \ldots$

Proof. Due to the equivalence of (3.12), (3.13) and the fact that $d_F(z_0) = 0$, the statements 1 and 2 are equivalent.

Let us now show that statements 2 and 3 are equivalent. In fact, condition 2 means that $d'_F(z_0; \bar z) = 0$, i.e. $d_F(z_0 + \varepsilon\bar z) = d_F(z_0) + o(\varepsilon) = o(\varepsilon)$, which implies 3.

Furthermore, condition 4 is an obvious consequence of condition 3. Finally, from 4. we get $d_F(z_0 + \varepsilon_k\bar z) = o(\varepsilon_k)$ for any sequence $\varepsilon_k \downarrow 0$, i.e. $d'_F(z_0; \bar z) = 0$. Hence statement 2 is valid. ∎

In a quite similar way the following lemma can be proved.

LEMMA 3.50 *Let $z_0 \in \mathrm{gr}\,F$. Then the following statements are equivalent:*

1. $\bar y \in \hat D_U F(z_0; \bar x)$;

2. $D_+ d_F(z_0; \bar z) = \liminf\limits_{\varepsilon \downarrow 0} \varepsilon^{-1}[d_F(z_0 + \varepsilon\bar z) - d_F(z_0)] = 0$;

3. there exist sequences $\varepsilon_k \downarrow 0$ and $\bar y_k \to \bar y$ such that $y_0 + \varepsilon_k\bar y_k \in F(x_0 + \varepsilon_k\bar x)$, $k = 1, 2, \ldots$

2.2   Description of Derivatives of Multivalued Mappings in Terms of the Distance Function

Let $X = R^n$, $Y = R^m$ and let $F : X \to 2^Y$ be a closed-valued mapping. We consider the lower and upper derivatives

$$D^+ d_F(z_0; \bar z) \triangleq \limsup_{\varepsilon \downarrow 0} \varepsilon^{-1}[d_F(z_0 + \varepsilon\bar z) - d_F(z_0)],$$

$$D_+ d_F(z_0; \bar z) \triangleq \liminf_{\varepsilon \downarrow 0} \varepsilon^{-1}[d_F(z_0 + \varepsilon\bar z) - d_F(z_0)]$$

of the function d_F at the point $z_0 \in \operatorname{gr} F$ in the direction $\bar{z} = (\bar{x}, \bar{y}) \in Z = X \times Y$.

LEMMA 3.51 *Let $\hat{D}_L F(z_0; \bar{x}) \neq \emptyset$. Then*

$$D_+ d_F(z_0; \bar{z}) = \rho(\bar{y}, \hat{D}_U F(z_0; \bar{x})).$$

Proof. Let $\varepsilon_k \downarrow 0$ be a sequence on which $D_+ d_F(z_0; \bar{z})$ is attained. For simplicity, we denote $y_k = y_0 + \varepsilon_k \bar{y}$, $x_k = x_0 + \varepsilon_k \bar{x}$, $z_k = (x_k, y_k)$. Let us fix an arbitrary vector $\bar{y}^* \in \hat{D}_L F(z_0; \bar{x})$. Then $y_0 + \varepsilon_k \bar{y}^* + o(\varepsilon_k) \in F(x_k)$, and, consequently,

$$d_F(z_k) = \rho(y_k, F(x_k)) \leq |y_0 + \varepsilon_k \bar{y}^* + o(\varepsilon_k) - y_k| \leq \varepsilon_k |\bar{y}^* - \bar{y}| + |o(\varepsilon_k)| \leq l \cdot \varepsilon_k,$$

where $l = \operatorname{const} > 0$.

Denote by v_k a point in $F(x_k)$ nearest to y_k. Then $\xi_k \overset{\triangle}{=} \varepsilon_k^{-1}(y_k - v_k)$ satisfies the inequality $|\xi_k| \leq l$, $k = 1, 2, \ldots$, i.e. the sequence $\{\xi_k\}$ is bounded and, without loss of generality, we can consider it to be convergent: $\xi_k \to \xi$. Comparing the equalities $y_k = y_0 + \varepsilon_k \bar{y}$ and $y_k = v_k + \varepsilon_k \xi_k$, we get

$$v_k = y_0 + \varepsilon_k(\bar{y} - \xi_k) \in F(x_k), \quad k = 1, 2, \ldots$$

Then due to Lemma 3.50, $\bar{y} - \xi \in \hat{D}_U F(z_0; \bar{x})$, i.e. $\xi = \bar{y} - \bar{y}_0$, where $\bar{y}_0 \in \hat{D}_U F(z_0; \bar{x})$. Thus $d_F(z_0 + \varepsilon_k \bar{z}) - d_F(z_0) = |y_k - v_k| = \varepsilon_k |\xi_k|$ and, therefore, $D_+ d_F(z_0; \bar{z}) = |\xi| = |\bar{y} - \bar{y}_0|$, i.e.

$$D_+ d_F(z_0; \bar{z}) \geq \rho(\bar{y}, \hat{D}_U F(z_0; \bar{x})). \tag{3.14}$$

Assume now $D_+ d_F(z_0; \bar{z}) > \rho(\bar{y}, \hat{D}_U F(z_0; \bar{x}))$. Let $\rho(\bar{y}, \hat{D}_U F(z_0; \bar{x})) = |\bar{y} - \hat{y}|$, where $\hat{y} \in \hat{D}_U F(z_0; \bar{x})$. Then we can find a sequence $\varepsilon_k \downarrow 0$ such that $v_k = y_0 + \varepsilon_k \hat{y} + o(\varepsilon_k) \in F(x_0 + \varepsilon_k \bar{x})$ for all $k = 1, 2, \ldots$ Denoting as above $y_k = y_0 + \varepsilon_k \bar{y}$, $x_k = x_0 + \varepsilon_k \bar{x}$, we can write

$$d_F(z_0 + \varepsilon_k \bar{z}) - d_F(z_0) = \rho(y_k, F(x_k)) \leq |v_k - y_k| \leq \varepsilon_k |\bar{y} - \hat{y}| + o(\varepsilon_k)$$

for $k = 1, 2, \ldots$ According to (3.14), from this we get

$$\lim_{\varepsilon_k \downarrow 0} \varepsilon_k^{-1}[d_F(z_0 + \varepsilon_k \bar{z}) - d_F(z_0)] \leq |\bar{y} - \hat{y}| < D_+ d_F(z_0; \bar{z}),$$

which contradicts the definition of $D_+ d_F(z_0; \bar{z})$. Therefore, the statement of the lemma is valid. ∎

COROLLARY 3.52 *If the function d_F is Lipschitz continuous in a neighbourhood of the point z_0, then*

$$D_+ d_F(z_0; \bar{z}) = \rho(\bar{y}, \hat{D}_U F(z_0; \bar{x})).$$

Proof. It is sufficient to repeat the proof of Lemma 3.51, assuming the vector $\bar{y}^*$ to be an element of the set $\limsup\limits_{k\to\infty} \varepsilon_k^{-1}[F(x_0 + \varepsilon_k\bar{x}) - y_0]$. ∎

COROLLARY 3.53 *Let C be a closed set in Y and $y \in C$. Then*

$$D_+\rho_C(y; \bar{y}) = \rho(\bar{y}, T_C^U(y)).$$

Proof. Let the set $F(x)$ be equal to C in Corollary 3.52. Then we have $\hat{D}_U F(z; \bar{x}) = T_C^U(y)$ for any $\bar{x} \in X$. On the other hand, the function $d_F(z) = \rho_C(y)$ is Lipschitz continuous and, due to Corollary 3.52, we get

$$D_+\rho_C(y; \bar{y}) = \rho(\bar{y}, T_C^U(y)). \quad ∎$$

REMARK. It is possible to show that $\hat{D}_U F(z_0; \bar{x}) = \emptyset$ if and only if $d_F'(z_0; \bar{z}) = +\infty$. Moreover, irrespective of the condition $\hat{D}_L F(z_0; \bar{x}) \neq \emptyset$, one always has

$$D^+ d_F(z_0; \bar{z}) \geq \rho(\bar{y}, \hat{D}_U F(z_0; \bar{x})).$$

LEMMA 3.54 $D^+ d_F(z_0; \bar{z}) \leq \rho(\bar{y}, \hat{D}_L F(z_0; \bar{x}))$.

Proof. If $\hat{D}_L F(z_0; \bar{x}) = \emptyset$, then $\rho(\bar{y}, \hat{D}_L F(z_0; \bar{x})) = +\infty$ and the inequality to be proved is valid.

Let $\hat{y}$ be an arbitrary vector from $\hat{D}_L F(z_0; \bar{x})$. Then according to Lemma 3.49, $y_0 + \varepsilon\hat{y} + o(\varepsilon) \in F(x_0 + \varepsilon\bar{x})$ for $\varepsilon \geq 0$, where $o(\varepsilon)/\varepsilon \to 0$ if $\varepsilon \downarrow 0$. Hence

$$d_F(z_0 + \varepsilon\bar{z}) - d_F(z_0) = \rho(y_0 + \varepsilon\bar{y}, F(x_0 + \varepsilon\bar{x}))$$
$$\leq |y_0 + \varepsilon\hat{y} + o(\varepsilon) - y_0 - \varepsilon\bar{y}| \leq \varepsilon|\bar{y} - \hat{y}| + |o(\varepsilon)|,$$

i.e., for all $\hat{y} \in \hat{D}_L F(z_0; \bar{x})$ the inequality $D^+ d_F(z_0; \bar{z}) \leq |\bar{y} - \hat{y}|$ holds, which is equivalent to the statement of the lemma. ∎

DEFINITION 3.55 *A multivalued mapping F is called differentiable at the point $z_0 = (x_0, y_0) \in \operatorname{gr} F$ in the direction $\bar{x} \in X$ if $\hat{D}_L F(z_0; \bar{x}) = \hat{D}_U F(z_0; \bar{x})$. In this case the common value $\hat{D}_L F(z_0; \bar{x}) = \hat{D}_U F(z_0; \bar{x})$ is denoted by $\hat{D}F(z_0; \bar{x})$, i.e. $\hat{D}F(z_0; \bar{x}) = \hat{D}_L F(z_0; \bar{x}) = \hat{D}_U F(z_0; \bar{x})$.*

THEOREM 3.56 *Let $z_0 = (x_0, y_0) \in \operatorname{gr} F$. Necessary and sufficient for the function d_F to be differentiable at the point z_0 in the direction $\bar{z} = (\bar{x}, \bar{y})$, where $\bar{y}$ is an arbitrary vector from Y, is the differentiability of the mapping F at the point z_0 in the direction $\bar{x}$, where*

$$d_F'(z_0; \bar{z}) = \rho(\bar{y}, \hat{D}_L F(z_0; \bar{x})). \tag{3.15}$$

Proof. The sufficiency follows directly from Lemma 3.51. Let us prove the necessity. Let $d'_F(z_0; \bar{z})$ exist for all $\bar{y} \in Y$. Suppose $\hat{D}_U F(z_0; \bar{x}) \neq \emptyset$ and take some element $\bar{y} \in \hat{D}_U F(z_0; \bar{x})$. Then there exist $\varepsilon_k \downarrow 0$ and $o(\varepsilon_k)$ such that $o(\varepsilon_k)/\varepsilon_k \to 0$ as $k \to \infty$ and

$$y_0 + \varepsilon_k \bar{y} + o(\varepsilon_k) \in F(x_0 + \varepsilon_k \bar{x}), \quad k = 1, 2, \ldots$$

From this we conclude $d_F(z_0 + \varepsilon_k \bar{z}) = \tilde{o}(\varepsilon_k)$, where $\tilde{o}(\varepsilon_k)/\varepsilon_k \to 0$ for $k \to \infty$. Hence, according to Lemma 3.49, $d'_F(z_0; \bar{z}) = 0$, i.e. $\bar{y} \in \hat{D}_L F(z_0; \bar{x})$. Since $\bar{y}$ is an arbitrary element from $\hat{D}_U F(z_0; \bar{x})$, one gets

$$\hat{D}_U F(z_0; \bar{x}) \subset \hat{D}_L F(z_0; \bar{x}).$$

Taking into account the inverse inclusion which always holds, we get that $\hat{D}_U F(z_0; \bar{x}) = \hat{D}_L F(z_0; \bar{x})$. Hence, by Lemma 3.51 relation (3.15) follows.

It remains to note that if $\hat{D}_U F(z_0; \bar{x})$ is empty, then $\hat{D}_L F(z_0; \bar{x})$ is empty as well and, according to Remark 2.2, relation (3.15) is also valid. ∎

The obtained results allow us quite easily to prove a well-known statement concerning the connection between $T_E^C(z)$ and $T_E^U(z)$.

LEMMA 3.57 *Let E be a closed set in Z, $z \in E$. Then*

$$\liminf_{z' \xrightarrow{E} z} T_E^U(z') = T_E^C(z).$$

Proof. Due to Lemma 3.48 the inclusion $\bar{z} \in T_E^C(z)$ is equivalent to the relation $\lim_{z' \xrightarrow{E} z} D_+ \rho_E(z'; \bar{z}) = 0$ or, what is the same, $\limsup_{z' \xrightarrow{E} z} D_+ \rho_E(z'; \bar{z}) = 0$. According to Corollary 3.53 this means $0 = \limsup_{z' \xrightarrow{E} z} \rho(z', T_E^U(z'))$. In this way, by the definition of the lower topological limit we obtain the relation $\bar{z} \in \liminf_{z' \xrightarrow{E} z} T_E^U(z')$. ∎

LEMMA 3.58 *Let $F : X \to 2^Y$, $z \in \mathrm{gr}\, F$ and $\bar{z} \in Z$. Then $d_F^\uparrow(z; \bar{z}) \leq \rho(\bar{y}, D_C F(z; \bar{x}))$.*

Proof. If $D_C F(z; \bar{x}) = \emptyset$, then $\rho(\bar{y}, D_C F(z; \bar{x})) = +\infty$ and the statement of the lemma is valid. Let $D_C F(z; \bar{x}) \neq \emptyset$ and

$$r_\delta(\tilde{z}, \varepsilon) \stackrel{\triangle}{=} \inf\{\varepsilon^{-1}[d_F(\tilde{z} + \varepsilon \hat{z}) - d_F(\tilde{z})] \mid |\hat{z} - \bar{z}| \leq \delta\}.$$

Then the limit $\limsup_{\tilde{z} \xrightarrow{d_F} z, \varepsilon \downarrow 0} r_\delta(\tilde{z}, \varepsilon)$ is attained on the sequence $z_k = (x_k, y_k) \to z$, $\varepsilon_k \downarrow 0$ depending on the value δ and such that $d_F(z_k) \to d_F(z)$. By y_{0k}

we denote a point from $F(x_k)$ such that $|y_k - y_{0k}| \leq \rho(y_k, F(x_k)) + \varepsilon_k^2$. Since $(x_k, y_{0k}) \in \operatorname{gr} F$, then for any $\tilde{y} \in D_C F(z; \bar{x})$ there exist sequences $\tilde{y}_k \to \tilde{y}$, $\tilde{x}_k \to \tilde{x}$ for which $y_{0k} + \varepsilon_k \tilde{y}_k \in F(x_k + \varepsilon_k \tilde{x}_k)$, $k = 1, 2, \ldots$ Hence, for all k beginning with some $k = k_0$, we get

$$r_\delta(z_k, \varepsilon_k) \leq d_F(z_k + \varepsilon_k(\tilde{x}_k, \bar{y})) - d_F(z_k)$$
$$\leq |y_k + \varepsilon_k \bar{y} - (y_{0k} + \varepsilon_k \tilde{y}_k)| - d_F(z_k)$$
$$= |y_{0k} + \varepsilon_k \bar{y} - (y_{0k} + \varepsilon_k \tilde{y}_k) + (y_k - y_{0k})| - d_F(z_k)$$
$$\leq \varepsilon_k |\bar{y} - \tilde{y}_k| + |y_k - y_{0k}| - d_F(z_k) = \varepsilon_k |\bar{y} - \tilde{y}_k| + \varepsilon_k^2.$$

From this inequality it immediately follows that $d_F^\uparrow(z; \bar{z}) \leq |\bar{y} - \tilde{y}|$ for all $\tilde{y} \in D_C F(z; \bar{x})$. This is equivalent to the statement of the lemma. ∎

2.3   First-order Approximations of Multivalued Mappings

For studying differential properties of multivalued mappings in [59]–[61] Demyanov and Rubinov introduced the concept of first-order approximation to a mapping. We consider this concept in connection with derivatives of distance functions.

DEFINITION 3.59 *The multivalued mapping F is said to have a first-order approximation at the point $z_0 = (x_0, y_0) \in \operatorname{gr} F$ in the direction $\bar{x} \in X$ if for any sequence $\{y_k\}$ such that $y_k \in F(x_0 + \varepsilon_k \bar{x})$, $k = 1, 2, \ldots$, $\varepsilon_k \downarrow 0$, $y_k \to y_0 \in F(x_0)$ if $k \to \infty$ the representation*

$$y_k = y_0 + \varepsilon_k \hat{y}_k + o(\varepsilon_k) \tag{3.16}$$

holds, where $\hat{y}_k \in \hat{D}_L F(z_0; \bar{x})$, $\varepsilon_k \hat{y}_k \to 0$ as $k \to \infty$.

Let us show that the mapping F having a first-order approximation at the point $z_0 \in \operatorname{gr} F$ in the direction $\bar{x}$ is differentiable at the point z_0 in the direction $\bar{x}$.

In fact, from the existence of a first-order approximation it follows that $\hat{D}_L F(z_0; \bar{x}) \neq \emptyset$. Let $\bar{y}$ be an arbitrary element from the set $\hat{D}_U F(z_0; \bar{x})$. Then we define

$$y_k \overset{\triangle}{=} y_0 + \varepsilon_k \bar{y} + o(\varepsilon_k) \in F(x_0 + \varepsilon_k \bar{x}), \quad k = 1, 2, \ldots,$$

where $\varepsilon_k \downarrow 0$, $o(\varepsilon_k)/\varepsilon_k \to 0$ for $k \to \infty$. According to relation (3.16), $y_k = y_0 + \varepsilon_k \hat{y}_k + o(\varepsilon_k)$ with $\hat{y}_k \in \hat{D}_L F(z_0; \bar{x})$. Hence, $\hat{y}_k \to \bar{y}$ and due to the closedness of $\hat{D}_L F(z_0; \bar{x})$ we conclude $\bar{y} \in \hat{D}_L F(z_0; \bar{x})$. In this way $\hat{D}_U F(z_0; \bar{x}) \subset \hat{D}_L F(z_0; \bar{x})$, which means that the equality $\hat{D}_U F(z_0; \bar{x}) = \hat{D}_L F(z_0; \bar{x})$ is true.

Let $\bar{x} \in X$. We choose arbitrary sequences $\{\varepsilon_k\}$ and $\{y_k\}$ such that $\varepsilon_k \downarrow 0$, $y_k \in F(x_0 + \varepsilon_k \bar{x})$, $k = 1, 2, \ldots$ and $y_k \to y_0 \in F(x_0)$. Furthermore, we denote $x_k = x_0 + \varepsilon_k \bar{x}$, $z_k = (x_k, y_k)$, $\bar{z}_k = \varepsilon_k^{-1}(z_k - z_0)$, where $z_0 = (x_0, y_0)$. From Theorem 3.56 the following lemma can be immediately derived.

LEMMA 3.60 *The following statements are equivalent:*

1. the mapping F has a first-order approximation at the point $z_0 \in$ gr F in the direction $\bar{x}$;

2. $d'_F(z_0; \bar{z}_k) \to 0$ for any sequence $z_k \xrightarrow{\text{gr } F} z_0 \in$ gr F;

3. for any sequence $z_k \xrightarrow{\text{gr } F} z_0 \in$ gr F the following inequality is valid:

$$d_F(z_k) - d_F(z_0) \geq \varepsilon_k d'_F(z_0; \bar{z}_k) + o(\varepsilon_k).$$

2.4 Some Properties of Derivatives of Multivalued Mappings

Let $X = R^n$, $Y = R^m$ and let $F : X \to 2^Y$ be a closed mapping.

LEMMA 3.61 *Let $z_0 = (x_0, y_0) \in$ gr F and the mapping F be convex-valued. Then $\hat{D}_L F(z_0; \bar{x})$ is a convex set for all $\bar{x} \in X$.*

Proof. If $\hat{D}_L F(z_0; \bar{x}) = \emptyset$, then this set is convex by definition. Assume $\hat{D}_L F(z_0; \bar{x}) \neq \emptyset$ and let $\bar{y}_1, \bar{y}_2 \in \hat{D}_L F(z_0; \bar{x})$. Then there exist vector functions $o_i(\varepsilon)$, $i = 1, 2$, such that $o_i(\varepsilon)/\varepsilon \to 0$ if $\varepsilon \downarrow 0$ and

$$y_0 + \varepsilon \bar{y}_1 + o_1(\varepsilon) \in F(x_0 + \varepsilon \bar{x}), \quad \varepsilon \geq 0,$$
$$y_0 + \varepsilon \bar{y}_2 + o_2(\varepsilon) \in F(x_0 + \varepsilon \bar{x}), \quad \varepsilon \geq 0.$$

Multiplying the first inclusion by $\lambda \in [0, 1]$ and the second by $(1 - \lambda)$ and adding both, we get

$$y_0 + \varepsilon(\lambda \bar{y}_1 + (1 - \lambda)\bar{y}_2) + o(\varepsilon) \in \lambda F(x_0 + \varepsilon \bar{x}) + (1 - \lambda)F(x_0 + \varepsilon \bar{x}),$$

from which, in view of the convexity of $F(x_0 + \varepsilon \bar{x})$ it follows that

$$y_0 + \varepsilon(\lambda \bar{y}_1 + (1 - \lambda)\bar{y}_2) + o(\varepsilon) \in F(x_0 + \varepsilon \bar{x}), \quad \varepsilon \geq 0.$$

(Here $o(\varepsilon) = \lambda o_1(\varepsilon) + (1 - \lambda)o_2(\varepsilon)$, $o(\varepsilon)/\varepsilon \to 0$ as $\varepsilon \downarrow 0$.) Thus

$$\lambda \bar{y}_1 + (1 - \lambda)\bar{y}_2 \in \hat{D}_L F(z_0; \bar{x}), \quad \lambda \in [0, 1],$$

which means the convexity of $\hat{D}_L F(z_0; \bar{x})$. ∎

Let us denote by $f_F(x, y)$ the set of points in $F(x)$ nearest to $y \in Y$. It is obvious that if F is a mapping with non-empty and closed values, then $f_F : X \times Y \to 2^Y$ is a multivalued mapping defined on the set $Z = X \times Y$. Assuming, in addition, the convex-valuedness of F, f_F will be a singlevalued function from Z to Y. Note that in any case $f_F(x, y) = y$ whenever $y \in F(x)$.

For the mapping f_F we consider the upper and lower derivatives

$$\hat{D}_L f_F(z_0; \bar{z}) = \liminf_{\varepsilon \downarrow 0} \varepsilon^{-1}[f_F(z_0 + \varepsilon \bar{z}) - f_F(z_0)],$$

$$\hat{D}_U f_F(z_0; \bar{z}) = \limsup_{\varepsilon \downarrow 0} \varepsilon^{-1}[f_F(z_0 + \varepsilon \bar{z}) - f_F(z_0)]$$

at the point $z_0 \in \operatorname{gr} F$ in the direction $\bar{z} = (\bar{x}, \bar{y})$. If these derivatives coincide, their common value is called the *derivative in the direction $\bar{z}$* and is denoted by $\hat{D} f_F(z_0; \bar{z})$. If this set $\hat{D} f_F(z_0; \bar{z})$ is singlevalued, then as a special notation we use

$$f'_F(z_0; \bar{z}) = \hat{D} f_F(z_0; \bar{z}).$$

LEMMA 3.62 *Let the closed mapping F be pseudolipschitz continuous and differentiable in the direction $\bar{x}$ at the point $z_0 = (x_0, y_0) \in \operatorname{gr} F$. If the set $\hat{D} F(z_0; \bar{x})$ is non-empty and convex, then for every $\bar{y} \in Y$ there exists the derivative $f'_F(z_0; \bar{z})$.*

Proof. For any point $\bar{y} \in Y$ and any $v \in f_F(x_0 + \varepsilon \bar{x}, y_0 + \varepsilon \bar{y})$ the relation $d_F(x_0 + \varepsilon \bar{x}, y_0 + \varepsilon \bar{y}) = |v - y_0 - \varepsilon \bar{y}|$ holds, so that, due to Lemma 3.26 and Remark 1.2, one gets

$$\left| \frac{1}{\varepsilon}(v - y_0) - \bar{y} \right| = \varepsilon^{-1}[d_F(z_0 + \varepsilon \bar{z}) - d_F(z_0)] \leq l|\bar{x}| + |\bar{y}| \qquad (3.17)$$

for $\varepsilon \in (0, \varepsilon_0]$, where ε_0 is some positive number sufficiently small. Consequently, the expression $\varepsilon^{-1}|v - y_0|$ is bounded. Let us denote by $W^*(z_0; \bar{z})$ the set of its partial limits under $\varepsilon \downarrow 0$. It is quite obvious that $W^*(z_0; \bar{z}) \subset \hat{D}_U F(z_0; \bar{x}) = \hat{D} F(z_0; \bar{x})$. Using Theorem 3.56, the passage to the limit in (3.17) yields

$$|w^* - \bar{y}| = d'_F(z_0; \bar{z}) = \rho(\bar{y}, \hat{D} F(z_0; \bar{x}))$$

for all $w^* \in W^*(z_0; \bar{z})$. Due to the convexity of $\hat{D} F(z_0; \bar{x})$ it can be easily seen that $W^*(z_0; \bar{z}) = \{w^*\}$, i.e. $\hat{D}_U f_F(z_0; \bar{z}) = w^* - \bar{y}$, where w^* is a point in $\hat{D} F(z_0; \bar{x})$ closest to $\bar{y}$. It is easy to see that in this case $\hat{D}_L f_F(z_0; \bar{z}) = \hat{D}_U f_F(z_0; \bar{z})$, i.e. the derivative $f'_F(z_0; \bar{z}) = w^* - \bar{y}$ exists. ∎

COROLLARY 3.63 *The value of $f'_F(z_0; \bar{z})$ in Lemma 3.62 coincides with $w^* - \bar{y}$, where w^* is a point in $\hat{D} F(z_0; \bar{x})$ closest to $\bar{y}$.*

COROLLARY 3.64 *If the mapping F is pseudolipschitz continuous at the point $z_0 = (x_0, y_0) \in \operatorname{gr} F$, then $\hat{D}_U F(z_0; \bar{x}) \neq \emptyset$ for all $\bar{x} \in X$.*

Proof. Indeed, the non-emptiness of $W^*(z_0; \bar{z}) \subset \hat{D}_U F(z_0; \bar{x})$ follows from the pseudolipschitz continuity of F.

LEMMA 3.65 *Let the closed-valued mapping F be pseudolipschitz continuous at the point $z_0 = (x_0, y_0) \in \operatorname{gr} F$ (with Lipschitz constant $l > 0$). Then the mapping $\hat{D}_U F(z_0; \cdot)$ is Lipschitz continuous on the set X (with the same constant l).*

Proof. Let us take arbitrary $\bar{x}_1, \bar{x}_2 \in X$. Then due to Corollary 3.64, we have $\hat{D}_U F(z_0; \bar{x}_1) \neq \emptyset$. Let $\bar{y}_1 \in \hat{D}_U F(z_0; \bar{x}_1)$. By definition of the upper derivative one can find a sequence $\{\varepsilon_k\}$, $\varepsilon_k \downarrow 0$ such that

$$y_k^0 \overset{\triangle}{=} y_0 + \varepsilon_k \bar{y}_1 + o(\varepsilon_k) \in F(x_0 + \varepsilon_k \bar{x}_1)$$

for all $k = 1, 2, \ldots$ Denote by y_k the projection of y_k^0 on the set $F(x_0 + \varepsilon_k \bar{x}_2)$. Then in view of the pseudolipschitz continuity of F, without loss of generality, we can assume that for all $k = 1, 2, \ldots$ the inequality

$$|y_k - y_k^0| \leq l\varepsilon_k |\bar{x}_2 - \bar{x}_1|, \tag{3.18}$$

is true, i.e., the sequence $\{\varepsilon_k^{-1}(y_k - y_k^0)\}$ is bounded and converges to some vector ξ. This means that $y_k = y_k^0 + \varepsilon_k \xi + o(\varepsilon_k)$, $k = 1, 2, \ldots$, and, hence,

$$y_k = y_0 + \varepsilon_k(\xi + \bar{y}_1) + o(\varepsilon_k) \in F(x_0 + \varepsilon_k \bar{x}_2), \quad k = 1, 2, \ldots$$

From this it immediately follows that $w \overset{\triangle}{=} \xi + \bar{y}_1 \in \hat{D}_U F(z_0; \bar{x}_2)$. Owing to relation (3.18), we then have $|w - \bar{y}_1| = |\xi| \leq l|\bar{x}_2 - \bar{x}_1|$, which means $\rho(\bar{y}_1, \hat{D}_U F(z_0; \bar{x}_2)) \leq |\bar{x}_2 - \bar{x}_1|$ for any $\bar{y}_1 \in \hat{D}_U F(z_0; \bar{x}_1)$. Therefore, for all $\bar{x}_1, \bar{x}_2$, we obtain the inclusion

$$\hat{D}_U F(z_0; \bar{x}_1) \subset \hat{D}_U F(z_0; \bar{x}_2) + l|\bar{x}_2 - \bar{x}_1| B. \quad \blacksquare$$

For a convex set C there exist several equivalent definitions of the recession cone $0^+ C$. In the case when C is not convex, these definitions fail to be equal. In particular, for the set $\hat{D}_U F(z_0; \bar{x})$ we can establish two forms of the recession cone:

$$0_p^+ \hat{D}_U F(z_0; \bar{x}) \overset{\triangle}{=} \limsup_{\lambda \downarrow 0} \lambda \hat{D}_U F(z_0; \bar{x})$$

and

$$0_a^+ \hat{D}_U F(z_0; \bar{x}) \overset{\triangle}{=} \{\bar{y} \mid \hat{D}_U F(z_0; \bar{x}) + \bar{y} \subset \hat{D}_U F(z_0; \bar{x})\}.$$

They coincide if $\hat{D}_U F(z_0; \bar{x})$ is convex. From Lemma 3.65 the following statement immediately follows.

COROLLARY 3.66 *Let the mapping F be pseudolipschitz continuous at the point $z_0 = (x_0, y_0) \in \operatorname{gr} F$. Then for any $\bar{x} \in X$*

$$0_p^+ \hat{D}_U F(z_0; \bar{x}) = \hat{D}_U F(z_0; 0).$$

Proof. According to the assumption of pseudolipschitz continuity of F, the set $\hat{D}_U F(z_0; \bar{x})$ is non-empty. Moreover, since the graph of the mapping $\hat{D}_U F(z_0; \cdot)$ is a cone, then for $\lambda > 0$ one has

$$\lambda \hat{D}_U F(z_0; \bar{x}) = \hat{D}_U F(z_0; \lambda \bar{x}).$$

Due to the Lipschitz continuity, which implies continuity of $\hat{D}_U F(z_0; \cdot)$, the passage to the limit in this equality yields

$$0_p^+ \hat{D}_U F(z_0; \bar{x}) = \limsup_{\lambda \downarrow 0} \hat{D}_U F(z_0; \lambda \bar{x})$$
$$= \lim_{\lambda \downarrow 0} \hat{D}_U F(z_0; \lambda \bar{x}) = \hat{D}_U F(z_0; \bar{x}). \quad \blacksquare$$

LEMMA 3.67 *Let the assumptions of Lemma 3.62 be satisfied. Then*

$$0^+ \hat{D} F(z_0; \bar{x}) = \hat{D} F(z_0; 0).$$

Proof. This statement is a direct consequence of Corollary 3.66 and the convexity of $\hat{D} F(z_0; \bar{x})$. $\blacksquare$

Finally, we would like to emphasize that if the mapping F is pseudolipschitz continuous at the point $z_0 \in \operatorname{gr} F$, then the derivatives $\hat{D}_U F(z_0; \bar{x})$, $\hat{D}_L F(z_0; \bar{x})$ and $\hat{D} F(z_0; \bar{x})$ coincide with the derivatives in the sense of Hadamard $D_U F(z_0; \bar{x})$, $D_L F(z_0; \bar{x})$ and $D F(z_0; \bar{x})$.

3.   Lemma About the Removal of Constraints

Let $X = R^n$, $Y = R^m$. We consider the multivalued mapping $F : X \to 2^Y$ and suppose it to be closed-valued and uniformly bounded at some point $x_0 \in \operatorname{dom} F$. Remind that uniform boundedness of F at a point x_0 means that there exist a neighbourhood X_0 in X of the point x_0 and a bounded set $Y_0 \subset Y$ such that $F(X_0) \subset Y_0$.

Denote as above

$$\varphi(x) = \inf\{f(x, y) \mid y \in F(x)\},$$
$$\omega(x) = \{y \in F(x) \mid f(x, y) = \varphi(x)\},$$

where the function $f : X \times Y \to R$ is Lipschitz continuous on the set $X_0 \times [Y_0 + \varepsilon_0 B]$ with Lipschitz constant $l_0 > 0$. Moreover, let $\varepsilon_0 > 2 \operatorname{diam} Y_0$.

Setting $z = (x, y)$, we introduce the function $L_\beta(z) = f(z) + \beta d_F(z)$, where β is an arbitrary number greater than l_0.

LEMMA 3.68 (**About the removal of constraints**) *For all* $x \in X_0$ *we have*

$$\varphi(x) = \inf\{L_\beta(z) \mid y \in Y_0 + \frac{\varepsilon_0}{2}B\}. \tag{3.19}$$

Proof. If $F(x) = \emptyset$, then the validity of (3.19) is trivial. Let $F(x) \neq \emptyset$. Take any number $\varepsilon \in (0, \varepsilon_0]$ and set

$$\varphi_\varepsilon(x) \overset{\triangle}{=} \inf_y \{f(z) \mid d_F(z) \leq \varepsilon\}.$$

Let $y \in F(x) + \varepsilon B$ and $\tilde{y}(x)$ be a point in $F(x)$ closest to y. Then, according to the Lipschitz continuity of f, we get

$$f(x, y) \geq f(x, \tilde{y}(x)) - l_0|y - \tilde{y}(x)| \geq \varphi(x) - l_0\varepsilon > \varphi(x) - \beta\varepsilon.$$

Consequently,

$$\varphi_\varepsilon(x) - \varphi(x) > -\varepsilon\beta, \quad \varepsilon \in (0, \varepsilon_0].$$

Then for any $y(x) \in \omega(x)$ we obtain

$$L_\beta(x, y(x)) = f(x, y(x)) = \varphi(x) \leq \inf_{\varepsilon \in (0, \varepsilon_0]} [\varphi_\varepsilon(x) + \varepsilon\beta] =$$

$$= \inf_{\varepsilon,\, y}\{f(x, y) + \varepsilon\beta \mid d_F(x, y) \leq \varepsilon \leq \varepsilon_0\} = \inf_{y \in F(x) + \varepsilon_0 B} L_\beta(z) \leq$$

$$\leq \inf_{y \in y(x) + \varepsilon_0 B} L_\beta(z) \leq \inf_{y \in Y_0 + \frac{\varepsilon_0}{2} B} L_\beta(z) \leq \varphi(x). \quad \blacksquare$$

COROLLARY 3.69 *If* $F(x) \neq \emptyset$, *then any value* $y(x) \in \omega(x)$ *yields the minimum in problem (3.19).*

In the case of Lipschitz continuity of f on the set $X_0 \times Y$ the lemma proved above admits a strengthening. Of course, the following statement holds (uniform boundedness is no longer required).

LEMMA 3.70 *Let the function* f *be Lipschitz continuous on the set* $X_0 \times Y$. *Then for all* $x \in X_0$ *the equality (3.19) holds on* Y.

Proof. It suffices to repeat the proof of Lemma 3.68 assuming $\varepsilon_0 = +\infty$. $\blacksquare$

Lemma 3.68 about the removal of constraints allows us to get a variety of statements about semicontinuity of φ and ω. Suppose that the above stated requirements concerning the mapping F and the function f hold.

LEMMA 3.71 *Let $z_0 = (x_0, y_0) \in \mathrm{gr}\, \omega$ and $\hat{D}_L F(z_0; \bar{x}) \neq \emptyset$. Then we have $D^+ \varphi(x_0; \bar{x}) < +\infty$ and the function $\varphi(x_0 + \varepsilon \bar{x})$ is upper semicontinuous with respect to ε at the point $\varepsilon = 0$ (for $\varepsilon \geq 0$).*

Proof. Since due to Lemma 3.68, for any $\bar{y} \in Y$, $y_0 \in \omega(x_0)$ and $\varepsilon \in (0, \varepsilon_0]$, we have

$$\varphi(x_0 + \varepsilon \bar{x}) - \varphi(x_0) \leq L_\beta(z_0 + \varepsilon \bar{z}) - L_\beta(z_0) \leq \varepsilon l_0 |\bar{z}| + \beta d_F(z_0 + \varepsilon \bar{z}),$$

then taking into account Lemma 3.54, we get

$$D^+ \varphi(x_0; \bar{x}) \leq l_0 |\bar{z}| + \beta D^+ d_F(z_0; \bar{z}) \leq l_0 |\bar{z}| + \beta \rho(\bar{y}, D_L F(z_0; \bar{x})) < +\infty.$$

From this relation and the previous inequality it follows that

$$\varphi(x_0 + \varepsilon \bar{x}) \leq \varphi(x_0) + M\varepsilon, \quad \varepsilon \in [0, \varepsilon_0'],$$

where $M = M(\bar{x}) > 0$. Here $0 < \varepsilon_0' \leq \varepsilon_0$ and M does not depend on ε. Thus

$$\limsup_{\varepsilon \downarrow 0} \varphi(x_0 + \varepsilon \bar{x}) \leq \varphi(x_0). \quad \blacksquare$$

LEMMA 3.72 *Let the assumptions of Lemma 3.71 hold and, in addition, the mapping F be upper semicontinuous at the point x_0. Then the multivalued mapping $\omega(x_0 + \varepsilon \bar{x})$ is upper semicontinuous with respect to ε at the point $\varepsilon = 0$ (for $\varepsilon \geq 0$).*

Proof. Let $\varepsilon_k \downarrow 0$, $y_k \in \omega(x_0 + \varepsilon_k \bar{x})$, $k = 1, 2, \ldots$, and $y_k \to y$. Then

$$f(x_0 + \varepsilon_k \bar{x}, y_k) = \varphi(x_0 + \varepsilon_k \bar{x}), \quad y_k \in F(x_0 + \varepsilon_k \bar{x}), \quad k = 1, 2, \ldots$$

Passing to the limit in these relations, according to the upper semicontinuity of F and φ (cf. Lemma 3.71), we get

$$f(x_0, y) \leq \varphi(x_0), \quad y \in F(x_0).$$

This means $y \in \omega(x_0)$. $\blacksquare$

REMARK. In Lemma 3.72 it is sufficient to require the upper semicontinuity of the mapping $F(x_0 + \varepsilon \bar{x})$ at the point $\varepsilon = 0$.

DEFINITION 3.73 *The multivalued mapping F is said to be pseudolipschitz continuous at the point $(x_0, y_0) \in \mathrm{gr}\, F$ in the direction $\bar{x}$ if there exist numbers $l > 0$, $\varepsilon_1 > 0$ and a neighbourhood $V(y_0)$ of the point y_0 such that*

$$F(x_0 + \varepsilon \bar{x}) \cap V(y_0) \subset F(x_0) + \varepsilon l |\bar{x}| B \qquad (3.20)$$

for all $\varepsilon \in [0, \varepsilon_1]$.

It is not hard to see that relation (3.20) is equivalent to the condition

$$d_F(x_0 + \varepsilon \bar{x}, y) - d_F(x_0, y) \geq -\varepsilon l |\bar{x}| \tag{3.21}$$

for all $\varepsilon \in [0, \varepsilon_1]$, $y \in V(y_0)$.

LEMMA 3.74 *Let F be pseudolipschitz continuous in the direction $\bar{x}$ at any point $z_0 = (x_0, y_0)$ such that $y_0 \in \omega(x_0)$. Then $D_+\varphi(x_0; \bar{x}) > -\infty$.*

Proof. Let $D_+\varphi(x_0; \bar{x})$ be attained on some sequence $\varepsilon_k \downarrow 0$. Choose a sequence $\{y_k\}$ such that $y_k \in \omega(x_0 + \varepsilon_k \bar{x})$, $k = 1, 2, \dots$ According to the uniform boundedness of F at x_0, the sequence $\{y_k\}$ is bounded. Therefore, without loss of generality, we can assume it to be convergent, i.e. $y_k \to y_0$. The pseudolipschitz continuity of the mapping F in the direction $\bar{x}$ supposes upper semicontinuity of $F(x_0 + \varepsilon \bar{x})$ at $\varepsilon = 0$. Due to Lemma 3.72 (and according to Remark 3) $y_0 \in \omega(x_0)$. Thus, in view of Lemma 3.68

$$\varphi(x_0 + \varepsilon_k \bar{x}) - \varphi(x_0) = L_\beta(x_0 + \varepsilon_k \bar{x}, y_k) - L_\beta(x_0, y_0)$$
$$\geq L_\beta(x_0 + \varepsilon_k \bar{x}, y_k) - L_\beta(x_0, y_k) \geq -\varepsilon_k l_0 |\bar{x}| - \beta \varepsilon_k l |\bar{x}|,$$

and, hence,
$$D_+\varphi(x_0; \bar{x}) \geq -(l_0 + \beta l)|\bar{x}| > -\infty. \quad \blacksquare$$

COROLLARY 3.75 *Let the assumptions of Lemma 3.74 hold. Then the function φ is locally Lipschitz continuous at the point x_0.*

Chapter 4

SUBDIFFERENTIALS OF MARGINAL FUNCTIONS

In this chapter we consider the marginal function

$$\varphi(x) = \inf \{f(x,y) \mid y \in F(x)\},$$

where $f : X \times Y \to R$, $F : X \to 2^Y$, $X = R^n$, $Y = R^m$. As above we denote

$$\omega(x) = \{y \in F(x) \mid f(x,y) = \varphi(x)\}.$$

Let $x_0 \in \operatorname{dom} F$. The following assumptions are supposed to hold throughout this chapter without special mentioning unless the opposite is said. These assumptions are:

(A1) – the multivalued mapping F is uniformly bounded at the point x_0, i.e., one can find a neighbourhood X_0 of the point x_0 and a bounded set $Y_0 \subset Y$ such that $F(X_0) \subset Y_0$;

(A2) – the multivalued mapping F is closed;

(A3) – the function f is Lipschitz continuous on the set $X_0 \times [Y_0 + \varepsilon_0 B]$ with $\varepsilon_0 > 2 \operatorname{diam} Y_0$, where X_0 and Y_0 satisfy assumption (A1).

1.   Clarke Subdifferentials of Marginal Functions
1.1   Estimates for Subdifferentials

To get estimates for the subdifferentials $\partial^0 \varphi(x_0)$ we need some auxiliary statements.

LEMMA 4.1 *Let the sequences $\{x_k\}$ and $\{y_k\}$ be such that $x_k \to x_0$, $\varphi(x_k) \to \varphi(x_0)$, $y_k \in \omega(x_k)$, $k = 1, 2, \ldots$, and $y_k \to y_0$. Then $y_0 \in \omega(x_0)$.*

Proof. Under the assumptions of the lemma we have $f(x_k, y_k) = \varphi(x_k)$. Passing to the limit, we get $f(x_0, y_0) = \varphi(x_0)$. Since, due to the assumption (A2), $y_0 \in F(x_0)$, then from the last equality we obtain $y_0 \in \omega(x_0)$. ∎

LEMMA 4.2 *The function φ is finite and lower semicontinuous in a neighbourhood X_0 of the point x_0.*

Proof. The validity of this statement follows directly from Lemma 3.18 and the assumption (A1). ∎

Let $g : X \to R \cup \{\pm\infty\}$ be a lower semicontinuous function finite at the point $\hat{x}$.

DEFINITION 4.3 *A vector $x^* \in X$ is called an ε-subgradient of the function g at the point $\hat{x}$ if the function $g(x) - \langle x^*, x \rangle + \varepsilon|x - \hat{x}|$ attains a local minimum at the point $\hat{x}$.*

The set of all ε-subgradients of the function g at a point x will be denoted by $\partial_\varepsilon g(x)$.

Furthermore, let us denote by $\hat{\partial} g(x)$ the set of elements $x^* \in X$ such that for every x^* there exist sequences $\{\varepsilon_k\}$, $\{x_k\}$ and $\{x_k^*\}$ with $\varepsilon_k \downarrow 0$, $x_k \to x$, $g(x_k) \to g(x)$, $x_k^* \to x^*$, where x_k^* is an ε_k-subgradient of the function g at the point x_k. Finally, by $\hat{\partial}^\infty g(x)$ we denote the set of elements $x^* \in X$ such that for every x^* there exist sequences $\{\varepsilon_k\}$, $\{\tau_k\}$, $\{x_k\}$ and $\{x_k^*\}$ with $\varepsilon_k \downarrow 0$, $\tau_k \downarrow 0$, $x_k \to x$, $g(x_k) \to g(x)$, $x_k^* \to x^*$, where x_k^* is an ε_k-subgradient of the function $\tau_k g$ at the point x_k. In doing so, we have

$$\hat{\partial} g(x) = \limsup_{x' \xrightarrow{g} x,\, \varepsilon \downarrow 0} \hat{\partial}_\varepsilon g(x'),$$

$$\hat{\partial}^\infty g(x) = \limsup_{x' \xrightarrow{g} x,\, \varepsilon \downarrow 0,\, \tau \downarrow 0} \tau \cdot \hat{\partial}_\varepsilon g(x').$$

The sets $\hat{\partial} g(x)$ and $\hat{\partial}^\infty g(x)$ are called the *Mordukhovich subdifferential* and the *Mordukhovich singular subdifferential*, respectively.

LEMMA 4.4 *([174]) If the function g is closed and finite at the point x, then the relation $\partial^0 g(x) = \operatorname{cl co} \{\hat{\partial} g(x) + \hat{\partial}^\infty g(x)\}$ holds.*

Let us introduce the sets

$$K(z_0) = \{x^* \in X \mid (x^*, 0) \in \partial^0 f(z_0) + N_F(z_0)\},$$
$$K_\infty(z_0) = \{x^* \in X \mid (x^*, 0) \in N_F(z_0)\},$$

where $N_F(z_0)$ is the Clarke tangent cone to the set $\operatorname{gr} F$ at the point $z_0 = (x_0, y_0) \in \operatorname{gr} F$.

THEOREM 4.5 *The following inclusion is valid:*

$$\partial^0 \varphi(x_0) \subset \mathrm{cl\,co} \left\{ \bigcup_{y_0 \in \omega(x_0)} K(z_0) + \bigcup_{y_0 \in \omega(x_0)} K_\infty(z_0) \right\}. \qquad (4.1)$$

Proof. Due to Lemma 4.4 we have

$$\partial^0 \varphi(x_0) \subset \mathrm{cl\,co} \{ \hat{\partial} \varphi(x_0) + \hat{\partial}^\infty \varphi(x_0) \}. \qquad (4.2)$$

Let $x^* \in \hat{\partial} \varphi(x_0)$. Then by the definition of $\hat{\partial} \varphi(x_0)$ there exist sequences $\{\varepsilon_k\}$, $\{x_k\}$ and $\{x_k^*\}$ such that $\varepsilon_k \downarrow 0$, $x_k \to x_0$, $\varphi(x_k) \to \varphi(x_0)$, $x_k^* \to x^*$, where x_k^* is a ε_k-subgradient of the function φ at the point x_k, i.e., the function $\varphi(x) - \langle x_k^*, x \rangle + \varepsilon_k |x - x_k|$ attains a local minimum at x_k. In view of the assumption (A2) there exists a sequence $\{y_k\}$ such that $y_k \in \omega(x_k) \subset Y_0$. Without loss of generality, we can assume $y_k \to y_0$, where $y_0 \in \omega(x_0)$. In this case $\varphi(x_k) = f(z_k)$, where $z_k = (x_k, y_k)$, and, consequently, the function

$$h_k(z) = f(z) - \langle x_k^*, x \rangle + \varepsilon_k |x - x_k|$$

attains a minimum at the point z_k on the set

$$G_k = \mathrm{gr}\, F \cap (V(x_k) \times V(y_k))$$

with $V(x_k)$ and $V(y_k)$ being some neighbourhoods of x_k and y_k.

Due to the assumption (A3) the function h_k satisfies a Lipschitz condition on some set Z_0 such that $G_k \subset Z_0$ for all $k = 1, 2, \ldots$, where the Lipschitz constant $l = l(x^*)$ depends only on x^* and not on k. Then the point z_k yields a local minimum for the function $h_k + l(x^*)\rho_F$, where $\rho_F(z) = \rho(z, \mathrm{gr}\, F)$. Therefore, due to the necessary optimality condition from Theorem 2.21, we have

$$0 \in \partial^0 (h_k + l(x^*)\rho_F)(z_k),$$

from which it follows that

$$(x^*, 0) \in \partial^0 f(z_k) + l(x^*)\partial^0 \rho_F(z_k) + \varepsilon_k [\bar{B} \times \{0\}],$$

where $\bar{B} = \{ x^* \in X | \; |x^*| \leq 1 \}$.

Passing to the limit in the latter inclusion and taking into account the closedness of the multivalued mappings $\partial^0 f$ and $\partial^0 \rho_F$, we get

$$(x^*, 0) \in \partial^0 f(z_0) + l(x^*)\partial^0 \rho_F(z_0).$$

Since $l(x^*)\partial^0 \rho_F(z_0) \subset N_F(z_0)$, then

$$(x^*, 0) \in \partial^0 f(z_0) + N_F(z_0).$$

Thus, $x^* \in K(z_0)$ for some $y_0 \in \omega(x_0)$ and therefore

$$\hat{\partial}\varphi(x_0) \subset \bigcup_{y_0 \in \omega(x_0)} K(z_0). \qquad (4.3)$$

In the same way, if $x_0^* \in \hat{\partial}^\infty \varphi(x_0)$, then x_0^* is the limit of the sequence $\{x_k^*\}$, where x_k^* is an ε_k-subgradient of the function $\tau_k \varphi$ at the point x_k, i. e., the function

$$\tau_k \varphi(x) - \langle x_k^*, x \rangle + \varepsilon_k |x - x_k|$$

attains a minimum at the point x_k, where $x_k \to x_0$, $\varphi(x_k) \to \varphi(x_0)$, $\varepsilon_k \downarrow 0$, and $\tau_k \downarrow 0$.

Replacing φ by $\tau_k \varphi$ and repeating the proof given above, we get $(x_0^*, 0) \in l_0 \partial^0 \rho_F(z_0)$, and, consequently,

$$\hat{\partial}^\infty \varphi(x_0) \subset \bigcup_{y_0 \in \omega(x_0)} K_\infty(z_0). \qquad (4.4)$$

Thus, from (4.2) and taking into account (4.3) and (4.4), the relationship (4.1) can be concluded. ∎

REMARK. The cone $N_F(z_0)$ in the definitions of the sets $K(z_0)$ and $K_\infty(z_0)$ can be replaced by the cone $R^+ \partial^0 \rho_F(z_0)$, where $R^+ = [0, +\infty]$.

Together with the approximating subdifferentials we shall consider the *Mordukhovich normal cone* to the graph gr F of the multivalued mapping F at the point $z_0 = (x_0, y_0) \in \text{gr } F$ defined as

$$N_F^M(z_0) = \limsup_{z \to z_0} [\text{cone}(z - W_F(z))],$$

where $W_F(z) = \{w \in \text{cl gr } F \mid |z - w| = \rho(z, \text{gr } F)\}$. There exist also another definition of the cone $N_F^M(z_0)$. In particular,

$$N_F^M(z_0) = \limsup_{z \xrightarrow{\text{cl gr } F} z_0} \{-[T_{\text{gr } F}^U(z)]^+\}.$$

We would also like to emphasize the close connection between the cone $N_F^M(z_0)$ and Clarke's normal cone $N_F(z_0)$: $N_F(z_0) = \text{cl co } N_F^M(z_0)$.

Using the Mordukhovich normal cone, it is possible to describe equivalent definitions of the subdifferentials $\hat{\partial}g(x)$ and $\hat{\partial}^\infty g(x)$ for a lower semicontinuous function g:

$$\hat{\partial}g(x) = \left\{ x^* \in X \mid (x^*, -1) \in N_{\text{epi } g}^M(x, g(x)) \right\},$$

$$\hat{\partial}^\infty g(x) = \left\{ x^* \in X \mid (x^*, 0) \in N_{\text{epi } g}^M(x, g(x)) \right\}.$$

The validity of the following assertion follows directly from the proof of Theorem 4.5 replacing the Clarke subdifferentials by the Mordukhovich subdifferentials and using corresponding subdifferential calculus from [132].

COROLLARY 4.6 *The following two inclusions are valid:*

$$\hat{\partial}\varphi(x_0) \subset \bigcup_{y_0 \in \omega(x_0)} \pi_0(\partial f(z_0) + N_F^M(z_0)), \quad \hat{\partial}^\infty \varphi(x_0) \subset \bigcup_{y_0 \in \omega(x_0)} \pi_0 N_F^M(z_0).$$

Here $\pi_0 C = \{x \in X \mid (x, 0) \in C\}$ *for* $C \subset X \times Y$.

Let us consider the set $\text{dom}\, D_C F(z_0; \cdot)$ with $z_0 \in \text{gr}\, F$. It is not hard to see that $\text{dom}\, D_C F(z_0; \cdot)$ is a convex cone in X containing 0. The following lemma is a direct consequence of Lemma 3.38.

LEMMA 4.7 $[\text{dom}\, D_C F(z_0; \cdot)]^+ = -K_\infty(z_0).$

From this lemma one recognizes that $K_\infty(z_0)$ is a convex closed cone in X and
$$\text{cl}\, \text{dom}\, D_C F(z_0; \cdot) = -[K_\infty(z_0)]^+.$$

LEMMA 4.8 *Let* $K(z_0) \neq \emptyset$. *Then* $0^+ K(z_0) = K_\infty(z_0)$.

Proof. According to the definition of $K(z_0)$ we have

$$K(z_0) \times \{0\} = [\partial^0 f(z_0) + N_F(z_0)] \cap (X \times \{0\}).$$

From this the relationship

$$0^+[K(z_0) \times \{0\}] = N_F(z_0) \cap (X \times \{0\}) = K_\infty(z_0) \times \{0\}$$

follows (see Section 1). Therefore $0^+ K(z_0) = K_\infty(z_0)$. ∎

THEOREM 4.9 *Let* $K(z_0) \neq \emptyset$ *for all* $y_0 \in \omega(x_0)$. *Then*

$$\partial\varphi(x_0) \subset \text{cl}\, \text{co} \bigcup_{y_0 \in \omega(x_0)} K(z_0).$$

Proof. We denote

$$H = \text{cl}\, \text{co} \left\{ \bigcup_{y_0 \in \omega(x_0)} K(z_0) + \bigcup_{y_0 \in \omega(x_0)} K_\infty(z_0) \right\},$$

$$H_0 = \text{cl}\, \text{co} \left\{ \bigcup_{y_0 \in \omega(x_0)} K(z_0) \right\}, \quad S = \bigcap_{y_0 \in \omega(x_0)} \text{cl}\, \text{co}\, \text{dom}\, D_C F(z_0; \cdot).$$

Due to the properties of support functions we get

$$\delta^*(\bar{x} \mid H_0) = \sup_{y_0 \in \omega(x_0)} \delta^*(\bar{x} \mid K(z_0)),$$

$$\delta^*(\bar{x} \mid H) = \delta^*(\bar{x} \mid H_0) + \sup_{y_0 \in \omega(x_0)} \delta^*(\bar{x} \mid K_\infty(z_0)).$$

If $\bar{x} \in S$, then according to Lemma 4.2, one has $\delta^*(\bar{x} \mid K_\infty(z_0)) = 0$ for all $y_0 \in \omega(x_0)$ and, therefore, $\delta^*(\bar{x} \mid H) = \delta^*(\bar{x} \mid H_0)$.

Let $\bar{x} \notin S$. Then there exists a point $y_0 \in \omega(x_0)$ for which $\bar{x} \in \mathrm{cl\,dom}\, D_C F(z_0; \cdot)$. According to Lemma 4.7 and taking into account the properties of conjugate cones (see Section 1) we can find an element $x_0^* \in K_\infty(z_0)$ such that $\langle x_0^*, \bar{x} \rangle > 0$. The latter means that $\delta^*(\bar{x} \mid K_\infty(z_0)) = +\infty$ and, consequently, $\delta^*(\bar{x} \mid H) = +\infty$.

On the other hand, $K(z_0) \neq \emptyset$ and, due to Lemma 4.8, for some $x^* \in K(z_0)$ and all $\lambda \geq 0$ the inclusion $x^* + \lambda x_0^* \in K(z_0)$ holds. From this it follows that $\delta^*(\bar{x} \mid K(z_0)) = +\infty$, i.e. $\delta^*(\bar{x} \mid H) = +\infty$. Thus, in any case the equality $\delta^*(\bar{x} \mid H) = \delta^*(\bar{x} \mid H_0)$ is valid. Due to the properties of support functions this means the coincidence of the sets H and H_0. Replacing in Theorem 4.5 and in relation (4.1) H by H_0, we obtain the statement of the theorem. ∎

We study now properties of the function $L_\beta(z) = f(z) + \beta d_F(z)$, $\beta > l$, where l is the Lipschitz constant of the function f on $X_0 \times [Y_0 + \varepsilon_0 B]$.

THEOREM 4.10 *Let the assumptions (A1)–(A3) hold. Then for any $\bar{x} \in X$ the following inequality is valid, where $\bar{z} = (\bar{x}, \bar{y})$:*

$$\varphi^\uparrow(x_0; \bar{x}) \leq \sup_{y_0 \in \omega(x_0)} \inf_{\bar{y}} L_\beta^\uparrow(z_0; \bar{z}).$$

Proof. Let $\bar{x} \in X$. We take an arbitrary $\delta > 0$ and denote

$$r_\delta(x, \varepsilon) = \inf_{|\hat{x} - \bar{x}| \leq \delta} \varepsilon^{-1}[\varphi(x + \varepsilon \hat{x}) - \varphi(x)],$$

$$R_\delta(z, \varepsilon) = \inf_{|\hat{z} - \bar{z}| \leq \frac{\delta}{2}} \varepsilon^{-1}[L_\beta(z + \varepsilon \hat{z}) - L_\beta(z)],$$

where $|z| = \sqrt{|x| + |y|}$. Let the limit $\limsup\limits_{x \xrightarrow{\varphi} x_0, \varepsilon \downarrow 0} r_\delta(x, \varepsilon)$ be attained on certain sequences $x_k = x_k(\delta) \to x_0$ and $\varepsilon_k = \varepsilon_k(\delta) \downarrow 0$ with $\varphi(x_k) \to \varphi(x_0)$. We choose an arbitrary sequence $y_k \in \omega(x_k)$. Without loss of generality, due to (A1) one can assume $y_k \to y_0$, where $y_0 \in \omega(x_0)$ according to Lemma 4.1. We denote $z_0 = (x_0, y_0)$, and $z_k = (x_k, y_k)$. It is obvious that $L_\beta(z_k) = \varphi(x_k)$, $L_\beta(z_0) = \varphi(x_0)$ and, therefore, $L_\beta(z_k) \to L_\beta(z_0)$.

Let $\bar{y}$ be an arbitrary element from Y and $\bar{z} = (\bar{x}, \bar{y})$. In view of Lemma 3.68, for sufficiently large k we have

$$r_\delta(x_k, \varepsilon_k) \leq \inf_{|\hat{x} - \bar{x}| \leq \delta} \varepsilon_k^{-1}[L_\beta(z_k + \varepsilon_k \hat{z}) - L_\beta(z_k)]$$

for all $\hat{y}$ such that $|\hat{y}| \leq \delta$. From this we conclude

$$r_\delta(x_k, \varepsilon_k) \leq \inf_{\substack{|\hat{x} - \bar{x}| \leq \delta \\ |\hat{y} - \bar{y}| \leq \delta}} \varepsilon_k^{-1}[L_\beta(z_k + \varepsilon_k \hat{z}) - L_\beta(z_k)] \leq R_\delta(z_k, \varepsilon_k),$$

and, therefore,

$$\varphi^\uparrow(x_0; \bar{x}) = \lim_{\delta \downarrow 0} \lim_{x_k \to x_0, \varepsilon_k \downarrow 0} r_\delta(x_k, \varepsilon_k) \leq \lim_{\delta \downarrow 0} \limsup_{z_k \to z_0, \varepsilon_k \downarrow 0} R_\delta(z_k, \varepsilon_k)$$

$$\leq \lim_{\delta \downarrow 0} \limsup_{z \xrightarrow{L_\beta} z_0, \varepsilon \downarrow 0} R_\delta(z, \varepsilon) = L_\beta^\uparrow(z_0; \bar{z})$$

for any $\bar{y} \in Y$, i.e. $\varphi^\uparrow(x_0; \bar{x}) \leq \inf_{\bar{y}} L_\beta^\uparrow(z_0; \bar{z})$. Thus, we finally get $\varphi^\uparrow(x_0; \bar{x}) \leq \inf_{\bar{y}} L_\beta(x_0; y(x_0))$, from which the statement of the theorem results immediately. ∎

COROLLARY 4.11 *If there exists a selector* $y(x) \in \omega(x)$ *continuous at the point* x_0, *then*

$$\varphi^\uparrow(x_0; \bar{x}) \leq \inf_{\bar{y}} L_\beta(x_0; y(x_0)).$$

We introduce the set

$$K_\beta(z_0) \triangleq \{x^* \in X \mid (x^*, 0) \in \partial^0 L_\beta(z_0)\}.$$

THEOREM 4.12 *Let the assumptions (A1)–(A3) hold and suppose the multivalued mapping* F *to be pseudolipschitz continuous at every point* $z_0 \in \omega\{x_0\} \times \omega(x_0)$. *Then*

$$\partial^0 \varphi(x_0) \subset \text{cl co} \bigcup_{y_0 \in \omega(x_0)} K_\beta(z_0). \tag{4.5}$$

Proof. Due to Theorem 4.10, the definition of $L_\beta^\uparrow(z_0; \bar{z})$ and the Minimax Theorem, we get for all $\bar{x} \in X$

$$\varphi^\uparrow(x_0; \bar{x}) \leq \sup_{y_0 \in \omega(x_0)} \inf_{\bar{y}} L_\beta^\uparrow(z_0; \bar{z}) = \sup_{y_0 \in \omega(x_0)} \inf_{\bar{y}} \sup_{z^* \in \partial^0 L_\beta(z_0)} \langle z^*, \bar{z} \rangle =$$

$$\sup_{y_0\in\omega(x_0)}\ \sup_{z^*\in\partial^0 L_\beta(z_0)}\ \inf_{\bar y}\ \langle z^*,\bar z\rangle = \sup_{y_0\in\omega(x_0)}\ \sup_{x^*\in K_\beta(z_0)}\ \langle x^*,\bar x\rangle =$$

$$\sup_{y_0\in\omega(x_0)}\ \delta^*(\bar x\,|\,K_\beta(z_0)) = \delta^*(\bar x\,|\,\mathrm{cl\,co}\ \bigcup_{y_0\in\omega(x_0)}\ K_\beta(z_0)).$$

Since for $\partial^0\varphi(x_0)\neq\emptyset$ the equality $\varphi^\uparrow(x_0;\bar x) = \delta^*(\bar x\,|\,\partial^0\varphi(x_0))$ holds, then in view of the inequality obtained above and the properties of support functions inclusion (4.5) results. If $\partial^0\varphi(x_0) = \emptyset$, then the validity of (4.5) is trivial. ∎

THEOREM 4.13 *Let the assumptions (A1)–(A3) hold and suppose the multivalued mapping F to be pseudolipschitz continuous at every point $z_0 \in \omega\{x_0\} \times \omega(x_0)$. Then*

$$\partial^0\varphi(x_0) \cap K_\beta(z_0) \neq \emptyset \tag{4.6}$$

for every $y_0 \in \omega(x_0)$.

Proof. Let $z_0 \in \{x_0\} \times \omega(x_0)$. Due to Lemma 3.68 the function $\varphi(x) - L_\beta(z)$ attains a local minimum at the point z_0. Hence, in accordance with the necessary optimality condition we have $0 \in \partial^0(\varphi - L_\beta)(z_0)$. Consequently, in view of the local Lipschitz continuity of L_β, we get the inclusion $0 \in \partial^0\varphi(x_0) - \partial^0 L_\beta(z_0)$, which is equivalent to (4.6). ∎

Let us now compare estimate (4.6) with the estimates from Theorems 4.5 and 4.9.

LEMMA 4.14 *Let the multivalued mapping F be pseudolipschitz continuous at the point $x_0 \in \mathrm{gr}\,F$. Then the relations $\bar z \in T^C_{\mathrm{gr}\,F}(z_0)$ and $d^0_F(z_0;\bar z) = 0$ are equivalent.*

Proof. Let $\bar z \in T^C_{\mathrm{gr}\,F}(z_0)$. Then for any sequences $z_k \xrightarrow{\mathrm{gr}\,F} z_0$, $\varepsilon_k \downarrow 0$ we can find a sequence $\bar z_k \to \bar z$ such that $z_k + \varepsilon_k\bar z_k \in \mathrm{gr}\,F$, $k = 1, 2, \ldots$ This is equivalent to the condition $d_F(z_k + \varepsilon_k\bar z_k) = 0$, $k = 1, 2, \ldots$ Therefore, in view of the pseudolipschitz continuity of d_F, we get

$$0 \le d_F(z_k + \varepsilon_k\bar z) - d_F(\bar z_k) \le d_F(z_k + \varepsilon_k\bar z_k) - d_F(z_k) + l_{d_F}|\bar z_k - z|\varepsilon_k.$$

From this inequality it follows that

$$d^0_F(z_0;\bar z) = \limsup_{z\to z_0,\,\varepsilon\downarrow 0}\ \varepsilon^{-1}[d_F(z + \varepsilon\bar z) - d_F(z)] = 0.$$

Now let $d^0_F(z_0;\bar z) = 0$. Then $d_F(z_k + \varepsilon_k\bar z) = o(\varepsilon_k)$ for all sequences $\varepsilon_k \downarrow 0$, $z_k \to z_0$, i.e., there exists a sequence $\bar y_k \to \bar y$ such that $y_k + \varepsilon_k\bar y_k \in F(x_k + \varepsilon_k\bar x)$, $k = 1, 2, \ldots$ Hence $\bar z \in T^C_{\mathrm{gr}\,F}(z_0)$. ∎

LEMMA 4.15 *Let the multivalued mapping F be pseudolipschitz contin-uous at the point $z_0 \in \mathrm{gr}\, F$. Then*

$$\mathrm{cl\,cone}\, \partial^0 d_F(z_0) = N_F(z_0),$$

where $N_F(z_0)$ is the Clarke normal cone to $\mathrm{gr}\, F$ at the point z_0.

Proof. Let $\bar{z} \in T^C_{\mathrm{gr}\, F}(z_0)$. Due to Lemma 4.14, we have $d^0_F(z_0; \bar{z}) = 0$. Since $d^0_F(z_0; \bar{z})$ is the support function of the set $\partial^0 d_F(z_0)$, then $\langle z^*, \bar{z} \rangle \leq 0$ for all $z^* \in \partial^0 d_F(z_0)$. Thus $\bar{z} \in -[\partial^0 d_F(z_0)]^+$, i.e.

$$T^C_{\mathrm{gr}\, F}(z_0) \subset -[\partial^0 d_F(z_0)]^+.$$

Vice versa, let $\bar{z} \in X \times Y$, $\bar{z} \in -[\partial^0 d_F(z_0)]^+$. Then $\langle z^*, \bar{z} \rangle \leq 0$ for all $z^* \in \partial^0 d_F(z_0)$ and the support function $d^0_F(z_0; \bar{z})$ to the set $\partial^0 d_F(z_0)$ is non-positive, what means $d^0_F(z_0; \bar{z}) = 0$. Then due to Lemma 4.14, $\bar{z} \in T^C_{\mathrm{gr}\, F}(z_0)$. In this way

$$T^C_{\mathrm{gr}\, F}(z_0) = -[\partial^0 d_F(z_0)]^+,$$

and, consequently,

$$N_F(z_0) = -[T^C_{\mathrm{gr}\, F}(z_0)]^+ = \mathrm{cl\,cone}\, \partial^0 d_F(z_0). \quad \blacksquare$$

Using Clarke's subdifferential calculus as well as Lemma 4.15, we get

$$\partial^0 L_\beta(z_0) = \partial^0 (f + \beta d_F)(z_0) \subset \partial^0 f(z_0) + \beta \partial^0 d_F(z_0) \subset \partial^0 f(z_0) + N_F(z_0),$$

from which it immediately follows that

$$K_\beta(z_0) \subset K(z_0). \tag{4.7}$$

Thus, Theorem 4.12 gives a more exact estimate of Clarke's subdifferential $\partial^0 \varphi(x)$.

The following example shows that the inclusion (4.7) can be strict.

EXAMPLE 4.16 *Let $X = Y = R$, $f(z) = |y|$, $x_0 = 0$, $F(x) = \{y \mid 0 \leq y \leq x \text{ if } x \geq 0 \text{ and } x \leq y \leq 0 \text{ if } x < 0\}$. It is easy to see that $\omega(x_0) = \{0\}$, $z_0 = (0,0)$, $\varphi(x) \equiv 0$ and, therefore, $\partial^0 \varphi(x_0) = \{0\}$. At the same time, the function $d_F(z) = \frac{1}{2}(|y - x| + |y| - |x|)$ is Lipschitz continuous and*

$$\partial^0 d_F(z_0) = \mathrm{co}\, \{(0,0), (0,1), (0,-1), (-1,1), (1,-1)\}.$$

Hence, for any $\beta > 1$ we get the representation $\partial^0 L_\beta(z_0) = \mathrm{co}\, \{(0, \beta + 1), (-\beta, \beta + 1), (-\beta, \beta - 1), (\beta, -\beta + 1), (0, -\beta - 1), (\beta, \beta - 1)\}$ as well as $K_\beta(z_0) = [-\frac{1}{2}(\beta + 1), \frac{1}{2}(\beta + 1)]$. But $N_F(z_0) = X \times Y$ and, consequently, $K(z_0) = X$.

1.2 Pseudolipschitz Continuity and Metrical Regularity

Let $F : X \to 2^Y$, $f : X \times Y \to R$, $\rho_F(z) = \rho(z, \operatorname{gr} F)$.

DEFINITION 4.17 *A multivalued mapping F is called metrically regular (with respect to its graph) at the point z_0 if there exist a constant $k > 0$ and a neighbourhood $V(z_0)$ such that $d_F(z) \le k\rho_F(z)$ for all $z \in V(z_0)$.*

If the inequality $d_F(z) \le k\rho_F(z)$ holds for all points from a set Z_0, then the mapping F is said to be *metrically regular on the set Z_0.*

LEMMA 4.18 *Let the multivalued mapping F be metrically regular at every point of the compact set $Z_0 \subset \operatorname{gr} F$. Then F is metrically regular on some neighbourhood of Z_0.*

LEMMA 4.19 *Let the closed mapping F be pseudolipschitz continuous at the point $z_0 = (x_0, y_0) \in \operatorname{gr} F$. Then F is metrically regular at z_0.*

Proof. From the pseudolipschitz continuity it follows that

$$F(\bar{x}) \cap V(y_0) \subset F(x) + l|\bar{x} - x|B \qquad (4.8)$$

for all $x, \bar{x} \in V(x_0)$, where $V(x_0) = x_0 + \delta_0 B$, $V(y_0) = y_0 + \delta_1 B$. We choose positive numbers δ_0' and δ_1' such that

$$\sqrt{(\delta_0')^2 + (\delta_1')^2} < \min\{\delta_0/2,\ \delta_1/2\}, \quad \delta_1' > l\delta_0'.$$

Denote $V'(x_0) = x_0 + \delta_0' B$, $V'(y_0) = y_0 + \delta_1' B$. We take an arbitrary point $z = (x, y) \in V'(x_0) \times V'(y_0)$ and denote by $\bar{z} = (\bar{x}, \bar{y})$ a point in the closed set $\operatorname{gr} F$ nearest to z. Then

$$|\bar{z} - z| \le |z - z_0| \le \sqrt{(\delta_0')^2 + (\delta_1')^2}$$

and, hence,

$$|\bar{x} - x| < \delta_0/2, \quad |\bar{y} - y| < \delta_1/2.$$

From this it follows that

$$|\bar{x} - x_0| = |(\bar{x} - x) + (x - x_0)| \le |\bar{x} - x| + |x - x_0| < \delta_0.$$

In the same way we can prove that $|\bar{y} - y_0| \le \delta_1$. Thus, (4.8) implies the existence of a point $v \in F(x)$ such that $|v - \bar{y}| \le l|x - \bar{x}|$. Therefore

$$\rho(y, F(x)) \le |y - v| \le |y - \bar{y}| + |\bar{y} - v| \le |y - \bar{y}| + l|x - \bar{x}|$$
$$\le (1 + l)|y - \bar{y}| + (1 + l)|x - \bar{x}| \le 2(1 + l)\max\{|x - \bar{x}|,\ |y - \bar{y}|\}$$
$$\le 2(1 + l)(\sqrt{|x - \bar{x}|^2 + |y - \bar{y}|^2}) = 2(1 + l)\rho_F(z). \quad \blacksquare$$

An effective tool for the investigation of multivalued mappings with respect to their pseudolipschitz continuity is the co-derivative introduced by Mordukhovich. Let

$$D_M^* F(z_0; y^*) \triangleq \{x^* \mid (x^*, -y^*) \in N_F(z_0)\}.$$

The mapping $D_M^* F(z_0; \cdot)$ is called the *Mordukhovich co-derivative (conjugate derivative)* of the multivalued mapping F at the point $z_0 = (x_0, y_0) \in \operatorname{gr} F$. It is well-known [132] that the co-derivative of a mapping F being pseudolipschitz continuous at z_0 satisfies the condition

$$\sup\{|x^*| \mid x^* \in D_M^* F(z_0; y^*)\} \le l|y^*|,$$

from which the equality

$$D_M^* F(z_0; 0) = \{0\} \tag{4.9}$$

results. On the other hand, due to Corollary 4.6,

$$\hat{\partial}^\infty d_F(z_0) \subset D_M^* F(z_0; 0).$$

If condition (4.9) holds, then obviously $\hat{\partial}^\infty d_F(z_0) = \{0\}$ (note that $\hat{\partial}^\infty g(x)$ always contains 0).

But the property of the singular subdifferential $\hat{\partial}^\infty g(x)$ to consist of only the null element is equivalent to the Lipschitz continuity of g in a neighbourhood of x (cf. [132]). Thus, condition (4.9) implies the Lipschitz continuity of the function d_F in a neighbourhood of z_0, which due to Lemma 3.26 means the pseudolipschitz continuity of F at z_0. Consequently, the following assertion is true.

THEOREM 4.20 *([134]) Let $z_0 \in \operatorname{gr} F$ and let the mapping F be closed and uniformly bounded at the point x_0. Then the following statements are equivalent:*

1. F is pseudolipschitz continuous at the point z_0;
2. $D_M^* F(z_0; 0) = \{0\}$.

We introduce the function $\bar{L}_\beta(z) = f(z) + \beta \rho_F(z)$ and denote by $V(\omega(x_0))$ some neighbourhood of the set $\omega(x_0)$.

LEMMA 4.21 *Let the multivalued mapping F and the function f satisfy the assumptions (A1)–(A3) and let, in addition, F be metrically regular at all points $z_0 \in \{x_0\} \times \omega(x_0)$. Then there exist a constant $\beta_1 > 0$ as well as neighbourhoods $V(x_0)$ and $V(\omega(x_0))$ such that*

$$\varphi(x) = \inf_y \{\bar{L}_\beta(z) \mid y \in V(\omega(x_0))\}$$

for all $x \in V(x_0)$ *and* $\beta > \beta_1$.

Proof. Due to Lemma 4.18, the mapping F is metrically regular on some set $V'(x_0) \times V'(\omega(x_0))$ with a certain constant $k > 0$. Consequently, $L_\beta(z) \leq \bar{L}_{k\beta}(z)$ and, with regard to Lemma 3.68, there exists β_0 such that for $\beta > \beta_0$ and $x \in V'(x_0) \cap X_0$ we have

$$\varphi(x) = \inf\{L_\beta(z) \mid y \in V'(\omega(x_0)) \cap (Y_0 + \varepsilon_0/2B)\}$$
$$\leq \inf\{\bar{L}_{k\beta}(z) \mid y \in V'(\omega(x_0)) \cap (Y_0 + \varepsilon_0/2B)\} = \varphi(x).$$

Therefore, using the notation $\beta_1 = k\beta$, $V(x_0) = V'(x_0) \cap X_0$ and $V(\omega(x_0)) = V'(\omega(x_0)) \cap [Y_0 + \varepsilon_0/2B]$, we get the desired statement. ∎

If we now repeat the proof of Theorem 6.3, replacing the reference to Lemma 3.68 by the reference to Lemma 4.21, under the assumptions of Lemma 4.21, we can deduce the inequality

$$\varphi^\uparrow(x_0; \bar{x}) \leq \sup_{y_0 \in \omega(x_0)} \inf_{\bar{y}} \bar{L}_\beta^\uparrow(z_0; \bar{z}) \qquad (4.10)$$

for all $\bar{x} \in X$, $\beta > \beta_1$.

Moreover, arguing as in the proof of Theorem 4.12, using inequality (4.10) and taking into account that the function $\bar{L}_\beta(z)$ is Lipschitz continuous, we recognize that the Clarke subdifferential $\partial^0 \bar{L}_\beta(z)$ is compact. In this way, the validity of the following theorem has been proved.

THEOREM 4.22 *Let the multivalued mapping F and the function f satisfy the assumptions (A1)–(A3) and, in addition, F be metrically regular at all points $z_0 \in \{x_0\} \times \omega(x_0)$. Then*

$$\partial^0 \varphi(x_0) \subset \operatorname{cl} \operatorname{co} \bigcup_{y_0 \in \omega(x_0)} \pi_0 \partial^0 \bar{L}_\beta(z_0), \qquad (4.11)$$

where $\pi_0 \partial^0 \bar{L}_\beta(z_0) = \{x^* \mid (x^*, 0) \in \partial^0 \bar{L}_\beta(z_0)\}$.

COROLLARY 4.23 *Let the closed mapping F be pseudolipschitz continuous at every point $z_0 \in \{x_0\} \times \omega(x_0)$. If the assumptions (A1)–(A3) hold, then the function φ is Lipschitz continuous in a neighbourhood of the point x_0.*

Proof. Due to Lemma 4.19 and Theorem 4.22, the inclusion (4.11) holds. Since the function $\bar{L}_\beta(z)$ is Lipschitz continuous on some set $Z_0 = V(x_0) \times V(\omega(x_0))$, then $\partial^0 \bar{L}_\beta(\cdot)$ is a multivalued mapping being compact-valued, upper semicontinuous and uniformly bounded on Z_0. Therefore, due to Lemma 3.2, $\partial^0 \bar{L}_\beta(x_0, \omega(x_0))$ is compact. Thus, the right-hand side of inclusion (4.11) is bounded, which implies the boundedness of $\partial^0\varphi(x_0)$. The non-emptiness of $\partial^0\varphi(x_0)$ follows from Theorem 4.13. Now, applying Theorem 2.20, we get the Lipschitz continuity of φ. ∎

THEOREM 4.24 *Let $z_0 = (x_0, y_0) \in \text{gr}\, F$ and the mapping F be closed and uniformly bounded at the point x_0. Then the following statements are equivalent:*
1. *F is metrically regular at the point z_0;*
2. *F is pseudolipschitz continuous at the point z_0.*

Proof. 1. $\Rightarrow$ 2. We intend to show that the function d_F is Lipschitz continuous in a neighbourhood of z_0. To this aim, we denote $\tilde{x} = (\tilde{x}_1, \tilde{x}_2)$, $\tilde{y} = (\tilde{y}_1, \tilde{y}_2)$, where $\tilde{x}_1 = x$, $\tilde{x}_2 = y$, $\tilde{y}_1 = v$, $\tilde{y}_2 = y$, $\tilde{F}(\tilde{x}) = F(x) \times \{y\}$. Consequently, as can be easily seen, $\rho_{\tilde{F}}((\tilde{x}, \tilde{y})) = \rho_F(x, v)$, $\rho(\tilde{y}, \tilde{F}(\tilde{x})) = \rho(v, F(x)) = d_F(x, v)$. One thus gets that the mapping $\tilde{F}$ is metrically regular at the point $(\tilde{x}_0, \tilde{y}_0)$, where $\tilde{x}_0 = (x_0, y_0)$, $\tilde{y}_0 = (v, y_0)$.

Let $\tilde{f}(\tilde{x}, \tilde{y}) = |y - v| = |\tilde{x}_2 - \tilde{y}_1|$. The function $\tilde{f}$ is Lipschitz continuous on the whole space. Consequently, the assumptions (A2) and (A3) are true for $\tilde{f}$ and $\tilde{F}$.

Applying Corollary 4.23 and taking into account Lemma 4.19 as well as the relation $d_F(x, y) = \inf\{\tilde{f}(\tilde{x}, \tilde{y}) \,|\, \tilde{y} \in \tilde{F}(\tilde{x})\}$, we deduce the Lipschitz continuity of d_F in a neighbourhood of z_0. According to Lemma 3.26, this implies the pseudolipschitz continuity of F at the point z_0.

2. $\Leftarrow$ 1. This follows directly from Lemma 4.19. ∎

With the help of the following generalization of metrical regularity, where ν is assumed to be a positive constant, an analogue of Theorem 4.24 for pseudohölder continuous mappings was proved in [128].

DEFINITION 4.25 *A mapping F is called metrically regular (with respect to its graph) of order ν at the point z_0 if there exist a constant $k > 0$ and a neighbourhood $V(z_0)$ such that $d_F(z) \le k(\rho_F(z))^\nu$ for all $z \in V(z_0)$.*

THEOREM 4.26 *Let $z_0 = (x_0, y_0) \in \text{gr}\, F$ and the mapping F be closed and uniformly bounded at the point x_0. Then for $0 < \nu \le 1$ the following statements are equivalent:*
1. *F is metrically regular of order ν at the point z_0;*
2. *F is pseudohölder continuous of order ν at the point z_0.*

2.   Locally Convex Mappings

2.1   Weakly Pseudoconvexity and Directional Derivatives of Marginal Functions

Let $X = R^n$, $Y = R^m$, $F : X \to 2^Y$, $x_0 \in \text{dom}\, F$, $f : X \times Y \to R$, and suppose that assumptions (A1)–(A3) hold.

DEFINITION 4.27 *A function* $g : X \times Y \to R$ *is called weakly pseudo-convex with respect to* y *at the point* z_0 *on the set* $E \subset X \times Y$ *in the direction* $\bar{x}$ *if* g *is directionally differentiable and*

$$g(z_k) - g(z_0) \geq g'(z_0; z_k - z_0) + o(|x_k - x_0|)$$

for all $z_k \xrightarrow{E} z_0$ *such that* $x_k = x_0 + \varepsilon_k \bar{x}$, *where* $\varepsilon_k \downarrow 0$.

Let $g : X \times Y \to R$ be a smooth function and $g(x_0, y)$ be convex in a neighbourhood of y_0. Then, due to Definition 4.27, it is not hard to see that the function g is weakly pseudoconvex with respect to y at z_0 for any $E \subset X \times Y$ and $\bar{x} \in X$.

A convex proper function f provides another example of a weakly pseudoconvex function.

From Lemma 3.60 it follows that the mapping F has a first-order approximation at the point z_0 in the direction $\bar{x}$ if and only if the function d_F is weakly pseudoconvex with respect to y at the point z_0 on the set gr F in the direction $\bar{x}$.

THEOREM 4.28 *Let there exist a constant* $\beta_0 > 0$ *such that for all* $\beta > \beta_0$ *the function* $L_\beta(z) = f(z) + \beta d_F(z)$ *is weakly pseudoconvex with respect to* y *at all points* $z_0 = \{x_0\} \times \omega(x_0)$ *on the set* gr F *in the direction* $\bar{x}$. *Then the function* φ *is directionally differentiable at* x_0 *in the direction* $\bar{x}$ *and*

$$\varphi'(x_0; \bar{x}) = \inf_{y_0 \in \omega(x_0)} \inf_{\bar{y}} L_\beta'(z_0; \bar{z}), \qquad (4.12)$$

where $\bar{z} = (\bar{x}, \bar{y})$.

Proof. To start with, let us note that under the assumptions of the theorem the mapping F is directionally differentiable at the points z_0 in the direction $\bar{x}$ and $\hat{D}_L F(z_0; \bar{x}) \neq \emptyset$. Let $D_+\varphi(x_0; \bar{x})$ be attained on the sequence $\varepsilon_k \downarrow 0$. We choose an arbitrary sequence $y_k \in \omega(x_0 + \varepsilon_k \bar{x})$. Due to Lemma 3.72 we can, without loss of generality, assume that $y_k \to y_0 \in \omega(x_0)$. Let us denote $\bar{y}_k = \varepsilon_k^{-1}(y_k - y_0)$, $\bar{z}_k = (\bar{x}, \bar{y}_k)$. Then, in view of Lemma 3.68 and the weakly pseudoconvexity of $L_\beta(z)$ with respect to y, we have

$$\varphi(x_0 + \varepsilon_k \bar{x}) - \varphi(x_0) = L_\beta(x_0 + \varepsilon_k \bar{x}, y_k) - L_\beta(x_0, y_0) \geq \varepsilon_k L_\beta'(z_0; \bar{z}_k) + o(\varepsilon_k),$$

and, consequently,

$$\varphi(x_0 + \varepsilon_k \bar{x}) - \varphi(x_0) \geq \varepsilon_k \left[\inf_{\bar{y}} L_\beta'(z_0; \bar{z}) + o(\varepsilon_k) \cdot \varepsilon_k^{-1} \right].$$

From the last inequality we easily obtain

$$D_+\varphi(x_0; \bar{x}) \geq \inf_{\bar{x}} L'_\beta(z_0; \bar{z}) \geq \inf_{y_0 \in \omega(x_0)} \inf_{\bar{y}} L'_\beta(z_0; \bar{z}). \qquad (4.13)$$

On the other hand, due to Lemma 3.68, for every $y_0 \in \omega(x_0)$ the inequality

$$\varphi(x_0 + \varepsilon\bar{x}) - \varphi(x_0) \leq L_\beta(x_0 + \varepsilon\bar{x}, y_0 + \varepsilon\bar{y}) - L_\beta(x_0, y_0)$$

holds for any vector $\bar{y} \in Y$. Therefore

$$D^+\varphi(x_0; \bar{x}) \leq \inf_{y_0 \in \omega(x_0)} \inf_{\bar{y}} L'_\beta(z_0; \bar{z}). \qquad (4.14)$$

Comparing (4.13) and (4.14) we get (4.12). ∎

COROLLARY 4.29 *Let f be a differentiable function, $f(x_0, y)$ be convex in a neighbourhood of the set $\omega(x_0)$ and let the mapping F have a first-order approximation at every point $z_0 \in \{x_0\} \times \omega(x_0)$ in the direction $\bar{x}$. Then the function φ is directionally differentiable at x_0 in the direction $\bar{x}$ and*

$$\varphi'(x_0; \bar{x}) = \inf_{y_0 \in \omega(x_0)} \inf_{\bar{y}} L'_\beta(z_0; \bar{z}) = \inf_{y_0 \in \omega(x_0)} \inf_{\bar{y} \in \hat{D}F(z_0;\bar{x})} \langle \nabla f(z_0), \bar{z} \rangle.$$

Proof. Under the assumptions of the corollary the function $L_\beta(z)$ is weakly pseudoconvex with respect to y in the direction $\bar{x}$ at all points $z_0 \in \{x_0\} \times \omega(x_0)$. Thus, the first equality follows from Theorem 4.28. The second one was proved in [55]. ∎

2.2 Subdifferentials of Marginal Functions for Locally Convex Multivalued Mappings

In addition to convex mappings, in the following we consider a broader class of mappings, so-called locally convex mappings.

DEFINITION 4.30 *A mapping $F : X \to 2^Y$ is said to be locally convex at a point $z_0 = (x_0, y_0) \in \operatorname{gr} F$ if there exists a neighbourhood $V(z_0)$ of the point z_0 such that $\operatorname{gr} F \cap V(z_0)$ is a convex set.*

Below it is assumed that the multivalued mapping F is locally convex at every point $z_0 \in \{x_0\} \times \omega(x_0)$ and the function f is continuous and convex in a neighbourhood of the set $\{x_0\} \times \omega(x_0)$. Then the function $L_\beta(z)$ is also convex in a neighbourhood of the set $\{x_0\} \times \omega(x_0)$, which we denote by $Z_0 = V(x_0) \times V(\omega(x_0))$. Under these assumptions, we have

$$\varphi(x) = \inf\{f(x, y) \mid y \in \tilde{F}(x)\},$$

where $\tilde{F}$ is a convex multivalued mapping with the graph gr $F \cap Z_0$. Thus, due to Lemma 3.33, the function φ is convex.

We denote $\pi_0 \partial L_\beta(z_0) = \{x^* \in X \mid (x^*, 0) \in \partial L_\beta(z_0)\}$. It is easy to see that if the convex function $L_\beta(z)$ is continuous (and, therefore, Lipschitz continuous) in a neighbourhood of z_0, then $\pi_0 \partial L_\beta(z_0) = K_\beta(z_0)$, where $K_\beta(z_0)$ was defined in Section 1.1.

LEMMA 4.31 *For any* $y_0, y_1 \in \omega(x_0)$ *the following relation holds:*

$$\pi_0 \partial L_\beta(x_0, y_0) = \pi_0 \partial L_\beta(x_0, y_1).$$

Proof. According to the definitions of $\partial L_\beta(z)$ and $\pi_0 \partial L_\beta(z)$, we have

$$\pi_0 \partial L_\beta(x_0, y_0) = \{x^* \mid L_\beta(x, y) - L_\beta(x_0, y_0) \geq \langle x^*, x - x_0 \rangle \ \forall (x, y) \in Z_0\}.$$

However, $L_\beta(x_0, y_0) = L_\beta(x_0, y_1) = \varphi(x_0)$. Consequently,

$$\pi_0 \partial L_\beta(x_0, y_0) = \{x^* \mid L_\beta(x, y) - L_\beta(x_0, y_1)$$
$$\geq \langle x^*, x - x_0 \rangle \ \ \forall (x, y) \in Z_0\} = \pi_0 \partial L_\beta(x_0, y_1). \ \blacksquare$$

LEMMA 4.32 *For any* $\bar{x} \in X$ *and* $y_0 \in \omega(x_0)$ *the inequality*

$$\varphi'(x_0; \bar{x}) \geq \delta^*(\bar{x} \mid \pi_0 \partial L_\beta(z_0))$$

is valid.

Proof. Under the assumptions described above the function φ is a proper convex function finite at x_0. This means that $\varphi'(x_0; \bar{x})$ exists for any $\bar{x} \in X$. Since the function $L_\beta(z)$ is weakly pseudoconvex with respect to y at any point $z_0 = (x_0, y_0)$, $y_0 \in \omega(x_0)$ in any direction $\bar{x}$, then due to Theorem 4.28, we get

$$\varphi'(x_0; \bar{x}) = \inf_{y_0 \in \omega(x_0)} \inf_{\bar{y}} L'_\beta(z_0; \bar{z}).$$

From this in view of Lemma 4.31 we conclude

$$\varphi'(x_0; \bar{x}) \geq \inf_{y_0 \in \omega(x_0)} \inf_{\bar{y}} \operatorname*{cl}_{\bar{z}} L'_\beta(z_0; \bar{z}) = \inf_{y_0 \in \omega(x_0)} \inf_{\bar{y}} \sup_{z^* \in \partial L_\beta(z_0)} \langle z^*, \bar{z} \rangle$$

$$\geq \inf_{y_0 \in \omega(x_0)} \sup_{z^* \in \partial L_\beta(z_0)} \inf_{\bar{y}} \langle z^*, \bar{z} \rangle = \inf_{y_0 \in \omega(x_0)} \sup_{x^* \in \pi_0 \partial L_\beta(z_0)} \langle x^*, \bar{x} \rangle$$

$$= \inf_{y_0 \in \omega(x_0)} \delta^*(\bar{x} \mid \pi_0 \partial L_\beta(z_0)) = \delta^*(\bar{x} \mid \pi_0 \partial L_\beta(z_0)),$$

where the notation $\operatorname*{cl}_{\bar{z}}$ means the closure with respect to $\bar{z}$. $\blacksquare$

COROLLARY 4.33 *For all $y_0 \in \omega(x_0)$ the following inclusion is valid:*

$$\pi_0 \partial L_\beta(z_0) \subset \partial \varphi(x_0).$$

THEOREM 4.34 *For any $y_0 \in \omega(x_0)$ the following equality holds:*

$$\partial \varphi(x_0) = K(z_0).$$

Proof. The function f is continuous and convex and, therefore, Lipschitz continuous on Z_0. Hence $\partial^0 f(z_0) = \partial f(z_0)$. Moreover, $\rho_{\mathrm{gr}\bar{F}}(z) \le d_F(z)$ and, consequently, $\rho'_{\bar{F}}(z_0; \bar{z}) \le d'_F(z_0; \bar{z})$ and $\partial \rho_{\bar{F}}(z_0) \subset \partial d_F(z_0)$. Taking into account Remark 1.1 and the Moreau-Rockafellar Theorem, we get

$$K(z_0) = \pi_0(\partial f(z_0) + R^+ \partial \rho_{\bar{F}}(z_0)) = \bigcup_{\beta > 0} \pi_0 \partial L_\beta(z_0).$$

Obviously, due to Lemma 4.31, $K(z_0)$ does not depend on the choice of y_0 from the set $\omega(x_0)$, i. e., $K(x_0, y_0) = K(x_0, y_1)$ for any y_0, $y_1 \in \omega(x_0)$.

Two cases are possible:

1) $K(z_0) = \emptyset$. Then $K(z_0) = \emptyset$ for all $y_0 \in \omega(x_0)$ and, consequently, owing to Theorem 4.5, $\partial \varphi(x_0) = \emptyset$. In this case, in view of Corollary 4.33, we can find a number β_0 such that

$$\partial \varphi(x_0) = \pi_0 \partial L_\beta(z_0) = \emptyset$$

for all $\beta > \beta_0$, $y_0 \in \omega(x_0)$.

2) $K(z_0) \ne \emptyset$. Applying Theorem 4.9, we obtain $\partial \varphi(x_0) \subset K(z_0)$ for every $y_0 \in \omega(x_0)$. On the other hand, by Corollary 4.33, we have

$$\pi_0 \partial L_\beta(z_0) \subset \partial \varphi(x_0)$$

for all $\beta > \beta_0$. From this it follows that

$$\partial \varphi(x_0) \supset \bigcup_{\beta > \beta_0} \pi_0 \partial L_\beta(z_0) = K(z_0).$$

Thus, in any case $\partial \varphi(x_0) = K(z_0)$. ∎

Let us introduce the notation

$$D^* F(z_0; y^*) \overset{\triangle}{=} \{x^* \in X \mid (-x^*, y^*) \in [T_{\mathrm{gr}\, F}(z_0)]^+\}.$$

The multivalued mapping $D^* F(z_0; \cdot)$ is called the *co-derivative* (or *conjugate derivative*) of the mapping F at the point $z_0 \in \mathrm{gr}\, F$. With the help of the co-derivative $D^* F(z_0; \cdot)$ we can reformulate Theorem 4.34.

THEOREM 4.35 *For all $y_0 \in \omega(x_0)$ the following equality holds:*

$$\partial \varphi(x_0) = \{x^* + D^* F(z_0; y^*) \mid (x^*, y^*) \in \partial f(z_0)\}.$$

In particular, if f is differentiable, then

$$\partial \varphi(x_0) = \nabla_x f(z_0) + D^* F(z_0; \nabla_y f(z_0)).$$

Chapter 5

DIRECTIONAL DERIVATIVES OF MARGINAL FUNCTIONS

1. Weakly Uniformly Differentiable Functions

Let $X = R^n$, $Y = R^m$ and let U be a compact set in Y. We consider the functions

$$\varphi(x) = \inf \{f(x,y) \mid y \in U\},$$
$$\Phi(x) = \sup \{f(x,y) \mid y \in U\},$$

where $f : X \times Y \to R$ is certain function. It is of considerable interest to find out sufficient assumptions to be imposed on the function f that are minimal in a sense and ensure directional differentiability of the functions φ and Φ. Moreover, our aim is to obtain formulas for the calculation of derivatives $\varphi'(x; \bar{x})$ and $\Phi'(x; \bar{x})$.

As above we denote

$$\omega(x) = \{y \in U \mid f(x,y) = \varphi(x)\},$$
$$\Omega(x) = \{y \in U \mid f(x,y) = \Phi(x)\}.$$

It is known that the maximum function Φ is differentiable at a point x_0 in the direction $\bar{x} \in X$ if f is continuous and uniformly differentiable at the point x_0 in the direction $\bar{x}$, i.e.

$$f(x_0 + \varepsilon\bar{x}, y) = f(x_0, y) + \varepsilon f'(x_0, y; \bar{x}) + \varepsilon\gamma(\varepsilon, y)$$

for all $\varepsilon > 0$, where $\gamma(\varepsilon, y)$ converges to 0 for $\varepsilon \downarrow 0$ uniformly with respect to y. The notation $f'(x_0, y; \bar{x})$ means the directional derivative of f at the point (x_0, y) with respect to x in the direction $\bar{x}$.

Let us consider a class of functions broader than that of continuous and uniformly differentiable functions f, for which the function Φ is directionally differentiable. This class will include also another functions

f for which $\Phi'(x; \bar{x})$ exists, in particular, functions being continuously differentiable and convex with respect to x.

DEFINITION 5.1 *1. A function f is called upper weakly uniformly differentiable (upper w.u.d.) with respect to U at a point (x_0, y_0) in the direction $\bar{x} \in X$ if there exists the finite derivative $f'(x_0, y; \bar{x})$ for all y from some neighbourhood $V(y_0) \cap U$ and*

$$\limsup_{\varepsilon \downarrow 0, \, y \xrightarrow{U} y_0} \varepsilon^{-1}[f(x_0 + \varepsilon\bar{x}, y) - f(x_0, y)] \le f'(x_0, y_0; \bar{x}). \qquad (5.1)$$

2. A function f is called lower weakly uniformly differentiable (lower w.u.d.) with respect to U at a point (x_0, y_0) in the direction $\bar{x}$ if there exists the finite derivative $f'(x_0, y; \bar{x})$ for all y from some neighbourhood $V(y_0) \cap U$ and

$$\liminf_{\varepsilon \downarrow 0, \, y \xrightarrow{U} y_0} \varepsilon^{-1}[f(x_0 + \varepsilon\bar{x}, y) - f(x_0, y)] \ge f'(x_0, y_0; \bar{x}).$$

3. A function f is called weakly uniformly differentiable (w.u.d.) with respect to U at a point (x_0, y_0) in the direction $\bar{x}$ if f is upper and lower w.u.d. at this point in the direction $\bar{x}$.

In the following, the words "with respect to U" will be omitted. It is obvious that a function f is upper w.u.d. if and only if the function $-f$ is lower w.u.d. Thus, we can restrict ourselves to the study of properties of upper w.u.d. functions.

To determine the borders of the class of upper w.u.d. functions let us investigate necessary features of such functions.

LEMMA 5.2 *Let the function f be upper w.u.d. at the point (x_0, y_0) in the direction $\bar{x}$. Then $f'(x_0, y; \bar{x})$ is upper semicontinuous at the point y_0.*

Proof. Due to (5.1), we get

$$f'(x_0, y; \bar{x}) \ge \limsup_{\varepsilon \downarrow 0, \, y \xrightarrow{U} y_0} \varepsilon^{-1}[f(x_0 + \varepsilon\bar{x}, y) - f(x_0, y)]$$

$$\ge \limsup_{y \xrightarrow{U} y_0} \{\lim_{\varepsilon \downarrow 0} \varepsilon^{-1}[f(x_0 + \varepsilon\bar{x}, y) - f(x_0, y)]\} = \limsup_{y \xrightarrow{U} y_0} f'(x_0, y_0; \bar{x}). \quad \blacksquare$$

Let us now show that the class of upper w.u.d. functions is sufficiently broad.

LEMMA 5.3 *Let the function f be continuously differentiable with respect to x. Then f is w.u.d. at every point (x, y) in any direction $\bar{x}$.*

Proof. It follows immediately from the definition of a w.u.d. function. ∎

LEMMA 5.4 *Let the function f be Lipschitz continuous with respect to x in a neighbourhood of x_0 for every $y \in Y$ and let it have a derivative $f'(x, y; \bar{x})$ being upper semicontinuous with respect to x and y at the point (x_0, y_0). Then f is upper w.u.d. at the point (x_0, y_0) in the direction $\bar{x}$.*

Proof. From the Lipschitz continuity of $f(x_0 + \varepsilon\bar{x}, y)$ with respect to ε it follows that almost everywhere there exists the derivative

$$\frac{d}{d\varepsilon} f(x_0 + \varepsilon\bar{x}, y) = f'(x_0 + \varepsilon\bar{x}, y; \bar{x}).$$

Due to the upper semicontinuity of $f'(x_0 + \varepsilon\bar{x}, y; \bar{x})$ with respect to ε and y for any $\mu > 0$ one can find a number $\delta > 0$ and a neighbourhood $V(y_0)$ such that for all $\tau \in (0, \delta)$ and $y \in V(y_0)$

$$\varepsilon^{-1}[f(x_0 + \varepsilon\bar{x}, y) - f(x_0, y)] = \varepsilon^{-1} \int_0^\varepsilon f'(x_0 + \tau\bar{x}, y; \bar{x}) \, d\tau$$

$$\leq \varepsilon^{-1} \int_0^\varepsilon [f'(x_0, y_0; \bar{x}) + \mu] \, d\tau = f'(x_0, y_0; \bar{x}) + \mu. \quad \blacksquare$$

REMARK. In the statement of Lemma 5.4 it suffices to assume Lipschitz continuity of $f(x_0 + \tau\bar{x}, y)$ with respect to τ in a neighbourhood of $\tau = 0$ and the upper semicontinuity of $f'(x_0 + \tau\bar{x}, y; \bar{x})$ with respect to τ and y at the point $(0, y_0)$.

LEMMA 5.5 *Let the function $f : X \times Y \to R$ be convex with respect to x for every $y \in Y$. Then f is upper w.u.d. at every point $(x, y) \in X \times U$ in any direction $\bar{x}$.*

Proof. The function $f(x + \lambda\bar{x}, y)$ is continuous and convex with respect to λ. Applying Theorem 2.17, we obtain that there exists the derivative

$$\frac{d}{d\varepsilon} f(x + \varepsilon\bar{x}, y)|_{\varepsilon=+0} = f'(x, y; \bar{x})$$

being upper semicontinuous with respect to x and y. Applying Lemma 5.4 and taking into account Remark 1, we get the statement to be proved. ∎

LEMMA 5.6 *Let the function $f(x, y)$ be continuous with respect to y for every x and uniformly differentiable at the point x_0 in the direction $\bar{x}$. Then f is w.u.d. at all points (x_0, y) for any $y \in U$ in the direction $\bar{x}$.*

Proof. Under the assumptions of the lemma the derivative $f'(x_0, y; \bar{x})$ is continuous with respect to y at the point y_0. According to this fact and with regard to Definition 5.1, we obtain the statement of the lemma. ∎

LEMMA 5.7 *Let $\alpha, \beta \geq 0$ and the functions f and g be upper w.u.d. at the point (x_0, y_0) in the direction $\bar{x}$. Then the function $\alpha f + \beta g$ is upper w.u.d. at the point (x_0, y_0) in the direction $\bar{x}$.*

Proof. It is a straightforward consequence of the definition of an upper w.u.d. function. ∎

LEMMA 5.8 *Let the functions $h(x, y)$ and $g(x, y)$ be w.u.d. at the point (x_0, y_0) in the direction $\bar{x}$. Then their sum, difference, product and quotient g/h for $h(x, y) \neq 0$ are w.u.d. at the point (x_0, y_0) in the direction $\bar{x}$.*

Proof. This assertion results immediately from the definition of a w.u.d. function. ∎

THEOREM 5.9 *1. Let the upper semicontinuous function f be upper w.u.d. in the direction $\bar{x}$ at every point of the set $\{x_0\} \times \Omega(x_0)$. Then the function Φ is directionally differentiable at the point x_0 in the direction $\bar{x}$ and*

$$\Phi'(x_0; \bar{x}) = \max_{y_0 \in \Omega(x_0)} f'(x_0, y_0; \bar{x}). \tag{5.2}$$

2. Let the lower semicontinuous function f be lower w.u.d. in the direction $\bar{x}$ at every point of the set $\{x_0\} \times \omega(x_0)$. Then the function φ is directionally differentiable at the point x_0 in the direction $\bar{x}$ and

$$\varphi'(x_0; \bar{x}) = \min_{y_0 \in \omega(x_0)} f'(x_0, y_0; \bar{x}). \tag{5.3}$$

Proof. 1. Let us take an arbitrary point $y_0 \in \Omega(x_0)$. Then

$$\Phi(x_0 + \varepsilon \bar{x}) - \Phi(x_0) \geq f(x_0 + \varepsilon \bar{x}, y_0) - f(x_0, y_0).$$

Dividing this inequality by $\varepsilon > 0$ and passing to the limit, due to Lemma 5.5, we get

$$D_+\Phi(x_0; \bar{x}) = \liminf_{\varepsilon \downarrow 0} \varepsilon^{-1}[\Phi(x_0 + \varepsilon \bar{x}) - \Phi(x_0)]$$
$$\geq f'(x_0, y_0; \bar{x}) \geq \sup_{y_0 \in \Omega(x_0)} f'(x_0, y_0; \bar{x}) = \max_{y_0 \in \Omega(x_0)} f'(x_0, y_0; \bar{x}). \tag{5.4}$$

Note that from (5.4) we can derive the inequality

$$\liminf_{\varepsilon \downarrow 0} \Phi(x_0 + \varepsilon \bar{x}) \geq \Phi(x_0). \tag{5.5}$$

Let us now take a sequence $\varepsilon_k \downarrow 0$ on which

$$D_+\Phi(x_0; \bar{x}) = \liminf_{\varepsilon \downarrow 0} \varepsilon^{-1}[\Phi(x_0 + \varepsilon\bar{x}) - \Phi(x_0)]$$

is attained. We denote $x_k = x_0 + \varepsilon_k\bar{x}$ and choose an arbitrary sequence $y_k \in \Omega(x_k)$, $k = 1, 2, \ldots$ Due to the compactness of U, without loss of generality, we can assume that $y_k \to \tilde{y}$. Let us show that $\tilde{y} \in \Omega(x_0)$. Indeed, owing to the upper semicontinuity of the function f and inequality (5.5), the passage to the limit in the equality $\Phi(x_k) = f(x_k, y_k)$ implies

$$\Phi(x_0) \leq \liminf_{k\to\infty} \Phi(x_k) \leq \limsup_{k\to\infty} f(x_k, y_k) \leq f(x_0, \tilde{y}),$$

where $\tilde{y} \in U$. This means that $\tilde{y} \in \Omega(x_0)$. In this way

$$\Phi(x_k) - \Phi(x_0) \leq f(x_k, y_k) - f(x_0, y_k),$$

and, according to the upper weakly uniform differentiability of the function f, we get

$$D^+\Phi(x_0; \bar{x}) \leq f'(x_0, \tilde{y}; \bar{x}).$$

Hence

$$D^+\Phi(x_0; \bar{x}) \leq \max_{y_0 \in \Omega(x_0)} f'(x_0, y_0; \bar{x}).$$

Taking into account the last inequality and (5.4), we get (5.2).

2. Since $\varphi(x) = -\max\{-f(x, y) \mid y \in U\}$ and $-f$ is upper semicontinuous and upper w.u.d. in the direction $\bar{x}$ at all points of the set $\{x_0\} \times \omega(x_0)$, then we can use the first part of the theorem to obtain (5.3). ∎

COROLLARY 5.10 *Under the assumptions of the theorem the multivalued mappings $\Omega(x + \varepsilon\bar{x})$ and $\omega(x + \varepsilon\bar{x})$ are upper semicontinuous at the point $\varepsilon = 0$ on the set of positive values of ε.*

We introduce the notation

$$\tilde{\Omega}(x_0, \bar{x}) = \limsup_{\varepsilon \downarrow 0} \Omega(x_0 + \varepsilon\bar{x}),$$
$$\tilde{\omega}(x_0, \bar{x}) = \liminf_{\varepsilon \downarrow 0} \omega(x_0 + \varepsilon\bar{x}).$$

COROLLARY 5.11 *1. Let the upper semicontinuous function f be upper w.u.d. in the direction $\bar{x}$ at all points of the set $\{x_0\} \times \tilde{\Omega}(x_0, \bar{x})$. Then*

$$\Phi'(x_0; \bar{x}) = f'(x_0, y_0; \bar{x}) \quad \text{for all} \quad y_0 \in \tilde{\Omega}(x_0, \bar{x}). \tag{5.6}$$

2. Let the lower semicontinuous function f be lower w.u.d. in the direction $\bar{x}$ at all points of the set $\{x_0\} \times \tilde{\omega}(x_0, \bar{x})$. Then

$$\varphi'(x_0; \bar{x}) = f'(x_0, y_0; \bar{x}) \quad \text{for all} \quad y_0 \in \tilde{\omega}(x_0, \bar{x}). \tag{5.7}$$

Proof. 1. From the closedness of the upper topological limit, the compactness of U as well as the upper semicontinuity of $\Omega(x_0 + \varepsilon\bar{x})$ (see Corollary 5.10) it follows that $\tilde{\Omega}(x_0, \bar{x})$ is a non-empty compact subset in $\Omega(x_0)$. Due to Theorem 5.9, the equality (5.2) holds. Now we want to estimate $\Phi'(x_0; \bar{x})$. Let $y_0 \in \tilde{\Omega}(x_0, \bar{x})$. Then there exist sequences $\varepsilon_k \downarrow 0$ and $y_k \in \Omega(x_0 + \varepsilon_k\bar{x})$, $k = 1, 2, \ldots$, such that $y_k \to y_0$. Since f is upper w.u.d. at (x_0, y_0), then we get

$$\Phi'(x_0; \bar{x}) = \lim_{k \to \infty} \varepsilon_k^{-1}[f(x_0 + \varepsilon_k\bar{x}, y_k) - \Phi(x_0)]$$

$$\leq \lim_{k \to \infty} \varepsilon_k^{-1}[f(x_0 + \varepsilon_k\bar{x}, y_k) - f(x_0, y_k)] \leq f'(x_0, y_0; \bar{x}).$$

Comparing the last relation with (5.3), we obtain (5.6).

2. The second statement of the lemma can be proved in the same way. ∎

It is possible to prove still another version of Theorem 5.9, which does not require upper semicontinuity of the function f with respect to both variables, but only with respect to y.

THEOREM 5.12 *1. Let the function f be upper semicontinuous with respect to y and upper w.u.d. in the direction $\bar{x}$ at all points of the set $\{x_0\} \times U$. Then the function Φ is directionally differentiable at the point x_0 in the direction $\bar{x}$ and formula (5.2) holds.*
2. Let the function f be lower semicontinuous with respect to y and lower w.u.d. in the direction $\bar{x}$ at all points of the set $\{x_0\} \times U$. Then the function φ is directionally differentiable at the point x_0 in the direction $\bar{x}$ and formula (5.3) holds.

Proof. 1. We repeat the proof of Theorem 5.9, but in contrast to it, the membership $\tilde{y} \in \Omega(x_0)$ is justified by the upper weakly uniform differentiability of the function f. Of course, since f is upper w.u.d. at the point $(x_0, \tilde{y})$, then due to (5.1)

$$f'(x_0, \tilde{y}; \bar{x}) \geq \limsup_{\substack{\varepsilon_k \downarrow 0 \\ y_k \to \tilde{y}}} \varepsilon_k^{-1}[f(x_k, y_k) - f(x_0, y_k)].$$

Therefore

$$\limsup_{k \to \infty} f(x_k, y_k) \leq \limsup_{k \to \infty} f(x_0, y_k) \leq f(x_0, \tilde{y}).$$

Hence, because of relation (5.5), the passage to the limit in the equality $\Phi(x_k) = f(x_k, y_k)$ yields

$$\Phi(x_0) \leq \liminf_{k \to \infty} \Phi(x_k) \leq \limsup_{k \to \infty} f(x_k, y_k) \leq f(x_0, \tilde{y}),$$

where $\tilde{y} \in U$. Therefore $\tilde{y} \in \Omega(x_0)$. The remaining part of the proof is the same as in Theorem 5.9.

2. It can be proved analogously. ∎

Lemmas 5.3–5.8 allow us to describe simple examples of w.u.d. functions. More complicated examples can be constructed with the help of the following lemma.

LEMMA 5.13 *Let U and V be compact sets in R^m and R^p, respectively, and the function $f : R^n \times U \times V \to R$ be continuous with respect to y and v on $U \times V$. Then the following assertions are true:*

1. If the function f is upper w.u.d. at all points from $\{x_0\} \times U \times V$ (with respect to $U \times V$) in the direction $\bar{x}$, then the function

$$g(x,y) = \max\{ f(x,y,v) \,|\, v \in V \}$$

is upper w.u.d. at the points $\{x_0\} \times U$ in the direction $\bar{x}$.

2. If the function f is lower w.u.d. at all points from $\{x_0\} \times U \times V$ (with respect to $U \times V$) in the direction $\bar{x}$, then the function

$$h(x,y) = \min\{ f(x,y,v) \,|\, v \in V \}$$

is lower w.u.d. at the points $\{x_0\} \times U$ in the direction $\bar{x}$.

Proof. 1. Due to Theorem 5.12 there exists the derivative

$$g'(x_0, y; \bar{x}) = \max\{ f'(x_0, y, v; \bar{x}) \,|\, v \in \Omega_v(x_0, y)\},$$

where $\Omega_v(x_0, y) = \{v \in V \,|\, f(x_0, y, v) = g(x_0, y)\}$ is a multivalued mapping upper semicontinuous with respect to y (see Lemma 3.23). In addition, due to Definition 5.1, $f'(x_0, y, v; \bar{x})$ is upper semicontinuous with respect to y and v. Therefore, from Lemma 3.23 it follows that the function $g'(x_0, y; \bar{x})$ is upper semicontinuous with respect to y. The latter is equivalent to the fact that the function g is w.u.d. at the point (x_0, y) in the direction $\bar{x}$ for any $y \in U$.

2. The second part of the theorem can be proved by using an argument similar to the one used in the first part. ∎

Let us now illustrate the property of weakly uniform differentiability on the following examples.

EXAMPLE 5.14 *Let $X = R$, $Y = R$, and $f(x,y) = -x + 2|y - x|$. It is easy to see that*

$$f(x,y) = \max\{x - 2y, 2y - 3x\}$$

Therefore, according to Lemma 5.13, one recognizes that the function f studied here is upper w.u.d.

EXAMPLE 5.15 *Let $X = R$, $Y = R$, $f(x, y) = x - 2|y - x|$, and $\Phi(x) = \max\{f(x, y) \mid |y| \leq 2\}$. In this case the function f reduces to*

$$f(x, y) = \min\{-x + 2y, 3x - 2y\}.$$

Due to Lemma 5.13, f is lower w.u.d. Let us study whether this function is upper w.u.d. Direct calculations yield

$$f'(x, y; \bar{x}) = \begin{cases} -\bar{x} & \text{if } y > x, \\ \min\{3\bar{x}, -\bar{x}\} & \text{if } y = x, \\ 3\bar{x} & \text{if } y < x. \end{cases}$$

It is not hard to see that the function $f'(x, y; \bar{x})$ is not upper semicontinuous with respect to y at the point $y = x$, i. e., the function f is not upper w.u.d. at the point (x, x) in the direction $\bar{x}$. Note that the absence of weakly uniform differentiability in this example implies that formula (5.2) does not hold. Indeed, if $|x| < 2$ we get $\Phi(x) = x$ and $\Omega(x) = \{x\}$. Consequently, for $\bar{x} = 1$ we have $\Phi'(x; \bar{x}) = \bar{x} = 1$. However, at the same time

$$\max\{f'(x, y; \bar{x}) \mid y \in \Omega(x)\} = \min\{3\bar{x}, -\bar{x}\} = \min\{3, -1\} = -1.$$

2.   Weakly Uniformly Differentiable Multivalued Mappings

Throughout this section we suppose that $X = R^n$, $Y = R^m$, and $F : X \to 2^Y$ is a closed convex-valued mapping uniformly bounded at the point $x_0 \in \operatorname{dom} F$. Using the results obtained in the last section, we want to describe a class of multivalued mappings F for which a constructive description of derivatives $\hat{D}_L F(z; \bar{x})$ exists and the maximum function

$$\Phi(x) = \sup\{f(x, y) \mid y \in F(x)\}$$

is directionally differentiable.

DEFINITION 5.16 *A mapping F is called weakly uniformly differentiable (w.u.d.) at the point x_0 in the direction $\bar{x} \in X$ if its support function S_F is lower w.u.d. at all points (x_0, p), $p \in Y$, in the direction $\bar{x}$.*

From this definition it immediately follows that the mapping F cannot be w.u.d. at the point x_0 in directions $\bar{x} \notin \gamma_{\operatorname{dom} F}(x_0)$.

Let us show that the class of multivalued mapping introduced above is rather broad.

LEMMA 5.17 *Let the support function $S_F(x, p)$ of the mapping F be uniformly differentiable at the point x_0 in the direction $\bar{x}$. Then F is w.u.d. at the point x_0 in the direction $\bar{x}$.*

Proof. Due to Lemmas 5.6 and 3.18, the function S_F is w.u.d. at all points (x_0, p) with $p \in Y$ in the direction $\bar{x}$. This means that F is w.u.d. at x_0 in the direction $\bar{x}$. ∎

LEMMA 5.18 *Let the mapping F be convex. Then F is w.u.d. at any point x_0 in any direction* $\bar{x} \in \gamma_{\mathrm{dom}\,F}(x_0)$.

Proof. Let $\bar{x} \in \gamma_{\mathrm{dom}\,F}(x_0)$. Under the assumptions made above the support function $S_F(x_0 + \varepsilon \bar{x}, p)$ is concave with respect to ε on $[0, \varepsilon_0]$ for some $\varepsilon_0 > 0$. Then due to Lemma 5.6, S_F is lower w.u.d. at x_0 in the direction $\bar{x}$, i.e., F is w.u.d. at x_0 in the direction $\bar{x}$. ∎

Note that multivalued mappings with uniformly differentiable support functions were studied in [35, 122, 150].

LEMMA 5.19 *Let the multivalued mappings F_1 and F_2 be w.u.d. at the point x_0 in the direction $\bar{x}$ and let $\lambda_1, \lambda_2 \in R^+$. Then the multivalued mapping $\lambda_1 F_1 + \lambda_2 F_2$ is w.u.d. at the point x_0 in the direction $\bar{x}$.*

Proof. It results from Lemma 5.7. ∎

Let $y \in F(x)$, $z = (x, y)$. We denote

$$M(z) = \{p \in Y \mid \langle p, y \rangle = S_F(x, p)\}, \quad \bar{B} = \{p \in Y \mid |p| \le 1\},$$

$$\Gamma'(z; \bar{x}) = \{\bar{y} \in Y \mid \langle p, \bar{y} \rangle \le S_F'(x, p; \bar{x}) \quad \forall p \in M(z) \cap \bar{B}\}.$$

LEMMA 5.20 *Let $y \in F(x)$. Then $M(z) = -[\mathrm{cone}\,(F(x) - y)]^+$.*

Proof. The condition $p \in M(z)$ is equivalent to $\langle p, v - y \rangle \le 0$ for all $v \in F(x)$, i.e. $p \in -[\mathrm{cone}\,(F(x) - y)]^+$. ∎

COROLLARY 5.21 $M(z)$ *is a closed convex cone.*

COROLLARY 5.22 $\hat{D}_L F(z_0; 0) = \overline{\mathrm{cone}}\,(F(x_0) - y_0) = \Gamma'(z_0; 0)$.

LEMMA 5.23 *Let $z_0 = (x_0, y_0)$, $y_0 \in F(x_0)$, and $S_F'(x_0, p; \bar{x}) > -\infty$. Then the derivative $S_F'(x_0, p; \bar{x})$ is positive homogeneous and convex on the cone $M(z_0)$.*

Proof. Let $\lambda > 0$. Then due to the properties of support functions, we have $S_F(x, \lambda p) = \lambda S_F(x, p)$ for any $p \in Y$ and $x \in X$. Therefore,

$$S_F'(x_0, \lambda p; \bar{x}) = \lim_{\varepsilon \downarrow 0} \varepsilon^{-1}[S_F(x_0 + \varepsilon \bar{x}, \lambda p) - S_F(x_0, \lambda p)]$$

$$= \lambda \lim_{\varepsilon \downarrow 0} \varepsilon^{-1}[S_F(x_0 + \varepsilon \bar{x}, p) - S_F(x_0, p)] = \lambda S_F'(x_0, \lambda p; \bar{x}),$$

i.e. $S_F'(x_0, \lambda p; \bar{x})$ is positive homogeneous with respect to p.

Let us now check the convexity of $S'_F(x_0, p; \bar{x})$ with respect to p. In view of the positive homogeneity of this function it is sufficient to show that

$$S'_F(x_0, p_1 + p_2; \bar{x}) \leq S'_F(x_0, p_1; \bar{x}) + S'_F(x_0, p_2; \bar{x})$$

for any p_1, $p_2 \in M(z_0)$. Indeed,

$$S'_F(x_0, p_1 + p_2; \bar{x}) = \lim_{\varepsilon \downarrow 0} \varepsilon^{-1}[S_F(x_0 + \varepsilon \bar{x}, p_1 + p_2) - S_F(x_0, p_1 + p_2)],$$

where, in view of the convexity and positive homogeneity of S_F with respect to p,

$$S_F(x_0 + \varepsilon \bar{x}, p_1 + p_2) \leq S_F(x_0 + \varepsilon \bar{x}, p_1) + S_F(x_0 + \varepsilon \bar{x}, p_2).$$

On the other hand, since according to Corollary 5.21) $p_1 + p_2 \in M(z_0)$, then

$$S_F(x_0, p_1 + p_2) = \langle p_1 + p_2, y_0 \rangle = \langle p_1, y_0 \rangle + \langle p_2, y_0 \rangle = S_F(x_0, p_1) + S_F(x_0, p_2).$$

Thus we get

$$S'_F(x_0, p_1 + p_2; \bar{x}) \leq \lim_{\varepsilon \downarrow 0} \varepsilon^{-1}[S_F(x_0 + \varepsilon \bar{x}, p_1) - S_F(x_0, p_1)]$$
$$+ \lim_{\varepsilon \downarrow 0} \varepsilon^{-1}[S_F(x_0 + \varepsilon \bar{x}, p_2) - S_F(x_0, p_2)] = S'_F(x_0, p_1; \bar{x}) + S'_F(x_0, p_2; \bar{x}). \quad \blacksquare$$

THEOREM 5.24 *Let* $z_0 = (x_0, y_0)$, $y_0 \in F(x_0)$, *and let* F *be w.u.d. at the point* x_0 *in the direction* $\bar{x}$. *Then*

$$\hat{D}_L F(z_0; \bar{x}) = \Gamma'(z_0; \bar{x}).$$

Proof. Due to Lemma 5.7, the function $\langle p, y \rangle - S_F(x, p)$ is upper w.u.d. in the direction $\bar{z} = (\bar{x}, \bar{y})$ at the point (z_0, p) for all $\bar{y} \in Y$, $p \in Y$. Since (see Section 1)

$$d_F(z) = \max \{\langle p, y \rangle - S_F(x, p) \,|\, p \in \bar{B}\},$$

then applying Theorem 5.9 and taking into account Lemma 3.18, we get

$$d'_F(z_0; \bar{z}) = \sup\{\langle p, \bar{y} \rangle - S'_F(x_0, p; \bar{x}) \,|\, p \in M(z_0) \cap \bar{B}\}.$$

In view of the definition of $\hat{D}_L F(z_0; \bar{x})$, we thus get the statement of the theorem. $\blacksquare$

Let $f : X \times Y \to R$ be a smooth function. We consider the function

$$\Phi(x) = \sup\{f(x, y) \,|\, y \in F(x)\}$$

and the set
$$\Omega(x) = \{y \in F(x) \mid f(x, y) = \Phi(x)\}.$$

LEMMA 5.25 *Let* $\bar{x} \in X$. *Then*

$$D_+\Phi(x_0; \bar{x}) \geq \sup_{y_0 \in \Omega(x_0)} \sup_{\bar{y} \in \hat{D}_L F(z_0; \bar{x})} \langle \nabla f(z_0), \bar{z} \rangle. \qquad (5.8)$$

Proof. Let $y_0 \in \Omega(x_0)$, $\bar{y} \in \hat{D}_L F(z_0; \bar{x})$. Then according to the definition of $\hat{D}_L F(z_0; \bar{x})$, there exists a function $o(\varepsilon)$ such that

$$y_0 + \varepsilon\bar{y} + o(\varepsilon) \in F(x_0 + \varepsilon\bar{x}) \qquad \forall \varepsilon \geq 0.$$

Consequently, we get

$$\Delta\Phi = \Phi(x_0 + \varepsilon\bar{x}) - \Phi(x_0) \geq f(x_0 + \varepsilon\bar{x}, y_0 + \varepsilon\bar{y} + o(\varepsilon)) - f(x_0, y_0),$$

and, therefore,

$$D_+\Phi(x_0; \bar{x}) = \liminf_{\varepsilon \downarrow 0} \varepsilon^{-1}\Delta\Phi \geq \langle \nabla f(z_0), \bar{z} \rangle$$

for all $\bar{y} \in \hat{D}_L F(z_0; \bar{x})$, what is equivalent to (5.8). ∎

LEMMA 5.26 *Let* $y_0 \in \Omega(x_0)$. *Then* $\nabla_y f(z_0) \in M(z_0)$.

Proof. Since $y_0 \in \Omega(x_0)$, then $f(x_0, y)$ attains a maximum on the set $F(x_0)$ at the point y_0. Due to the necessary optimality condition (see, e.g., Lemma 3.23), we get $\langle \nabla_y f(z_0), y - y_0 \rangle \leq 0$ for all $y \in F(x_0)$ and, according to Lemma 5.20, it follows that $\nabla_y f(z_0) \in M(z_0)$. ∎

LEMMA 5.27 *Let* F *be w.u.d. at the point* x_0 *in the direction* $\bar{x}$. *Then for all* $y_0 \in \Omega(x_0)$ *the following inequality is true:*

$$D_+\Phi(x_0; \bar{x}) \geq \langle \nabla_x f(z_0), \bar{x} \rangle + S'_F(x_0, \nabla_y f(z_0); \bar{x}).$$

Proof. Due to Lemma 5.26, we have $\nabla_y f(z_0) \in M(z_0)$, and therefore $\langle \nabla_y f(z_0), y_0 \rangle = S_F(x_0, \nabla_y f(z_0))$. Then, according to Lemma 5.25 and Theorem 5.24, we get the desired statement. ∎

LEMMA 5.28 *Let* F *be w.u.d. at the point* x_0 *in the direction* $\bar{x}$. *Then* $\Omega(x_0 + \varepsilon\bar{x})$ *is upper semicontinuous with respect to* ε *at the point* $\varepsilon = +0$.

Proof. With regard to Lemma 5.27

$$\liminf_{\varepsilon \downarrow 0} \Phi(x_0 + \varepsilon\bar{x}) \geq \Phi(x_0),$$

i.e., $\Phi(x_0 + \varepsilon\bar{x})$ is lower semicontinuous at the point $\varepsilon = +0$. In view of the closedness of F (cf. Lemma 3.23), we thus obtain the upper semicontinuity of $\Omega(x_0 + \varepsilon\bar{x})$ at the point $\varepsilon = +0$. ∎

THEOREM 5.29 *Let the mapping F be w.u.d. at the point x_0 in the direction $\bar{x}$, and let the smooth function f satisfy the condition*

$$f(x_0, y) - f(x_0, y_0) \leq \langle \nabla_y f(z_0), y - y_0 \rangle \qquad (5.9)$$

for all $y_0 \in \Omega(x_0)$ and all y from a neighbourhood of $\Omega(x_0)$. Then there exists $\Phi'(x_0; \bar{x})$, where

$$\Phi'(x_0; \bar{x}) = \sup_{y_0 \in \Omega(x_0)} \{ \langle \nabla_x f(z_0), \bar{x} \rangle + S'_F(x_0, \nabla_y f(z_0); \bar{x}) \}.$$

Proof. Let

$$D^+ \Phi(x_0; \bar{x}) \overset{\triangle}{=} \limsup_{\varepsilon \downarrow 0} \varepsilon^{-1} [\Phi(x_0 + \varepsilon\bar{x}) - \Phi(x_0)]$$

be attained on some sequence $\varepsilon_k \downarrow 0$. We choose a sequence $y_k \in \Omega(x_0 + \varepsilon_k \bar{x})$, $k = 1, 2, \ldots$ Due to the uniform boundedness of F at the point x_0, this sequence is bounded. Hence, without loss of generality, we can assume that $y_k \to y_0$, where $y_0 \in \Omega(x_0)$ according to Lemma 5.28. Applying relation (5.9) and Lemma 5.26, we get

$$\Delta\Phi(\varepsilon_k) \overset{\triangle}{=} \Phi(x_0 + \varepsilon_k \bar{x}) - \Phi(x_0) = f(x_0 + \varepsilon_k \bar{x}, y_k) - f(x_0, y_0)$$
$$\leq [f(x_0 + \varepsilon_k \bar{x}, y_k) - f(x_0, y_k)] + [f(x_0, y_k) - f(x_0, y_0)]$$
$$\leq \varepsilon_k \langle \nabla_x f(x_0 + \tau_k \varepsilon_k \bar{x}, y_k), \bar{x} \rangle + \langle \nabla_y f(x_0, y_0), y_k - y_0 \rangle$$
$$\leq \varepsilon_k \langle \nabla_x f(x_0 + \tau_k \varepsilon_k \bar{x}, y_k), \bar{x} \rangle + \max \{ \langle \nabla_y f(z_0), y - y_0 \rangle \mid y \in F(x_0 + \varepsilon_k \bar{x}) \}$$
$$= \varepsilon_k \langle \nabla_x f(x_0 + \tau_k \varepsilon_k \bar{x}, y_k), \bar{x} \rangle + S_F(x_0 + \varepsilon_k \bar{x}, \nabla_y f(z_0)) - S_F(x_0, \nabla_y f(z_0)),$$

where $\tau_k \in (0, 1)$. Dividing this inequality on $\varepsilon_k > 0$ and passing to the limit for $\varepsilon_k \downarrow 0$, we get

$$D^+ \Phi(x_0; \bar{x}) \leq \langle \nabla_x f(z_0), \bar{x} \rangle + S'_F(x_0, \nabla_y f(z_0); \bar{x})$$
$$\leq \sup_{y_0 \in \Omega(x_0)} \{ \langle \nabla_x f(z_0), \bar{x} \rangle + S'_F(x_0, \nabla_y f(z_0); \bar{x}) \}.$$

From this in view of Lemma 5.27 we obtain the statement of the theorem. ∎

REMARK. Condition (5.9) always holds true if the function $f(x_0, y)$ is concave in a neighbourhood of the set $\Omega(x_0)$.

We can also obtain another version of Theorem 5.29 more suitable for calculations by imposing more stringent conditions on the function f.

THEOREM 5.30 *Let the mapping F be w.u.d. at the point x_0 in the direction $\bar{x}$, and let the function f satisfy the condition*

$$f(z) - f(z_0) \leq \langle \nabla f(z_0), z - z_0 \rangle \qquad (5.10)$$

at every point $z_0 \in \{x_0\} \times \Omega(x_0)$ and for all z from a neighbourhood of the set $\{x_0\} \times \Omega(x_0)$. Then for all $y_0 \in \Omega(x_0)$,

$$\Phi'(x_0; \bar{x}) = \langle \nabla_x f(z_0), \bar{x} \rangle + S'_F(x_0, \nabla_y f(z_0); \bar{x}).$$

Proof. Let $y_0 \in \Omega(x_0)$. Applying Lemma 5.26 and condition (5.10), we get

$$\Delta\Phi(\varepsilon) \overset{\triangle}{=} \Phi(x_0 + \varepsilon\bar{x}) - \Phi(x_0) = \sup_{y \in \Omega(x_0 + \varepsilon\bar{x})} [f(x_0 + \varepsilon\bar{x}, y) - f(x_0, y_0)]$$

$$\leq \sup_{y \in F(x_0 + \varepsilon\bar{x})} [\varepsilon\langle \nabla_x f(z_0), \bar{x} \rangle + \langle \nabla_y f(z_0), y_k - y_0 \rangle]$$

$$= \varepsilon\langle \nabla_x f(z_0), \bar{x} \rangle + S_F(x_0 + \varepsilon\bar{x}, \nabla_y f(z_0)) - S_F(x_0, \nabla_y f(z_0)).$$

Hence

$$D^+\Phi(x_0; \bar{x}) \overset{\triangle}{=} \limsup_{\varepsilon \downarrow 0} \varepsilon^{-1}\Delta\Phi(\varepsilon) \leq \langle \nabla_x f(z_0), \bar{x} \rangle + S'_F(x_0, \nabla_y f(z_0); \bar{x}).$$

Uniting the last inequality and (5.8), we get the statement of the theorem. ∎

REMARK. Condition (5.10) always holds if f is a concave function.

Let us note that Theorems 5.29 and 5.30 generalize results from [35, 55, 122, 149, 150]. The formulas obtained for the derivative $\Phi'(x_0; \bar{x})$ can be used if one knows the derivative $S'_F(x_0, p; \bar{x})$. On the other hand, if F is w.u.d. at x_0 in the direction $\bar{x}$, then

$$S'_F(x_0, p; \bar{x}) = \sup\{\langle p, \bar{y} \rangle \mid \bar{y} \in \hat{D}_L F(z_0; \bar{x})\}.$$

Thus, due to this fact we can rewrite the formulas for $\Phi'(x_0; \bar{x})$ from Theorems 5.29 and 5.30.

THEOREM 5.31 *Let the mapping F be w.u.d. at the point x_0 in the direction $\bar{x}$ and the function $f(x_0, y)$ be concave in a neighbourhood of the set $\Omega(x_0)$. Then*

$$\Phi'(x_0; \bar{x}) = \sup_{y_0 \in \Omega(x_0)} \sup_{\bar{y} \in \hat{D}_L F(z_0; \bar{x})} \langle \nabla f(z_0), \bar{z} \rangle.$$

If the function f is concave in a neighbourhood of $\{x_0\} \times \Omega(x_0)$, then

$$\Phi'(x_0; \bar{x}) = \sup_{\bar{y} \in \hat{D}_L F(z_0; \bar{x})} \langle \nabla f(z_0), \bar{z} \rangle$$

for any $y_0 \in \Omega(x_0)$.

Let us note that similar theorems can be proved for the function φ.

3.   Strongly Differentiable Mappings and Directional Differentiability of Marginal Functions

As has been shown in the last section, uniformly differentiable and convex multivalued mappings belong to the class of w.u.d. mappings. For this class Theorems 5.29–5.31 are valid which allow us to calculate directional derivatives of marginal functions. In the present section we consider another well-known class of multivalued mappings, so-called strongly differentiable mappings. The concept of strong differentiability has been introduced by Tiurin [173] and Banks and Jacobs [14]. Later on this type of differentiability was studied by Pecherskaya [140].

Below it will be shown that the class of strongly differentiable multivalued mapping is embedded in the considerably broader class of w.u.d. mappings. It should be mentioned, however, that the assumption of strong differentiability of a multivalued mapping requires stronger conditions concerning the multivalued mappings under study. On the other hand, it allows to obtain more powerful results about directional differentiability of marginal functions.

3.1   Strong differentiability of multivalued mappings

As in Chapter 2, we denote by $CS(Y)$ (resp. $CCS(Y)$) the set of all non-empty compact (non-empty convex compact) subsets in Y. Moreover, we consider multivalued mappings $F : X \to CS(Y)$ (resp. $F : X \to CCS(Y)$) with non-empty compact values (non-empty convex compact values) $F(x) \subset Y$, where as above $X = R^n$, $Y = R^m$.

Let $\bar{x} \in X$. The following notion of strong differentiability has been introduced in [14], [173].

DEFINITION 5.32 *A mapping $F: X \to CCS(Y)$ is called strongly differentiable at the point $x \in X$ in the direction $\bar{x}$ if there exist two non-empty convex compact subsets $G^+(x, \bar{x})$ and $G^-(x, \bar{x})$ in Y such that*

$$\rho_H(F(x + t\bar{x}) + tG^-(x, \bar{x}), F(x) + tG^+(x, \bar{x})) = o(t) \qquad (5.11)$$

where $o(t)/t \to 0$ *for* $t \downarrow 0$.

Along with this definition we introduce the following modified notion.

DEFINITION 5.33 *A multivalued mapping* $F : X \to CS(Y)$ *is called strongly differentiable (in a broad sense) at the point* x *in the direction* $\bar{x}$ *if there exist two non-empty compact subsets* $G^+(x, \bar{x})$ *and* $G^-(x, \bar{x})$ *such that (5.11) holds.*

Note that in general the pair $(G^+(x, \bar{x}), G^-(x, \bar{x}))$ in Definition 5.32 or Definition 5.33 is not uniquely defined. Furthermore, from 5.11 it follows that the mapping $F(x + \varepsilon\bar{x})$ is continuous in the sense of Hausdorff with respect to ε at the point $\varepsilon = +0$.

It is well-known that strong differentiability (in the sense of Definition 5.32) has the following simple interpretation (see [14], [60]).

Let us consider the set $CCS(Y) \times CCS(Y)$. The elements of this set are pairs (A, C) of non-empty convex compact sets A and C. For elements of $CCS(Y) \times CCS(Y)$ we define the addition as

$$(A_1, C_1) + (A_2, C_2) = (A_1 + A_2, C_1 + C_2)$$

and the multiplication by a number α via

$$\alpha(A, C) = \begin{cases} (\alpha A, \alpha C) & \text{if } \alpha \geq 0 \\ (\alpha C, \alpha A) & \text{if } \alpha < 0. \end{cases}$$

Introducing an equivalence relation on $CCS(Y) \times CCS(Y)$ by the relation

$$(A, C) \sim (B, D) \iff A + D = B + C,$$

we obtain the set $\mathcal{L}$ of equivalence classes of pairs of convex compact sets. It is known that this set $\mathcal{L}$ is a linear space in which the sum of two classes is defined as the class containing the sum of two arbitrary representatives of these classes. It is quite obvious that the zero element in $\mathcal{L}$ is the class of all pairs (A, A). The multiplication of a class by a number is defined in an analogous way.

Following [14] and [60], we define the distance between two classes containing (A, B) and (C, D):

$$d((A, B), (C, D)) = \rho_H(A + D, B + C).$$

In doing so, the space $\mathcal{L}$ becomes a metric space. From Definition 5.32 it follows that the mapping $F : X \to CCS(Y)$ is strongly differentiable at the point x in the direction $\bar{x}$ if and only if the mapping $\tilde{F} : X \to \mathcal{L}$ defined by $\tilde{F}(x) = (F(x), 0)$ is differentiable at the point x in the direction $\bar{x}$, i. e., there is an element $\tilde{F}'(x,\bar{x}) = (G^+(x,\bar{x}), G^-(x,\bar{x}))$ from $\mathcal{L}$ such that

$$\tilde{F}(x + t\bar{x}) = \tilde{F}(x) + t\tilde{F}'(x,\bar{x}) + \tilde{o}_{x,\bar{x}}(t), \quad t \geq 0,$$

where $d(0, \tilde{o}_{x,\bar{x}}(t)) \cdot t^{-1} \to 0$ for $t \downarrow 0$. Hence $\tilde{F}'(x,\bar{x}) = (G^+(x,\bar{x}), G^-(x,\bar{x}))$ is the ordinary derivative of $\tilde{F}$ at the point x in the direction $\bar{x}$.

Simple examples of strongly differentiable mappings are the following:

1) $F(x) = \sum\limits_{i=1}^{k} f_i(x) C_i$, where $f_i : X \to R$ are continuously differentiable functions, $C_i \subset Y$ are convex compact sets for $i = 1, \ldots, k$. Let us note that this simultaneously proves to be an example of strong differentiability in a broad sense if C_i are only compact sets.

2) $F(x) = \{y \in R^m \mid y = (y_1, \ldots, y_m), \ g_i(x) \leq y_i \leq h_i(x), \ i = 1, \ldots, m\}$, where $g_i(x)$ and $h_i(x)$ are differentiable functions for $i = 1, \ldots, m$.

Let $F : X \to CCS(Y)$. By $S_F(x,p)$ we denote the support function of the set $F(x)$ and consider the sets

$$M(x,y) = \{p \in R^m \mid \langle p, y \rangle = S_F(x,p)\}$$

with $y \in F(x)$ and

$$\Gamma'(x,y;\bar{x}) = \{\bar{y} \in Y \mid \langle p, \bar{y} \rangle \leq S'_F(x,p;\bar{x}) \quad \forall p \in M(x,y) \cap \bar{B}\}.$$

Let us recall that $M(x,y)$ is a closed convex cone. The set $\Gamma'(x,y;\bar{x})$ is convex as well and its support function $S_{\Gamma'}(x,y;\bar{x})$ is determined by

$$S_{\Gamma'}(x,y;\bar{x}) = \begin{cases} S'_F(x,p;\bar{x}) & \text{if } p \in M(x,y) \\ +\infty & \text{if } p \notin M(x,y). \end{cases}$$

Now we study several properties of strongly differentiable multivalued mappings.

Let $F : X \to CCS(Y)$ be a multivalued mapping strongly differentiable at the point x in the direction $\bar{x}$. We denote by $S^+(x,\bar{x},p)$ and $S^-(x,\bar{x},p)$ the support functions of the sets $G^+(x,\bar{x})$ and $G^-(x,\bar{x})$, respectively. Due to relation (2.7) and taking into account the properties of support functions, from (5.11) we get

$$\max_{|p| \leq 1} \{S_F(x + t\bar{x}, p) + tS^-(x,\bar{x},p) - S_F(x,p) - tS^+(x,\bar{x},p)\} = o(t). \quad (5.12)$$

This means that the derivative $S'_F(x,p;\bar{x})$ does exits, where

$$S'_F(x,p;\bar{x}) = S^+(x,\bar{x},p) - S^-(x,\bar{x},p) \quad (5.13)$$

for all $p \in Y$. Moreover, from (5.12) we deduce that the support function S_F is uniformly (with respect to p) differentiable in the direction

$\bar{x}$. Therefore, using Lemma 5.17 and Theorem 5.24, we can prove the following property of strongly differentiable mappings.

THEOREM 5.34 *Let* $F : X \to CCS(Y)$ *be strongly differentiable at* x *in the direction* $\bar{x}$. *Then for all* $y \in F(x)$, *we have*

$$\hat{D}_L F(x,y;\bar{x}) + G^-(x,\bar{x}) = G^+(x,\bar{x}) + \hat{D}_L F(x,y;0). \tag{5.14}$$

Proof. As a consequence of (5.13) the derivative $S'_F(x,p;\bar{x})$ is continuous with respect to p. On the other hand, the support functions of the sets $G^+(x,\bar{x}) + \Gamma'(x,y;0)$ and $G^-(x,\bar{x}) + \Gamma'(x,y;0)$ are equal to $S^+(x,\bar{x},p)$ and $S^-(x,\bar{x},p)$, respectively, if $p \in M(x,y)$ and equal to $+\infty$ provided that $p \notin M(x,y)$. Since the coincidence of support functions to convex sets implies the coincidence of the closures of these sets, then by virtue of (5.13) we get

$$\begin{aligned} \text{cl} \, [\Gamma'(x,y;\bar{x}) + G^-(x,\bar{x}) + \Gamma'(x,y;0)] \\ = \text{cl} \, [G^+(x,\bar{x}) + \Gamma'(x,y;0)]. \end{aligned} \tag{5.15}$$

It is not hard to see that $\Gamma'(x,y;\bar{x}) + \Gamma'(x,y;0) = \Gamma'(x,y;\bar{x})$, i.e., the recession cone $0^+\Gamma'(x,y;\bar{x})$ coincides with $\Gamma'(x,y;0)$. Thus the sets

$$\Gamma'(x,y;\bar{x}) + G^-(x,\bar{x}) + \Gamma'(x,y;0) = \Gamma'(x,y;\bar{x}) + G^-(x,\bar{x})$$

and

$$G^+(x,\bar{x}) + \Gamma'(x,y;0)$$

are closed (see Corollary 9.1.2 in [154]). Consequently, (5.15) can be written in the form

$$\Gamma'(x,y;\bar{x}) + G^-(x,\bar{x}) = G^+(x,\bar{x}) + \Gamma'(x,y;0).$$

Applying now Theorem 5.24, the statement has been proved. ■

3.2 Directional differentiability of marginal functions

Let $f : X \times Y \to R$ be a smooth function and assume $F : X \to CS(Y)$. We consider the marginal function $\Phi(x) = \max \{f(x,y) \,|\, y \in F(x)\}$.

THEOREM 5.35 *Let the multivalued mapping* F *be strongly differentiable at the point* x_0 *in the direction* $\bar{x}$. *Then the marginal function* Φ *is directionally differentiable at* x_0 *in the direction* $\bar{x}$ *and*

$$\Phi'(x_0;\bar{x}) =$$

$$\max_{y_0 \in \Omega(x_0)} \, \max_{y^+ \in G^+(\bar{x})} \, \min_{y^- \in G^-(\bar{x})} \{\langle \nabla_x f(z_0), \bar{x} \rangle + \langle \nabla_y f(z_0), y^+ - y^- \rangle\},$$

$$\tag{5.16}$$

where $z_0 = (x_0, y_0)$, $G^+(\bar{x}) = G^+(x, \bar{x})$, $G^-(\bar{x}) = G^-(x, \bar{x})$.

Proof. Let

$$\Delta\Phi(t) \overset{\triangle}{=} \Phi(x_0 + t\bar{x}) - \Phi(x_0),$$

$$D^+\Phi(x_0; \bar{x}) \overset{\triangle}{=} \limsup_{t\downarrow 0} t^{-1}\Delta\Phi(t),$$

$$D_+\Phi(x_0; \bar{x}) \overset{\triangle}{=} \liminf_{t\downarrow 0} \Delta\Phi(t).$$

Suppose that $D^+\Phi(x_0; \bar{x})$ is attained on the sequence $t_k \downarrow 0$. We choose an arbitrary sequence $y_k \in \Omega(x_0 + t_k\bar{x})$, $k = 1, 2, \dots$. Since the mapping $F(x_0 + t\bar{x})$ is continuous in the sense of Hausdorff with respect to t at the point $t = 0$ (for $t \geq 0$), then in view of Lemma 3.22 it follows that the mapping $\Omega(x_0 + t\bar{x})$ is upper semicontinuous with respect to ε at the point $\varepsilon = 0$. Furthermore, due to the continuity of $F(x_0 + t\bar{x})$ and the compactness of the sets $F(x)$ we get the boundedness of the sequence $\{y_k\}$ (see Lemma 3.2). Thus, without loss of generality we can assume that $y_k \to y_0$, where $y_0 \in \Omega(x_0)$.

Let us take an arbitrary element $y^- \in G^-(\bar{x})$. Then due to Definition 5.33, for every $k = 1, 2, \dots$ there exist $v_k \in F(x_0)$ and $a_k \in G^+(\bar{x})$ such that $y_k + t_k y^- = v_k + t_k a_k + o(t_k)$, $k = 1, 2, \dots$ Then

$$y_k = v_k + t_k(a_k - y^-) + o(t_k), \quad k = 1, 2, \dots \tag{5.17}$$

Since $G^+(\bar{x})$ is a compact set and since $y_k \to y_0$, we get $v_k \to y_0$.

On the other hand, without loss of generality it can be assumed that $a_k \to a^+ \in G^+(\bar{x})$. Thus, taking into account $v_k \in F(x_0)$ and $y_0 \in \Omega(x_0)$, we obtain

$$t_k^{-1}\Delta\Phi(t_k) = t_k^{-1}[f(x_0 + t_k\bar{x}, y_k) - f(x_0, y_0)]$$
$$\leq t_k^{-1}[f(x_0 + t_k\bar{x}, y_k) - f(x_0, y_k) + f(x_0, y_k) - f(x_0, v_k)]$$
$$= \langle \nabla_x f(x_0, y_k), \bar{x}\rangle + \langle \nabla_y f(x_0, v_k), y_k - v_k\rangle + o(t_k),$$

where $o(t_k)/t_k \to 0$ for $t_k \downarrow 0$. Dividing this inequality by t_k and passing to the limit for $t_k \downarrow 0$, in view of (5.17) we get

$$D^+\Phi(x_0; \bar{x}) \leq \langle \nabla_x f(z_0), \bar{x}\rangle + \langle \nabla_y f(z_0), a^+ - y^-\rangle$$

for any $y^- \in G^-(\bar{x})$. Therefore

$$D^+\Phi(x_0; \bar{x}) \leq \langle \nabla_x f(z_0), \bar{x}\rangle + \max_{y^+ \in G^+(\bar{x})} \langle \nabla_y f(z_0), y^+ - y^-\rangle,$$

i. e.

$$D^+\Phi(x_0;\bar{x}) \leq \langle \nabla_x f(z_0), \bar{x} \rangle + \min_{y^- \in G^-(\bar{x})} \max_{y^+ \in G^+(\bar{x})} \langle \nabla_y f(z_0), y^+ - y^- \rangle$$

$$= \langle \nabla_x f(z_0), \bar{x} \rangle + \max_{y^+ \in G^+(\bar{x})} \min_{y^- \in G^-(\bar{x})} \langle \nabla_y f(z_0), y^+ - y^- \rangle.$$

We finally get

$$D^+\Phi(x_0;\bar{x})$$
$$\leq \max_{y_0 \in \Omega(x_0)} \{ \langle \nabla_x f(z_0), \bar{x} \rangle + \max_{y^+ \in G^+(\bar{x})} \min_{y^- \in G^-(\bar{x})} \langle \nabla_y f(z_0), y^+ - y^- \rangle \}.$$
$$\tag{5.18}$$

Now we choose arbitrary elements $y_0 \in \Omega(x_0)$ and $y^+ \in G^+(\bar{x})$. Let $D_+\Phi(x_0;\bar{x})$ be attained on the sequence $t_k \downarrow 0$. Due to (5.11), there exist $w_k \in F(x_0 + t_k\bar{x})$ and $b_k \in G^-(\bar{x})$ such that

$$y_0 + t_k y^+ = w_k + t_k b_k + o(t_k), \tag{5.19}$$

where $o(t_k)/t_k \to 0$ for $k \to \infty$. Owing to the compactness of $G^-(\bar{x})$, the sequence $\{b_k\}$ is bounded. Therefore, without loss of generality we can assume that $b_k \to \tilde{y}^- \in G^-(\bar{x})$. In this case (5.19) can be written in the form

$$w_k = y_0 + t_k(y^+ - \tilde{y}^-) + o(t_k) \in F(x_0 + t_k\bar{x}), \quad k = 1,2,\dots \tag{5.20}$$

Hence $t_k^{-1}\Delta\bar{\Phi}(t_k) \geq t_k^{-1}[f(x_0 + t_k\bar{x}, w_k) - f(x_0, y_0)]$ which leads us to

$$D_+\Phi(x_0;\bar{x}) \geq \langle \nabla_x f(z_0), \bar{x} \rangle + \langle \nabla_y f(z_0), y^+ - \tilde{y}^- \rangle$$

for all $y_0 \in \Omega(x_0)$ and all $y^+ \in G^+(\bar{x})$. This means that

$$D_+\Phi(x_0;\bar{x}) \geq \langle \nabla_x f(z_0), \bar{x} \rangle + \min_{y^- \in G^-(\bar{x})} \langle \nabla_y f(z_0), y^+ - y^- \rangle$$

for all $y_0 \in \Omega(x_0)$ and all $y^+ \in G^+(\bar{x})$. Consequently

$$D_+\Phi(x_0;\bar{x})$$
$$\geq \max_{y_0 \in \Omega(x_0)} \max_{y^+ \in G^+(\bar{x})} \{ \langle \nabla_x f(z_0), \bar{x} \rangle + \min_{y^- \in G^-(\bar{x})} \langle \nabla_y f(z_0), y^+ - y^- \rangle \}.$$

From the last inequality and (5.18) the relation (5.16) follows. ∎

In the case of a compact-valued mapping $F : X \to CCS(Y)$ we can modify Theorem 5.35 in the following way.

THEOREM 5.36 *Suppose that the multivalued mapping $F\colon X \to CCS(Y)$ is strongly differentiable at the point x_0 in the direction $\bar{x}$. Then*

$$\Phi'(x_0; \bar{x}) = \max_{y_0 \in \Omega(x_0)} \max_{\bar{y} \in \hat{D}_L F(z_0; \bar{x})} \langle \nabla f(z_0), \bar{z} \rangle, \qquad (5.21)$$

where $z_0 = (x_0, y_0)$, $\bar{z} = (\bar{x}, \bar{y})$.

Proof. For shortness we denote $G^+ = G^+(x_0, \bar{x})$, $G^- = G^-(x_0, \bar{x})$, and $\tilde{G} = G^+(x, \bar{x}) + \hat{D}_L F(z_0; 0)$. Due to Lemma 5.26, we have $\nabla_y f(z_0) \in M(z_0)$. Applying now (5.14), we obtain

$$\max_{y^+ \in G^+} \langle \nabla_y f(z_0), y^+ \rangle = \max_{y^- \in G^-} \langle \nabla_y f(z_0), y^- \rangle + \max_{\bar{y} \in \hat{D}_L F(z_0; \bar{x})} \langle \nabla_y f(z_0), \bar{y} \rangle.$$

Consequently, one obtains

$$\max_{y^+ \in G^+} \min_{y^- \in G^-} \langle \nabla_y f(z_0), y^+ - y^- \rangle$$
$$= \max_{y^+ \in G^+} \langle \nabla_y f(z_0), y^+ \rangle - \max_{y^- \in G^-} \langle \nabla_y f(z_0), y^- \rangle = \max_{\bar{y} \in \hat{D}_L F(z_0; \bar{x})} \langle \nabla_y f(z_0), \bar{y} \rangle.$$

With regard to the last equality, we can write relation (5.16) in the form (5.21). ∎

Chapter 6

FIRST AND SECOND ORDER SENSITIVITY ANALYSIS OF PERTURBED MATHEMATICAL PROGRAMMING PROBLEMS

In this chapter we suppose that the function $f : X \times Y \to R$ is continuously differentiable, where $X = R^n$ and $Y = R^m$. We consider the mathematical programming problem

$$(\bar{P}_x) : \qquad \begin{cases} f(x,y) \to \inf_y \\ y \in F(x), \end{cases}$$

with $x \in X$ being a vector of parameters.

If the multivalued mapping F is defined with the help of equality and/or inequality constraints, i. e.

$$F(x) = \{\, y \in Y \mid h_i(x,y) \le 0, \ i=1,\ldots,r, \ h_i(x,y)=0, \ i=r+1,\ldots,p \,\},$$

then we deal with the classical problem of *nonlinear programming*

$$(P_x) : \qquad \begin{cases} f(x,y) \to \inf_y \\ h_i(x,y) \le 0, \quad i = 1,\ldots,r, \\ h_i(x,y) = 0, \quad i = r+1,\ldots,p. \end{cases}$$

(Here we assume that $1 \le r \le p$; the case $r = p$ means that there are only inequality constraints).

We consider *the optimal value function*

$$\varphi(x) = \inf\{f(x,y) \mid y \in F(x)\}$$

as well as the *set of optimal solutions*

$$\omega(x) = \{y \in F(x) \mid \varphi(x) = f(x,y)\}$$

associated with the problems $(\bar{P}_x)$ and (P_x).

Let $x_0 \in \text{dom}\, F$. Throughout this chapter we suppose that the following assumption holds:

(A1)   –   the multivalued mapping F is uniformly bounded at the point x_0, i.e., there exist a compact set $Y_0 \subset Y$ and a neighbourhood $X_0 \subset R^n$ of the point x_0 such that $F(X_0) \subset Y_0$.

1.   Stability Properties of Optimal Solutions in Mathematical Programming Problems

Let us consider the mathematical programming problem $(\bar{P}_x)$ introduced above.

Besides (A1) we suppose the following assumptions to be fulfilled:

(A2)   –   the multivalued mapping F is closed, which means that its graph $gr F = \{(x, y) \mid x \in X,\ y \in F(x)\}$ is a closed set in the space $X \times Y$;

(A3)   –   the function f is locally Lipschitz continuous on the set $X_0 \times [Y_0 + \varepsilon_0 B]$, where $\varepsilon_0 \geq 2\,\text{diam}\, Y_0$ and B is the open unit ball.

Let $y_0 \in \omega(x_0)$, $r > 0$. We denote

$$F_r(x, y_0) = F(x) \cap (y_0 + r\bar{B}),$$
$$\varphi_r(x, y_0) = \inf\{f(x, y) \mid y \in F_r(x, y_0)\},$$
$$\omega_r(x, y_0) = \{y \in F_r(x, y_0) \mid f(x, y) = \varphi_r(x, y_0)\}.$$

Note that for any $y_0 \in \omega(x_0)$ we always have $\varphi(x) \leq \varphi_r(x, y_0)$, $x \in V(x_0)$, and $\varphi(x_0) = \varphi_r(x_0, y_0)$.

DEFINITION 6.1 *The set of optimal solutions $\omega(x)$ in the problem $(\bar{P}_x)$ is referred to as locally Lipschitz (or Hölder) continuous at the point (x_0, y_0), $y_0 \in \omega(x_0)$, if there exist numbers $l_0 > 0$, $r > 0$ and a neighbourhood $V(x_0)$ of the point x_0 such that the inequality*

$$|y - y_0|^i \leq l_0 |x - x_0| \tag{6.1}$$

holds for all $x \in V(x_0)$, $y \in \omega_r(x, y_0)$, where $i = 1$ (or $i = 2$).

The set of optimal solutions $\omega(x)$ is said to be globally Lipschitz (or Hölder) continuous at the point (x_0, y_0) if the inequality (6.1) holds for $r = +\infty$.

We will make use of the following definition of a local isolated minimizer (see Auslender [6], Ward [176])

DEFINITION 6.2 *We say that the point $y_0 \in \omega(x_0)$ is a local isolated minimizer of order i ($i=1,\ 2$) for the problem $(\bar{P}_x)$ if there exist $\delta > 0$*

and $r > 0$ such that

$$f(x_0, y) > f(x_0, y_0) + \delta| y - y_0 |^i \qquad (6.2)$$

for all $y \in F_r(x_0, y_0)$, $y \neq y_0$.

Clearly, if $y_0 \in \omega(x_0)$ is a local isolated minimizer, then $\omega(x_0) = \{y_0\}$.

THEOREM 6.3 *Let the assumptions (A1)–(A3) hold and the multivalued mapping F be pseudo-Lipschitz continuous at the point (x_0, y_0), where $y_0 \in \omega(x_0)$. Furthermore, let the point y_0 be a local isolated minimizer of order i (i=1, 2) for the problem $(\bar{P}_x)$. Then one can find a number $l > 0$ and a neighbourhood $V(x_0)$ such that*

$$|y - y_0|^i \leq l|x - x_0|$$

for all $x \in V(x_0)$, $y \in \omega_r(x, y_0)$, i. e., the set of optimal solutions is locally Lipschitz continuous for $i = 1$ and locally Hölder continuous for $i = 2$.

Proof. We introduce the auxiliary objective function

$$g(x, y) = f(x, y) + \delta|y - y_0|^i,$$

where δ, i and y_0 are supposed to be fixed and defined as in Definition 6.2.

We make use of the notation

$$\varphi_f(x) = \inf\{f(x, y) \mid y \in F_r(x, y_0)\},$$
$$\varphi_g(x) = \inf\{g(x, y) \mid y \in F_r(x, y_0)\},$$
$$\omega_{gr}(x) = \{y \in F_r(x, y_0) \mid g(x, y) = \varphi_g(x)\}.$$

Let us show that $\omega_{gr}(x_0) = \{y_0\}$. In fact, $g(x_0, y_0) = f(x_0, y_0) = \varphi_f(x_0) \leq \varphi_g(x_0)$. Due to $y_0 \in F_r(x_0, y_0)$ the last inequality means $g(x_0, y_0) = \varphi_g(x_0)$ and $y_0 \in \omega_{gr}(x_0)$.

On the other hand, let $v \in F_r(x_0, y_0)\backslash\{y_0\}$. We get that

$$g(x_0, v) = f(x_0, v) + \delta|v - y_0|^i \geq f(x_0, y_0) + \delta|v - y_0|^i$$
$$> f(x_0, y_0) = g(x_0, y_0) = \varphi_g(x_0),$$

i. e. $v \notin \omega_{gr}(x_0)$. Hence $\omega_{gr}(x_0) = \{y_0\}$.

We take now arbitrary elements $x \in V(x_0)$, $y(x) \in \omega_{gr}(x)$. Then the following inequality is true

$$\varphi_g(x) - \varphi_g(x_0) = g(x, y(x)) - g(x_0, y_0)$$
$$= f(x, y(x)) + \delta|y(x) - y_0|^i - f(x_0, y_0)$$
$$\geq f(x, y(x)) - f(x_0, y_0) \geq \varphi_f(x) - \varphi_f(x_0). \qquad (6.3)$$

From the assumptions of the theorem it follows that the functions $\varphi_f(x)$ and $\varphi_g(x)$ are Lipschitz continuous in a sufficiently small neighbourhood $V(x_0)$ (see Corollary 4.23). Let l_1 and l_2 be the corresponding Lipschitz constants on $V(x_0)$. Without loss of generality, we assume that $l_1 = \max\{l_1, l_2\}$. Then

$$-l_1|x - x_0| \le \varphi_f(x) - \varphi_f(x_0) \le f(x, y(x)) - f(x_0, y_0)$$
$$\le \varphi_g(x) - \varphi_g(x_0) \le l_1|x - x_0|,$$

and, hence,

$$|f(x, y(x)) - f(x_0, y_0)| \le l_1|x - x_0|.$$

From (6.2) we get that

$$\delta|y(x) - y_0|^i = (\varphi_g(x) - \varphi_g(x_0)) - (f(x, y(x)) - f(x_0, y_0)) \le (l_1 + l_2)|x - x_0|,$$

i. e.

$$\delta|y - y_0|^i \le l_0|x - x_0| \tag{6.4}$$

for all $y \in \omega_{gr}(x)$, $x \in V(x_0)$.

Now we denote

$$p(x, y) = f(x, y) - \delta|y - y_0|^i,$$
$$\varphi_p(x) = \inf\{p(x, y) \mid y \in F_r(x, y_0)\},$$

so that

$$f(x, y) = p(x, y) + \delta|y - y_0|^i.$$

Since y_0 is a local isolated minimizer of order i, then $p(x_0, y) > p(x_0, y_0)$ for all $y \in F_r(x_0, y_0)$, i. e. $y_0 \in \omega_{pr}(x_0)$, where

$$\omega_{pr}(x) = \{y \in F_r(x, y_0) \mid p(x, y) = \varphi_p(x)\},$$

and $\omega_{pr}(x_0) = \{y_0\}$. Moreover, repeating the above arguments for the function $p(x, y)$, we obtain for any $y(x) \in \omega_r(x, y_0)$ that

$$l_1|x - x_0| \ge \varphi_f(x) - \varphi_f(x_0) = f(x, y(x)) - f(x_0, y_0)$$
$$= p(x, y(x)) + \delta|y(x) - y_0|^i - p(x_0, y_0) \ge p(x, y(x)) - p(x_0, y_0)$$
$$\ge \varphi_p(x) - \varphi_p(x_0) \ge -l_1|x - x_0|.$$

Therefore $|p(x, y(x)) - p(x_0, y_0)| \le l_1|x - x_0|$ and $\delta|y(x) - y_0|^i \le l_0|x - x_0|$ for all $y \in \omega_r(x, y_0)$, $x \in V(x_0)$, $i = 1$ or $i = 2$. ∎

We emphasize that local isolated minimizers play an important role in sensitivity studies. The attempts to extend this concept to problems with non-singleton $\omega(x_0)$ have led to a number of interesting generalizations. One of them is the concept of the growth condition (see [32]).

DEFINITION 6.4 *We say that a growth condition of order i (i=1 or 2) holds for problem* $(\bar{P}_x)$ *if there exist* $\delta > 0$ *and* $r > 0$ *such that*

$$(GC) \qquad f(x_0, y) > \varphi(x_0) + \delta \rho^i(y, \omega(x_0))$$

for all $y \in F(x_0) \cap V_r(\omega(x_0))$, $y \notin \omega(x_0)$, *where* $V_r(\omega(x_0))$ *is a neighbourhood of the set* $\omega(x_0)$ *with radius* r.

Necessary and sufficient second-order conditions for local growth can be found in [33].

In the case $\omega(x_0) = \{y_0\}$ the above definition reduces to the definition of a local isolated minimizers.

Using condition (GC), the following generalization of Theorem 6.3 can be proved.

THEOREM 6.5 *Let the assumptions (A1)–(A3) hold and let the multivalued mapping F be pseudo-Lipschitz continuous at the points* $z_0 \in \{x_0\} \times \omega(x_0)$. *Then condition (GC) implies that there exist* $l > 0$ *and a neighbourhood* $V(x_0)$ *such that for all* $x \in V(x_0)$ *and all* $y \in \omega(x)$ *we have*

$$\rho^i(y, \omega(x_0)) \leq l \mid x - x_0 \mid .$$

Proof. It is quite similar to the proof of Theorem 6.3. ∎

Note that a similar result on stability properties of optimal solutions has been proved in [33] assuming a weaker constraint qualification property of F.

Let us now fix $\bar{x} \in X$. Our aim consist in the study of sensitivity estimates of the optimal solution set $\omega(x)$ belonging to problem $(\bar{P}_x)$ if the parameter x changes in the direction $\bar{x}$.

DEFINITION 6.6 *The set of optimal solutions* $\omega(x)$ *of the problem* $(\bar{P}_x)$ *is said to be sequentially Lipschitz (or Hölder) continuous at the point* x_0 *in the direction* $\bar{x}$, *if for any sequences* $t_k \downarrow 0$ *and* $y_k \to y_0 \in \omega(x_0)$ *such that* $y_k \in \omega(x_0 + t_k \bar{x})$, $k = 1, 2, \ldots$, *the following inequality holds* $(i = 1$ *or* $i = 2)$:

$$\limsup_{k \to \infty} t_k^{-1} |y_k - y_0|^i < \infty. \tag{6.5}$$

DEFINITION 6.7 *The set of optimal solutions* $\omega(x)$ *of the problem* $(\bar{P}_x)$ *is said to be weakly Lipschitz (or Hölder) continuous at the point* x_0 *in the direction* $\bar{x}$, *if for any sequence* $t_k \downarrow 0$ *there is a point* $y_0 \in \omega(x_0)$ *such that*

$$\rho^i(y_0, \omega(x_0 + t_k \bar{x})) \leq lt_k, \quad k = 1, 2, \ldots \tag{6.6}$$

for some $l > 0$, *where* $i = 1$ *(or* $i = 2$).

It is obvious that the estimations (6.5) and (6.6) generalize locally Lipschitz (Hölder) continuity (Definition 6.1). Since from the condition

$$\limsup_{k \to \infty} t_k^{-1} |y_k - y_0|^i < \infty$$

the boundedness of the sequence $\{t_k^{-1} |y_k - y_0|^i\}$ results and, therefore, $lt_k \geq |y_k - y_0|^i \geq \rho^i(y_0, \omega(x_0 + t_k \bar{x}))$, then we get that sequentially Lipschitz (Hölder) stability of $\omega(x)$ implies the estimate (6.6).

2.   Regular Multivalued Mappings

In this section we investigate the nonlinear programming problem

$$f(x, y) \to \inf_y$$

$(P_x):$ $\qquad \begin{cases} h_i(x, y) \leq 0, & i = 1, \dots, r, \\ h_i(x, y) = 0, & i = r + 1, \dots, p, \end{cases}$

in which x is a vector of parameters, while the functions $f : X \times Y \to R$ and $h_i : X \times Y \to R$ are continuously differentiable.

Furthermore, we consider the set of feasible points $F : R^n \to 2^{R^m}$ defined by $F(x) = \{ y \in Y \mid h_i(x, y) \leq 0, \ i = 1, \dots, r, \ h_i(x, y) = 0, \ i = r + 1, \dots, p \}$, the optimal value function $\varphi(x) = \inf\{f(x, y) \mid y \in F(x)\}$ and the set of optimal solutions $\omega(x) = \{y \in F(x) \mid \varphi(x) = f(x, y)\}$ of the problem (P_x).

We denote

$$I = \{1, \dots, r\}, \ I_0 = \{r + 1, \dots, p\} \ (\text{ for } r = p, \text{ we assume } I_0 = \emptyset);$$
$$z = (x, y), \ \bar{z} = (\bar{x}, \bar{y}) \in X \times Y; \quad I(z) = \{i \in I \mid h_i(z) = 0\}.$$

In addition, we set $\lambda = (\lambda_1, \dots, \lambda_p)$, $h = (h_1, \dots, h_p) \in R^p$ and introduce the *Lagrangian* $L(z, \lambda) = f(z) + \langle \lambda, h(z) \rangle$ as well as the *set of Lagrange multipliers*

$$\Lambda(z) = \{\lambda \in R^p \mid \nabla_y L(z, \lambda) = 0, \ \lambda_i \geq 0, \ \lambda_i h_i(z) = 0, \ i \in I\}$$

of the problem (P_x) at the point $y \in \omega(x)$.

Let $x_0 \in \text{dom } F$. It is easy to see that, due to (A1) and the assumptions imposed on the functions h_i, $i = 1, \dots, p$, the multivalued mapping F is closed. Thus, in view of (A1), F and ω are compact-valued mappings in a neighbourhood of the point x_0.

The multivalued mapping F can also be described in another form. Let us denote

$$h_+(z) \overset{\triangle}{=} \max\{h_i(z), i = 1, \dots, r, \ |h_i(z)|, i = r + 1, \dots, p\},$$
$$h_0(z) \overset{\triangle}{=} \max\{0, h_i(z), i = 1, \dots, r, \ |h_i(z)|, i = r + 1, \dots, p\}.$$

The functions h_+ and h_0 are locally Lipschitz continuous and directionally differentiable (see Theorems 2.23 and 5.1). With their help we can describe the mapping F as

$$F(x) = \{y \mid h_+(z) \leq 0\} \quad \text{and} \quad F(x) = \{y \mid h_0(z) = 0\}$$

respectively.

2.1 Regularity Conditions

In order to obtain meaningful results concerning differential properties of optimal value functions associated with the problem (P_x) it is necessary to suppose some additional assumptions on the mapping F, namely regularity conditions (see e.g. [32], [69], [163]).

The simplest and most famous one is the following.

DEFINITION 6.8 *We shall say that the regularity condition (LI) holds at the point $z_0 \in \operatorname{gr} F$ or, equivalently, the multivalued mapping F is (LI)-regular at z_0 if the system of vectors $\{\nabla_y h_i(z_0), \; i \in I(z_0) \cup I_0\}$ is linearly independent.*

The Mangasarian-Fromowitz constraint qualification is another well-known regularity condition.

DEFINITION 6.9 *We shall say that the Mangasarian-Fromowitz regularity condition (MF) holds at the point $z_0 \in \operatorname{gr} F$ or, equivalently, the multivalued mapping F is (MF)-regular at z_0 if*
1. the vectors $\nabla_y h_i(z_0), \; i \in I_0$, are linearly independent;
2. one can find an element $\bar{y} \in Y$ such that

$$\langle \nabla_y h_i(z_0), \bar{y} \rangle = 0, \quad i \in I_0,$$
$$\langle \nabla_y h_i(z_0), \bar{y} \rangle < 0, \quad i \in I(z_0).$$

The (MF)-regularity condition is obviously fulfilled at points where the mapping F is (LI)-regular.

Indeed, let condition (LI) be valid at the point $z_0 = (x_0, y_0) \in \operatorname{gr} F$. We take an arbitrary $\delta > 0$ and consider the system

$$\begin{cases} \langle \nabla_y h_i(z_0), \bar{y} \rangle &= 0, \quad i \in I_0, \\ \langle \nabla_y h_i(z_0), \bar{y} \rangle &= -\delta, \quad i \in I(z_0), \end{cases}$$

which has full rank. Therefore, it is possible to solve it with respect to $\bar{y}$, which means that condition (MF) is satisfied.

For functions h_i convex with respect to y and $r = p$, condition (MF) is fulfilled at the point $z_0 = (x_0, y_0) \in \operatorname{gr} F$ if there exists an element $\tilde{y}_0 \in R^m$ such that $h_i(x_0, \tilde{y}_0) < 0, \; i \in I$.

Now we want investigate the connection between the (MF)-regularity condition and the set of Lagrange multipliers $\Lambda(z)$ of problem (P_x).

Let $\Lambda_0(z)$ be the set of *degenerate Lagrange multipliers*, i.e.

$$\Lambda_0(z) = \{\lambda \in R^p \mid \sum_{i=1}^{p} \lambda_i \nabla_y h_i(z) = 0,\ \lambda_i \geq 0,\ \lambda_i h_i(z) = 0,\ i \in I\}.$$

It is easy to see that $\Lambda_0(z) = 0^+ \Lambda(z)$, i.e., $\Lambda_0(z)$ is the recession cone of the polyhedral set $\Lambda(z)$.

LEMMA 6.10 *(***Gauvin, Tolle** [73]*) Let $z_0 = (x_0, y_0) \in \operatorname{gr} F$ and suppose $y_0 \in \omega(x_0)$. Then the following statements are equivalent:*
  1. the multivalued mapping F is (MF)-regular at the point z_0;
  2. $\Lambda(z_0) \neq \emptyset$ and bounded;
  3. $\Lambda_0(z_0) = \{0\}$.

Proof. 1. $\Rightarrow$ 2. Suppose that at the point z_0 condition (MF) holds. Then one can find a number $\delta > 0$ such that the system

$$\begin{cases} \langle \nabla_y h_i(z_0), \bar{y} \rangle = 0, & i \in I_0, \\ \langle \nabla_y h_i(z_0), \bar{y} \rangle + \delta \leq 0, & i \in I(z_0) \end{cases}$$

has a solution. Let $\bar{y}$ be such a solution. We introduce the function $\xi : R \to R^m$. If $I_0 = \emptyset$, we set $\xi = 0$. Otherwise, if $I_0 \neq \emptyset$, then we consider the system $h_i(x_0, y_0 + \varepsilon \bar{y} + \xi) = 0$, $i \in I_0$, supposing the components ξ_k of the vector-function ξ to be zero if the k-th row of the matrix $[\nabla_y h_i(z_0),\ i \in I_0]$ does not belong to the basis. Since the gradients $\nabla_y h_i(z_0)$, $i \in I_0$, are linearly independent, then, due to the implicit function theorem, this system has a continuously differentiable solution $\xi = \xi(\varepsilon)$ such that $\xi(0) = 0$.

Subtracting $h_i(x_0, y_0) = 0$, $i \in I_0$, from the system of equations

$$h_i(x_0, y_0 + \varepsilon \bar{y} + \xi(\varepsilon)) = 0,\ i \in I_0,$$

dividing then by $\varepsilon > 0$ and passing to the limit, we obtain

$$\langle \nabla_y h_i(z_0), \bar{y} \rangle + \langle \nabla_y h_i(z_0), \xi'(0) \rangle = 0,\ i \in I_0,$$

where $\xi'(0) = \lim_{\varepsilon \downarrow 0} \varepsilon^{-1} \xi(\varepsilon)$. Due to condition (MF), it follows that

$$\langle \nabla_y h_i(z_0), \xi'(0) \rangle = 0,\ i \in I_0.$$

Thus, in view of the (MF)-regularity and the definition of $\bar{y}$, we get $\xi'(0) = 0$.

Let $i \in I$. If $i \notin I(z_0)$, then $h_i(z_0) < 0$ and, due to the continuity of h_i and $\xi(\varepsilon)$, there exists an $\varepsilon_0' > 0$ such that

$$h_i(x_0, y_0 + \varepsilon \bar{y} + \xi(\varepsilon)) < 0, \quad \varepsilon \in [0, \varepsilon_0'], \quad i \in I \backslash I(z_0).$$

Let $i \in I(z_0)$. Owing to the continuity of $\nabla_y h_i(z)$ and $\xi(\varepsilon)$, there exists a number ε_0, $0 < \varepsilon_0 \leq \varepsilon_0'$, such that

$$\langle \nabla_y h_i(x_0, y_0 + \varepsilon \bar{y} + \xi(\varepsilon)), \bar{y} \rangle \leq -\delta/2$$

for any $i \in I(z_0)$, $\varepsilon \in [0, \varepsilon_0]$. In this case we obtain

$$h_i(x_0, y_0 + \varepsilon \bar{y} + \xi(\varepsilon))$$
$$= h_i(x_0, y_0) + \varepsilon \langle \nabla_y h_i(x_0, y_0 + \Theta_i \varepsilon \bar{y} + \Theta_i \xi(\varepsilon)), \bar{y} \rangle \leq -\delta/2,$$

where $0 < \Theta_i < 1$, $i \in I(z_0)$. Consequently, $y_0 + \varepsilon \bar{y} + \xi(\varepsilon) \in F(x_0)$, $\varepsilon \in [0, \varepsilon_0]$, where $\xi(\varepsilon)/\varepsilon \to 0$ for $\varepsilon \downarrow 0$.

This implies $f(x_0, y_0 + \varepsilon \bar{y} + \xi(\varepsilon)) - f(x_0, y_0) \geq 0$ so that, dividing by $\varepsilon > 0$ and passing to the limit for $\varepsilon \downarrow 0$, we obtain $\langle \nabla_y f(z_0), \bar{y} \rangle \geq 0$.

According to all what was said above, the linear programming problem

$$\langle \nabla_y f(z_0), \bar{y} \rangle \to \min$$
$$(P'): \quad \begin{cases} \langle \nabla_y h_i(z_0), \bar{y} \rangle = 0, & i \in I_0, \\ \langle \nabla_y h_i(z_0), \bar{y} \rangle \leq -\delta, & i \in I(z_0) \end{cases}$$

has a solution. Owing to the duality theorem of linear programming (see, e.g., [112]), the dual problem

$$(P_D'): \quad \delta \sum_{i \in I(z_0)} \lambda_i \to \max, \quad \lambda \in \Lambda(z_0)$$

has a solution as well. This means that $\Lambda(z_0) \neq \emptyset$ and the components λ_i, $i \in I(z_0)$, of any vector $\lambda \in \Lambda(z_0)$ are bounded. Taking into account the linear independence of the gradients $\nabla_y h_i(z_0)$, $i \in I_0$, as well as the equality

$$\nabla_y f(z_0) + \sum_{i=1}^{p} \lambda_i \nabla_y h_i(z_0) = 0,$$

we get that the components λ_i, $i \in I_0$, are bounded either. Thus, the set $\Lambda(z_0)$ is non-empty and bounded.

2. $\Rightarrow$ 3. Since $\Lambda(z_0)$ is non-empty, bounded and obviously closed, then $\Lambda_0(z_0) = 0^+ \Lambda(z_0) = \{0\}$ (cf. Section 1).

3. $\Rightarrow$ 1. Let $\Lambda_0(z_0) = \{0\}$. From this we immediately get that the gradients $\nabla_y h_i(z_0)$ are linearly independent. Since $\Lambda_0(z_0)$ does not

depend on the function f, but only on the constraints of the problem (P_x), there exists a vector $\eta \in R^m$ such that the set

$$\Lambda_\eta = \{\lambda \in R^p \mid \eta + \sum_{i=1}^p \lambda_i \nabla_y h_i(z_0) = 0, \ \lambda_i \geq 0, \ \lambda_i h_i(z_0) = 0, \ i \in I\}$$

is non-empty. (Otherwise, all vectors $\nabla_y h_i(z_0)$, $i \in I(z_0)$, would be equal to zero and the condition (MF) were trivially fulfilled). Due to the fact that $0^+ \Lambda_\eta(z_0) = \Lambda_0(z_0) = \{0\}$, the set Λ_η is bounded. Consequently, problem (P_D') with $\Lambda_\eta(z_0)$ instead of $\Lambda(z_0)$ has a solution and, according to the duality theorem in linear programming, the set of feasible points of the problem (P') dual to (P_D') is non-empty. In view of the linear independence of $\nabla_y h_i(z_0)$, $i \in I_0$, this means that condition (MF) holds at the point z_0. ∎

COROLLARY 6.11 *Let the multivalued mapping F be (LI)-regular at the point $z_0 \in \operatorname{gr} F$. Then the set $\Lambda(z_0)$ is single-pointed, i. e. $\Lambda(z_0) = \{\lambda\}$.*

Proof. Let the mapping F satisfy the (LI)-regularity condition at the point z_0, i.e., the gradients $\nabla_y h_i(z_0)$, $i \in I(z_0) \cup I_0$, are linearly independent. Then the system

$$\begin{cases} \nabla_y f(z_0) + \sum_{i=1}^p \lambda_i \nabla_y h_i(z_0) = 0, \\ \lambda_i = 0, \ i \notin I(z_0) \cup I_0 \end{cases}$$

has full rank and, therefore, not more than one solution. On the other hand, a (LI)-regular mapping is also (MF)-regular. From this we conclude that $\Lambda(z_0) \neq \emptyset$ and $\Lambda(z_0) = \{\lambda\}$. ∎

2.2   (R)-regular Mappings

We still introduce another definition of regularity of a multivalued mapping (see, e.g., [65], [152]).

DEFINITION 6.12 *The multivalued mapping F is said to be (R)-regular at the point z_0 if there exist positive numbers $\alpha > 0$, $\delta_1 > 0$, $\delta_2 > 0$ such that*

(R) $\rho(y, F(x)) \leq \alpha \max\{0; \ h_i(z), \ i = 1, \dots, r; \ |h_i(z)|, \ i = r+1, \dots, p\}$

holds for all $x \in x_0 + \delta_1 B$, $y \in y_0 + \delta_2 B$.

If the mapping F is (R)-regular at every point $z_0 \in \{x_0\} \times \omega(x_0)$, then the problem (P_x) is called (R)-regular at the point x_0. We say that the mapping F is (R)-*regular on the set* Z_0 if it is (R)-regular at every point $z \in Z_0$ with one the same values α, δ_1 and δ_2.

Using the function $h_0(z)$ (see p. 103), we can rewrite the (R)-regularity condition: the mapping F is (R)-regular at the point z_0 if there exist positive numbers α, δ_1 and δ_2 such that for every $x \in x_0 + \delta_1 B$, $y \in y_0 + \delta_2 B$ the inequality $d_F(z) \leq \alpha h_0(z)$ holds.

We want to investigate the connection between the condition (MF) and (R)-regularity. To this aim we need an auxiliary lemma based on the well-known Ekeland theorem.

THEOREM 6.13 *(Ekeland) Let* $g : R^n \to R$ *be a function lower semi-continuous and bounded below. Then for any point* x_ε *satisfying the condition*

$$\inf g \leq g(x_\varepsilon) \leq \inf g + \varepsilon$$

and for any $\lambda > 0$ *there exists a point* $\bar{x}_\varepsilon$ *such that*
  1. $g(\bar{x}_\varepsilon) \leq g(x_\varepsilon)$;
  2. $|\bar{x}_\varepsilon - x_\varepsilon| \leq \lambda$;
  3. $g(\bar{x}_\varepsilon) \leq g(x) + \frac{\varepsilon}{\lambda}|x - \bar{x}_\varepsilon|$ *for all* $x \in R^n$.

Proof. We introduce the auxiliary function $\tilde{g}(x) = g(x) + \frac{\varepsilon}{\lambda}|x - x_\varepsilon|$, which is lower semicontinuous and bounded from below. Therefore, there exists a point $\bar{x}_\varepsilon$ yielding the minimum of $\tilde{g}$ on R^n such that

$$g(\bar{x}_\varepsilon) + \frac{\varepsilon}{\lambda}|\bar{x}_\varepsilon - x_\varepsilon| \leq g(x) + \frac{\varepsilon}{\lambda}|x - x_\varepsilon| \qquad (6.7)$$

for all $x \in R^n$. Setting $x = x_\varepsilon$ we obtain

$$g(\bar{x}_\varepsilon) + \frac{\varepsilon}{\lambda}|\bar{x}_\varepsilon - x_\varepsilon| \leq g(x_\varepsilon).$$

This implies the validity of condition 1. Furthermore, since $g(x_\varepsilon) \leq \inf g + \varepsilon$, we get $|\bar{x}_\varepsilon - x_\varepsilon| \leq \lambda$, i.e. condition 2. Finally, from (6.7) it follows that

$$g(\bar{x}_\varepsilon) \leq g(x) + \frac{\varepsilon}{\lambda}\left[|x - \bar{x}_\varepsilon| - |\bar{x}_\varepsilon - x_\varepsilon|\right] \leq g(x) + \frac{\varepsilon}{\lambda}|x - \bar{x}_\varepsilon|$$

for any $x \in R^n$. ∎

Following Borwein [36], we shall prove the following assertion.

LEMMA 6.14 *Let* F *fail to be* (R)-*regular at the point* $z_0 = (x_0, y_0)$. *Then there exist sequences* $\delta_k \downarrow 0$, $y_k \to y_0$ *and* $x_k \to x_0$ *such that* $h_+(x_k, y_k) > 0$ *and*

$$|h_+(x_k, y)| - |h_+(x_k, y_k)| \geq -\delta_k|y - y_k| \qquad (6.8)$$

for all $y \in R^m$ *(for the definition of* h_+, *see p. 103).*

Proof. If the mapping F is not (R)-regular at the point z_0, then one can find sequences $a_k \to y_0$ and $x_k \to x_0$ such that

$$\rho(a_k, F(x_k)) > kh_0(x_k, a_k).$$

This means that $a_k \notin F(x_k)$ and, hence, $h_0(x_k, a_k) = h_+(x_k, a_k) > 0$ for all $k = 1, 2, \ldots$ Thus, we get

$$\rho(a_k, F(x_k)) > kh_+(x_k, a_k), \quad k = 1, 2, \ldots \tag{6.9}$$

Now we shall apply the Ekeland theorem to the function $g_k(y) = |h_+(x_k, y)|$ with $\varepsilon_k = g_k(a_k)$, $\lambda_k = \min\{k\varepsilon_k, \sqrt{\varepsilon_k}\}$, $y_{\varepsilon_k} = a_k$. We set $\delta_k = \varepsilon_k / \lambda_k = \max\left\{\frac{1}{k}, \sqrt{\varepsilon_k}\right\}$.

According to the Ekeland theorem there exists a point y_k such that

$$|a_k - y_k| \leq \lambda_k, \quad g_k(y) - g_k(y_k) \geq -\delta_k|y - y_k|. \tag{6.10}$$

The function h_+ is continuous. Therefore, we get that $\varepsilon_k \downarrow 0$, $\lambda_k \downarrow 0$, $\delta_k \downarrow 0$ and from (6.10) the inequality (6.8) results. Since $\lambda_k \downarrow 0$ and $a_k \to y_0$, then $y_k \to y_0$. Finally, from relations (6.9) and (6.10) we deduce

$$|y_k - a_k| \leq k\varepsilon_k \leq kh_+(x_k, a_k) < \rho(a_k, F(x_k)).$$

Therefore $y_k \notin F(x_k)$, which means $h_+(x_k, y_k) > 0$. ∎

THEOREM 6.15 *Let the multivalued mapping F be (MF)-regular at the point $z_0 = (x_0, y_0) \in \operatorname{gr} F$. Then the mapping F is (R)-regular at z_0.*

Proof. Let the condition (MF) hold at the point z_0. Suppose that F is not (R)-regular at this point. Then, according to Lemma 6.14, there exist sequences $\delta_k \downarrow 0$, $y_k \to y_0$ and $x_k \to x_0$ such that $h_+(x_k, y_k) > 0$ and (6.8) is true. We denote $g_k(y) = |h_+(x_k, y)|$. Due to (6.8), we have

$$g_k(y) - g_k(y_k) \geq -\delta_k|y - y_k|,$$

i. e., the function $\tilde{g}_k(y) = g_k(y) + \delta_k|y - y_k|$ attains its minimum at the point y_k. Since

$$|h_+(x_k, y_k)| = h_+(x_k, y_k) > 0,$$

then in a sufficiently small neighbourhood of the point y_k we also have

$$|h_+(x_k, y)| = h_+(x_k, y).$$

Consequently

$$\tilde{g}_k(y) = h_+(x_k, y) + \delta_k|y - y_k|.$$

Applying the necessary minimum condition (see Lemma 2.19) to $\tilde{g}_k(y)$ at the point y_k, we get $g'_k(y_k; \bar{y}) \geq 0$ for all $\bar{y} \in R^m$. This implies

$$0 \leq h'_+(x_k, y_k; \bar{y}) + \delta_k |\bar{y}|. \qquad (6.11)$$

Let us find the value of $h'_+(x_k, y_k; \bar{y})$. For this purpose we denote

$$\tilde{I}(z_k) = \{i \in I \mid h_i(z_k) = h_+(z_k)\},$$
$$\tilde{I}_0(z_k) = \{i \in I_0 \mid |h_i(z_k)| <= h_+(z_k)\}.$$

With regard to $|h_i(z_k)| = h_+(z_k) > 0$ we obviously have

$$h'_i(z_k; \bar{y}) = \langle \nabla_y h_i(z_k), \bar{y} \rangle, \quad i \in \tilde{I}(z_k),$$

$$|h_i(z_k)|'_{\bar{y}} = \frac{h_i(z_k)}{|h_i(z_k)|} \cdot \langle \nabla_y h_i(z_k), \bar{y} \rangle, \quad i \in \tilde{I}_0(z_k).$$

Thus, applying Theorem 5.9, we get

$$h'_+(z_k; \bar{y})$$
$$= \max \left\{ \langle \nabla_y h_i(z_k), \bar{y} \rangle, \; i \in \tilde{I}(z_k), \; \frac{h_i(z_k)}{|h_i(z_k)|} \langle \nabla_y h_i(z_k), \bar{y} \rangle, \; i \in \tilde{I}_0(z_k) \right\}$$
$$= \max_{\lambda \in A(z_k)} \left\langle \sum_{i=1}^{r} \lambda_i \nabla_y h_i(z_k) + \sum_{i=r+1}^{p} \sigma_i(z_k) \lambda_i \nabla_y h_i(z_k), \bar{y} \right\rangle,$$

where $A(z_k) = \{\lambda \in R^p \mid \lambda_i \geq 0, \; i = 1, \ldots, p; \; \sum_{i=1}^{p} \lambda_i = 1, \; \lambda_i h_i(z_k) = \lambda_i h_0(z_k), \; i = 1 \ldots, r; \; \lambda_i |h_i(z_k)| = \lambda_i h_0(z_k), \; i = r + 1, \ldots, p\}$ and $\sigma_i(z_k) = \frac{h_i(z_k)}{|h_i(z_k)|}, i = r + 1, \ldots, p$. Turning back to (6.11) from Theorem 2.14, we can rewrite this condition in the form

$$0 \in \left\{ \sum_{i=1}^{r} \lambda_i \nabla_y h_i(z_k) + \sum_{i=r+1}^{p} \lambda_i \sigma_i(z_k) \nabla_y h_i(z_k) \; \Big| \; \lambda \in A(z_k) \right\} + \delta_k \bar{B},$$

or

$$0 = \sum_{i=1}^{r} \lambda_i(z_k) \nabla_y h_i(z_k) + \sum_{i=r+1}^{p} \lambda_i(z_k) \sigma_i(z_k) \nabla_y h_i(z_k) + \delta_k \xi_k \qquad (6.12)$$

for some $\lambda(z_k) \in A(z_k)$, $\sigma_i(z_k)$, $\xi_k \in B$.

Since the sequences $\{\lambda(z_k)\}$, $\{\sigma_i(z_k)\}$ and $\{\xi_k\}$ are bounded, then without loss of generality we can assume them to be convergent, i.e. $\lambda(z_k) \to \lambda$, $\sigma_i(z_k) \to 1$, $\xi_k \to \xi$. Passing in (6.12) to the limit, we get

$$0 = \sum_{i=1}^{p} \lambda_i \nabla_y h_i(z_0),$$

where $\lambda = (\lambda_1, \ldots, \lambda_p) \neq 0$, $\lambda_i \geq 0$, $\lambda_i h_i(z_0) = 0$ for $i = 1, \ldots, r$.

The last condition means that $\Lambda_0(z_0) \neq \{0\}$, i.e., condition (MF) does not hold at the point z_0. The obtained contradiction shows that the mapping F is (R)-regular at z_0. ∎

Due to Theorem 6.15, we established the following connection between regularity conditions:

$$(LI) \Longrightarrow (MF) \Longrightarrow (R).$$

Notice that (R)-regularity does not imply (MF)-regularity in general. An important example of (R)-regular mappings which are not (MF)-regular in general, is given in the following lemma.

LEMMA 6.16 *(***Fedorov** [65]*) Let* $a_i \in R^m$ *and the functions* $b_i(x)$ *be continuous for* $i = 1, \ldots, p$. *Then the multivalued mapping* F *defined by*

$$F(x) = \{y \in R^m \mid \langle a_i, y \rangle + b_i(x) \leq 0, \ i = 1, \ldots, p\} \qquad (6.13)$$

is (R)-*regular at each point* $z_0 = (x_0, y_0) \in \mathrm{gr}\, F$ *such that* $x_0 \in \mathrm{int}\,\mathrm{dom}\, F$. *Moreover,* (R)-*regularity holds for all* $y \in Y$ *and* $x \in \mathrm{int}\,\mathrm{dom}\, F$.

Proof. Let $h_i(z) = \langle a_i, y \rangle + b_i(x)$, $h(z) = \max\{h_i(z) \mid i = 1, \ldots, p\}$. We take an arbitrary boundary point y from $F(x)$, where $x \in V(x_0) \subset \mathrm{int}\,\mathrm{dom}\, F$. The set of elements $\tilde{y} \notin F(x)$ such that $\rho(\tilde{y}, F(x)) = |\tilde{y} - y|$ coincides with the set

$$\{y = y(t) \mid y(t) = y + tl, \ l \in N_{F(x)}(y), \ |l| = 1\}$$

with $t = \rho(y(t), F(x))$. (Here $N_{F(x)}(y)$ is the normal cone to the set $F(x)$ at the point y). For such $y(t)$ we get

$h(x, y(t))$

$$\geq \max_{i \in I(x,y)} \{\langle a_i, y + tl \rangle + b_i(x)\} \geq t \max_{\substack{l \in N_{F(x)}(y) \\ |l|=1}} \max_{i \in I(x,y)} \langle a_i, l \rangle \overset{\triangle}{=} t\beta(x, y).$$

Assume that $\beta(x, y) \leq 0$. Then, for some $l_0 \in N_{F(x)}(y)$, $|l_0| = 1$, the inequality

$$\max_{i \in I(x,y)} \langle a_i, l_0 \rangle \leq 0$$

holds. From this inequality for sufficiently small $t > 0$ it follows that

$$0 < h(x, y + tl_0) = \max_{i \in I(x,y)} \{\langle a_i, y \rangle + b_i(x) + t\langle a_i, l_0 \rangle\}$$

$$= t \max_{i \in I(x,y)} \langle a_i, l_0 \rangle = t\beta(x, y) \leq 0.$$

The contradiction obtained shows that $\beta(x,y) > 0$. On the other hand

$$\beta(x,y) = \max_{\substack{l \in N_{F(x)}(y) \\ |l|=1}} \max_{i \in I(x,y)} \langle a_i, l \rangle = \max_{\lambda \in \bar{\Lambda}(x,y)} \max_{i \in I(x,y)} \left\langle a_i, \sum_{j \in I(x,y)} \lambda_j a_j \right\rangle,$$

where the set $\bar{\Lambda}(x,y)$ is defined as follows: $\bar{\Lambda}(x,y) = \{\lambda \in R^p \mid \lambda_j \geq 0, j \in I(x,y), \lambda_j = 0, j \notin I(x,y), \sum_{j=1}^{p} \lambda_j a_j = 1\}$.

Since $\beta(x,y)$ is defined by the set $I(x,y)$ and there exists only a finite number of subsets of the set $I = \{1,\ldots,p\}$, then $\beta(x,y) \geq \beta > 0$ for y belonging to the boundary of $F(x)$ and arbitrary $x \in V(x_0)$. Because any $\tilde{y} \notin F(x)$ can be represented as $\tilde{y} = y + lt$ with $t > 0$, $l \in N_{F(x)}(y)$, $|l| = 1$, where y is a boundary point from $F(x)$, then for any $y \notin F(x)$ the relation

$$h(x,\tilde{y}) \geq \beta t \geq \beta \rho(\tilde{y}, F(x))$$

is valid for all $x \in V(x_0)$. Thus, we obtained the statement of the lemma. ∎

REMARK. Lemma 6.16 is often referred to as a weak form of Hoffman's famous lemma about Lipschitz continuity of the feasible set for a parametric linear programming problem (see [79]).

The following example shows that the mapping F defined by (6.13) does not satisfy the (MF)-condition in general.

EXAMPLE 6.17 *Let* $F(x) = \{y \in R^2 \mid y_1 + y_2 \leq x_1, \ y_1 + y_2 \geq x_2, \ y_1 \geq 0, \ y_2 \geq 0\}$, $x \in X = R^2$. *We choose the points* $y_0 = \left(\frac{1}{2}, \frac{1}{2}\right)$, $x_0 = (1,1)$. *It is easy to calculate that*

$$\Lambda_0(z_0) = \{\lambda \in R^4 \mid \lambda_1 = \alpha, \ \lambda_2 = \alpha, \ \lambda_3 = \lambda_4 = 0, \ \alpha \geq 0\} \neq \{0\},$$

i.e., according to Lemma 2.1, the (MF)-condition does not hold at z_0.

Since (R)-regularity does not imply (MF)-regularity, then the set of Lagrange multipliers is not necessarily bounded at points $z_0 \in \{x_0\} \times \omega(x_0)$, at which the mapping F is (R)-regular. It can be shown, however, that the set $\Lambda(z_0)$ is non-empty at these points.

In fact, let $M = \alpha\beta$, where α is the positive constant from Definition 6.12, and let $\beta > l_0$, where l_0 is the Lipschitz constant of the function f on the set $X_0 \times [Y_0 + \varepsilon_0 B]$ with X_0 and Y_0 being the sets from assumption (A1). Finally, let $\varepsilon_0 > 2 \operatorname{diam} Y_0$. For $z_0 \in \operatorname{gr} F$ we denote

$$\tilde{\Lambda}_M(z_0) \stackrel{\triangle}{=} \left\{\lambda \in R^p \mid \lambda_i \geq 0, \ \lambda_i h_i(z_0) = 0, \ i = 1,\ldots,p, \ \sum_{i=1}^{p} |\lambda_i| \leq M\right\},$$

$$\Lambda_M(z_0) \stackrel{\triangle}{=} \Lambda(z_0) \cap \left\{\lambda \in R^p \mid \sum_{i=1}^{p} |\lambda_i| \leq M\right\}.$$

LEMMA 6.18 *Assume the multivalued mapping F to be (R)-regular at the point $z_0 \in \{x_0\} \times \omega(x_0)$. Then $\Lambda_M(z_0) \neq \emptyset$.*

Proof. Due to Lemma 3.68 and Definition 6.12, we get

$$
\varphi(x_0) \leq f(x_0, y) + M \cdot \max\{0, \ h_i(x_0, y), \ i \in I, \ |h_i(x_0, y)|, \ i \in I_0\}
$$
$$
= f(x_0, y) + M \cdot \max\{0, \ h_i(x_0, y), \ i \in I(z_0), \ |h_i(x_0, y)|, \ i \in I_0\}
$$
$$
= f(x_0, y) + \max_{\lambda \in \tilde{\Lambda}_M(z_0)} \sum_{i=1}^{p} \lambda_i h_i(x_0, y) = \max_{\lambda \in \tilde{\Lambda}_M(z_0)} L(x_0, y, \lambda)
$$

for all y from some neighbourhood of y_0. Since $L(x_0, y_0, \lambda) = f(x_0, y_0) = \varphi(x_0)$, then y_0 is a local minimizer of the function

$$
Q(y) \triangleq \max\left\{ L(x_0, y, \lambda) \,\middle|\, \lambda \in \tilde{\Lambda}_M(z_0) \right\}.
$$

Therefore, $Q'(y_0; \bar{y}) \geq 0$ for all $\bar{y} \in Y$. Due to Theorem 5.9 and taking into account Lemma 5.3, we have

$$
Q'(y_0; \bar{y}) = \max\{\langle \nabla_y L(z_0, \lambda), \bar{y} \rangle \mid \lambda \in \tilde{\Lambda}_M(z_0)\},
$$

and, consequently,

$$
0 \leq \delta^*(\bar{y} \mid \nabla_y L(z_0, \tilde{\Lambda}_M(z_0))),
$$

for all $\bar{y} \in Y$, which means

$$
0 \in \nabla_y L(z_0, \tilde{\Lambda}_M(z_0)),
$$

i.e. $\Lambda_M(z_0) \neq \emptyset$. ∎

The following property of (R)-regularity is of great importance.

LEMMA 6.19 *Let the multivalued mapping F be (R)-regular at every point of the compact set $Z_0 \subset \operatorname{gr} F$. Then F is (R)-regular on some neighbourhood of the set Z_0.*

Proof. Let the mapping F be (R)-regular at the point $z_0 \in Z_0$ with parameters $\alpha(z_0)$, $\delta_1(z_0)$ and $\delta_2(z_0)$. We make use of the notation $\delta(z_0) = \min\{\delta_1(z_0), \delta_2(z_0)\}$.

According to Definition 6.12, the mapping F is (R)-regular at any point $z \in z_0 + \frac{1}{2}\delta(z_0)B$ with parameters $\alpha(z_0)$ and $\frac{1}{2}\delta(z_0)$. Since

$$
Z_0 \subset \bigcup_{z_0 \in Z_0} \left(z_0 + \frac{1}{2}\delta(z_0)B \right)
$$

and Z_0 is a compact set, then there exist $z_i \in Z_0$, $i = 1, \ldots, N$, such that

$$Z_0 \subset \bigcup_{i=1}^{N} \left(z_i + \frac{1}{2}\delta(z_i)B \right) = V(Z_0).$$

Thus, every point from $V(Z_0)$ belongs to one of the sets $z_i + \frac{1}{2}\delta(z_i)B$ associated with points where F is (R)-regular with parameters $\alpha(z_i)$ and $\frac{1}{2}\delta(z_i)$. We denote

$$\alpha = \max\{\alpha(z_i) \,|\, i = 1, \ldots, N\}, \quad \delta = \min\left\{\frac{1}{2}\delta(z_i) \,\Big|\, i = 1, \ldots, N\right\}.$$

Then the mapping F is (R)-regular at any point of a neighbourhood $V(Z_0)$ of the set Z_0 with parameters α and δ. ∎

COROLLARY 6.20 *If the problem (P_x) is (R)-regular at the point x_0, then the multivalued mapping F is (R)-regular on the set $\{x_0\} \times \omega(x_0)$.*

LEMMA 6.21 *Let the multivalued mapping F be (R)-regular at the point $z_0 \in \mathrm{gr}\, F$. Then the function d_F is continuous at z_0.*

Proof. According to Definition 6.12, we have

$$d_F(z) \le \alpha \max\{0, h_i(z), i = 1, \ldots, r, \; |h_i(z)|, i = r+1, \ldots, p\}$$

in a neighbourhood of z_0.

Since the right-hand side of this relation is continuous (see Theorem 2.23), then passing to the limit for $z \to z_0$, we obtain

$$\lim_{z \to z_0} d_F(z) = 0, \quad \text{i.e. } \lim_{z \to z_0} d_F(z) = d_F(z_0). \quad \blacksquare$$

LEMMA 6.22 *Let there exist a point $z_0 \in \{x_0\} \times \omega(x_0)$, where the mapping F is (R)-regular. Then the function φ is upper semicontinuous at x_0.*

Proof. Due to Lemma 3.68, all assumptions of which are satisfied, we have

$$\varphi(x) \le f(x, y_0) + \beta d_F(x, y_0).$$

Therefore

$$\limsup_{x \to x_0} \varphi(x) \le f(x_0, y_0) = \varphi(x_0). \quad \blacksquare$$

COROLLARY 6.23 *If the multivalued mapping F is (R)-regular at every point $z_0 \in \{x_0\} \times \omega(x_0)$, then the function φ is continuous at x_0.*

Proof. Since F is closed and uniformly bounded at x_0, then, due to Lemma 3.18, φ is lower semicontinuous at x_0. In view of Lemma 6.22 we get the statement of the corollary. ∎

LEMMA 6.24 *Let the multivalued mapping F be (R)-regular at some point $z_0 \in \{x_0\} \times \omega(x_0)$. Then the multivalued mapping $\omega(\cdot)$ is upper semicontinuous at x_0.*

Proof. Let $x_k \to x_0$, $y_k \in \omega(x_k)$, $k = 1, 2, \ldots$, and $y_k \to y$. Then $\varphi(x_k) = f(x_k, y_k)$ and, according to the upper semicontinuity of φ (see Lemma 6.22), we get

$$f(x_0, y) = \lim_{k \to \infty} f(x_k, y_k) \le \limsup_{k \to \infty} \varphi(x_k) \le \varphi(x_0).$$

Due to the closedness of F, the inclusion $y \in F(x_0)$ holds. Thus, with regard to the condition $f(x_0, y) \le \varphi(x_0)$ it follows that $y \in \omega(x_0)$. ∎

2.3 The Linear Tangent Cone and Derivatives of Regular Multivalued Mappings

We introduce the *linearized tangent cone* $\Gamma_F(z)$ to the set $\operatorname{gr} F$ at some point $z \in \operatorname{gr} F$ by setting

$$\Gamma_F(z) = \{\bar{z} \in X \times Y \mid \langle \nabla h_i(z), \bar{z} \rangle \le 0,\ i \in I(z),\ \langle \nabla h_i(z), \bar{z} \rangle = 0,\ i \in I_0 \}.$$

Furthermore, for $\bar{x} \in X$ we consider the set

$$\Gamma_F(z; \bar{x}) = \{\bar{y} \in R^m \mid (\bar{x}, \bar{y}) \in \Gamma_F(z)\}.$$

LEMMA 6.25 *Let $z \in \operatorname{gr} F$. Then $T^U_{\operatorname{gr} F}(z) \subset \Gamma_F(z)$.*

Proof. Let $\bar{z} \in T^U_{\operatorname{gr} F}(z)$. Then, according to the definition of the upper tangent cone, we have

$$\liminf_{\varepsilon \downarrow 0} \varepsilon^{-1} \rho(z + \varepsilon \bar{z}, \operatorname{gr} F) = 0,$$

i. e., there exists a sequence $\varepsilon_k \downarrow 0$ such that

$$z + \varepsilon_k \bar{z} + o(\varepsilon_k) \in \operatorname{gr} F, \quad k = 1, 2, \ldots,$$

where $o(\varepsilon_k)/\varepsilon_k \to 0$ for $\varepsilon_k \downarrow 0$. From this it follows that

$$h_i(z + \varepsilon_k \bar{z} + o(\varepsilon_k)) \le 0,\ i \in I(z),\ k = 1, 2, \ldots,$$
$$h_i(z + \varepsilon_k \bar{z} + o(\varepsilon_k)) = 0,\ i \in I_0,\ k = 1, 2, \ldots$$

Subtracting $h_i(z)$, $i \in I \cup I_0$, dividing by ε_k and passing to the limit, in view of the continuous differentiability of the functions $h_i(z)$, $i \in I \cup I_0$, we obtain

$$\langle \nabla h_i(z), \bar{z} \rangle \leq 0, \ i \in I(z), \qquad \langle \nabla h_i(z), \bar{z} \rangle = 0, \ i \in I_0.$$

This means that $\bar{z} \in \Gamma_F(z)$. Therefore $T^U_{\text{gr} F}(z) \subset \Gamma_F(z)$. ∎

COROLLARY 6.26 *Let $\bar{x} \in X$ and $z \in \text{gr} F$. Then*

$$D_U F(z; \bar{x}) \subset \Gamma_F(z; \bar{x}).$$

LEMMA 6.27 *Let $z \in \text{gr} F$ and $\hat{D}_U F(z; \bar{x}) = \emptyset$. Then $D_+ d_F(z; \bar{z}) = +\infty$ for all $\bar{y} \in R^m$.*

Proof. Suppose the opposite, i.e., one can find an element $\bar{y} \in Y$ and a sequence $\varepsilon_k \downarrow 0$ such that

$$\lim_{k \to \infty} \varepsilon_k^{-1} d_F(z + \varepsilon_k \bar{z}) \leq \delta < +\infty.$$

Then there exists a number k_0 such that for $k > k_0$ the inclusion

$$y + \varepsilon_k \bar{y} \in F(x + \varepsilon_k \bar{x}) + 2\delta \varepsilon_k B$$

holds. Hence, one can indicate a vector $\xi_k \in 2\delta B$ such that

$$y + \varepsilon_k \bar{y} + \varepsilon_k \xi_k \in F(x + \varepsilon_k \bar{x}) \quad \text{for all} \ k > k_0.$$

Without loss of generality we can assume that $\xi_k \to \xi$. Then from the last inclusion we get that $y + \xi \in \hat{D}_U F(z; \bar{x})$. The contradiction obtained shows that $D_+ d_F(z; \bar{z}) = +\infty$. ∎

LEMMA 6.28 *Let the multivalued mapping F be (R)-regular at the point $z \in \text{gr} F$. Then $\text{dom} \Gamma_F(z; \cdot) = X$ and*

$$T^U_{\text{gr} F}(z) = \hat{T}^U_{\text{gr} F}(z) = T^L_{\text{gr} F}(z) = \hat{T}^L_{\text{gr} F}(z) = \Gamma_F(z).$$

Proof. Let $\bar{z} \in X \times Y$. We choose a number $\varepsilon_1 > 0$ such that $h_i(z + \varepsilon \bar{z}) < 0$ for all $\varepsilon \in (0, \varepsilon_1]$ and $i \in I \backslash I(z)$. According to the definition of (R)-regularity, there exist numbers $\alpha > 0$ and $\varepsilon_0 \in (0, \varepsilon_1]$ such that for all $\varepsilon \in (0, \varepsilon_0]$ we have

$$d_F(z + \varepsilon \bar{z}) \leq \alpha \max\{ 0, \ h_i(z + \varepsilon \bar{z}), i \in I, \ |h_i(z + \varepsilon \bar{z})|, i \in I_0\}$$
$$= \alpha \max\{ 0, \ h_i(z + \varepsilon \bar{z}), i \in I(z), \ |h_i(z + \varepsilon \bar{z})|, i \in I_0\}.$$

This implies

$$
\begin{aligned}
0 \;\; &\le d_F(z + \varepsilon\bar{z}) - d_F(z) \\
&\le \alpha \max\{0,\; h_i(z) + \varepsilon\langle\nabla h_i(z), \bar{z}\rangle + \varepsilon\gamma_i(\varepsilon),\; i \in I(z), \\
&\qquad |h_i(z) + \varepsilon\langle\nabla h_i(z), \bar{z}\rangle + \varepsilon\gamma_i(\varepsilon)|,\; i \in I_0\} \\
&\le \varepsilon\alpha \max\{0,\; \langle\nabla h_i(z), \bar{z}\rangle,\; i \in I(z),\; |\langle\nabla h_i(z), \bar{z}\rangle| i \in I_0\} + \varepsilon\alpha\eta(\varepsilon),
\end{aligned}
\tag{6.14}
$$

where
$$
\begin{aligned}
\eta(\varepsilon) &= \max\{0, |\gamma_i(\varepsilon)|\; i \in I(z) \cup I_0\} \\
\gamma_i(\varepsilon) &= \langle\nabla h_i(z + \varepsilon\Theta_i\bar{z}) - \nabla h_i(z), \bar{z}\rangle,\; i \in I(z), \\
\gamma_i(\varepsilon) &= |\langle\nabla h_i(z + \varepsilon\Theta_i\bar{z}) - \nabla h_i(z), \bar{z}\rangle|,\; i \in I_0, \\
0 &< \Theta_i < 1,\; i \in I(z) \cup I_0.
\end{aligned}
$$

Arguing from the opposite, it is not hard to prove that $\eta(\varepsilon) \to 0$ for $\varepsilon \downarrow 0$. Consequently, dividing inequality (6.14) by ε and passing to the limit for $\varepsilon \downarrow 0$, we get

$$
\begin{aligned}
0 \le D_+ d_F(z; \bar{z}) &\le D^+ d_F(z; \bar{z}) \\
&\le \alpha \max\{\, 0,\; \langle\nabla h_i(z), \bar{z}\rangle,\; i \in I(z),\; |\langle\nabla h_i(z), \bar{z}\rangle|,\; i \in I_0\}.
\end{aligned}
\tag{6.15}
$$

In view of Lemma 6.27, we thus have $\operatorname{dom} \hat{D}_U F(z; \cdot) = X$.

Let $\bar{z} \in \Gamma_F(z)$. Then from relation (6.15) the equality $D^+ d_F(z; \bar{z}) = 0$ results, which means that $\bar{z} \in \hat{T}^L_{\mathrm{gr}\,F}(z)$. In this way, the inclusion $\Gamma_F(z) \subset \hat{T}^L_{\mathrm{gr}\,F}(z)$ is true. Taking into account the inclusions $\hat{T}^L_{\mathrm{gr}\,F}(z) \subset \hat{T}^U_{\mathrm{gr}\,F}(z)$ and $\hat{T}_{\mathrm{gr}\,F}(z) \subset T^L_{\mathrm{gr}\,F}(z) \subset T^U_{\mathrm{gr}\,F}(z)$ as well as Lemma 6.25, we finally get the statement of the lemma. ∎

LEMMA 6.29 *Let the multivalued mapping F be (R)-regular at the point $z \in \operatorname{gr} F$. Then $T^C_{gr\,F}(z) = \Gamma_F(z)$.*

Proof. Let $\bar{z} = (\bar{x}, \bar{y}) \in X \times Y$, $\bar{z} \ne 0$. From the (R)-regularity of the mapping F at the point z and Lemma 6.19 it follows that F is (R)-regular at all points in some δ_0-neighbourhood of the point z. Let us choose $\delta_1 > 0$ such that $2\delta_1 < \delta_0$ and $h_i(z') < 0$ for all $z' \in z + 2\delta_1 B$, $i \notin I(z)$. Denoting $\varepsilon_0 = \delta_1 \cdot |\bar{z}|^{-1}$, we get $h_i(z' + \varepsilon\bar{z}) < 0$ for all $z' \in z + \delta_1 B$, $\varepsilon \in [0, \varepsilon_0]$, $i \notin I(z)$. Then the following inequalities are valid for $z' \in (z + \delta_1 B) \cap \operatorname{gr} F$ and $\varepsilon \in [0, \varepsilon_0]$:

$$
\begin{aligned}
d_F(z' + \varepsilon\bar{z}) - d_F(z') &= d_F(z' + \varepsilon\bar{z}) \\
&\le \alpha \max\{0,\; h_i(z' + \varepsilon\bar{z}),\; i \in I,\; |h_i(z' + \varepsilon\bar{z})|,\; i \in I_0\} \\
&= \alpha \max\{0,\; h_i(z' + \varepsilon\bar{z}),\; i \in I(z),\; |h_i(z' + \varepsilon\bar{z})|,\; i \in I_0\}
\end{aligned}
$$

$$= \alpha \max\{0, \; h_i(z') + \varepsilon\langle\nabla h_i(z'), \bar{z}\rangle + \varepsilon\gamma_i, \; i \in I(z),$$
$$|h_i(z') + \varepsilon\langle\nabla h_i(z'), \bar{z}\rangle + \varepsilon\gamma_i|, \; i \in I_0\}$$
$$\leq \varepsilon\alpha \max\{0, \; \langle\nabla h_i(z'), \bar{z}\rangle, \; i \in I(z), \; |\langle\nabla h_i(z'), \bar{z}\rangle|, \; i \in I_0\} + \varepsilon\gamma,$$

where $\gamma = \max\{|\gamma_i|, \; i \in I(z) \cup I_0\}$, $\gamma_i(\varepsilon) = \langle\nabla h_i(z' + \tau_i\varepsilon\bar{z}) - \nabla h_i(z'), \; \bar{z}\rangle$, $\tau_i \in (0,1)$, $i \in I(z) \cup I_0$. From the last inequality we get

$$\eta(\bar{z}) \triangleq \limsup_{z' \xrightarrow{\mathrm{gr}\,F} z, \varepsilon\downarrow 0} \varepsilon^{-1}[d_F(z' + \varepsilon\bar{z}) - d_F(z')]$$
$$\leq \alpha \max\{0, \; \langle\nabla h_i(z), \bar{z}\rangle, \; i \in I(z), \; |\langle\nabla h_i(z), \bar{z}\rangle|, \; i \in I_0\}.$$

Then, for $\bar{z} \in \Gamma_F(z)$ we have $\eta(\bar{z}) = 0$. This means

$$\lim_{k\to\infty} \varepsilon_k^{-1}\rho(y_k + \varepsilon_k\bar{y}, F(x_k + \varepsilon_k\bar{x})) = 0$$

for arbitrary sequences $\varepsilon_k \downarrow 0$ and $z_k \xrightarrow{\mathrm{gr}\,F} z$, i.e., one can indicate a sequence $\xi_k \to 0$, $\xi_k \in Y$, such that $y_k + \varepsilon_k\bar{y} + \varepsilon_k\xi_k \in F(x_k + \varepsilon_k\bar{x})$, $k = 1, 2, \ldots$, where $z_k = (x_k, y_k)$. But this is equivalent to $\bar{z} \in T^C_{\mathrm{gr}\,F}(z)$. Consequently $\Gamma_F(z) \subset T^C_{\mathrm{gr}\,F}(z)$. On the other hand, due to Lemma 6.25 and the inclusion $T^C_{\mathrm{gr}\,F}(z) \subset T^U_{\mathrm{gr}\,F}(z)$, the reverse inclusion holds. ∎

Generalizing Lemmas 6.28 and 6.29, we obtain the following result.

THEOREM 6.30 *Let the multivalued mapping F be (R)-regular at the point $z \in \mathrm{gr}\,F$. Then for any $\bar{x} \in X$ the following equalities are valid:*

$$D_S F(z; \bar{x}) = \Gamma_F(z; \bar{x}) \neq \emptyset, \quad S = U, L, C,$$
$$\hat{D}_U F(z; \bar{x}) = \hat{D}_L F(z; \bar{x}) = \Gamma_F(z; \bar{x}).$$

2.4 Subdifferentials of Marginal Functions in Regular Problems

Our next aim consists in estimating the subdifferential $\partial^0\varphi(x_0)$ in an (R)-regular problem (P_x).

LEMMA 6.31 *If $\mathrm{dom}\,D_C F(z; \cdot) = X$ at the point $z \in \mathrm{gr}\,F$, then the mapping F is pseudo-Lipschitz continuous at the point z.*

Proof. Let $\mathrm{dom}\,D_C F(z; \cdot) = X$. Then, due to Lemma 3.58, $d^\uparrow_F(z; \bar{z}) < +\infty$ for all $\bar{z} \in X \times Y$. Since $d_F(z)$ is lower semicontinuous (see Lemma 3.18), then, according to Theorem 2.20, the function $d_F(z)$ is Lipschitz continuous in a neighbourhood of z. Applying now Lemma 3.26, we get the statement of the lemma. ∎

THEOREM 6.32 *Let the multivalued mapping F be (R)-regular at the point $z_0 = (x_0, y_0) \in \mathrm{gr}\, F$. Then F is pseudo-Lipschitz continuous at the point z_0.*

Proof. The statement of the theorem follows directly from Lemmas 6.29 and 6.31. ∎

Now we suppose that the problem (P_x) is (R)-regular at the point x_0. According to Theorem 6.32, the mapping F is pseudo-Lipschitz continuous at every point $z_0 \in \{x_0\} \times \omega(x_0)$ and, therefore, due to Corollary 4.23, the function φ is Lipschitz continuous in a neighbourhood of x_0. Let us estimate the subdifferential $\partial^0 \varphi(x_0)$ making use of Theorem 4.9. In view of Lemma 6.29 and Theorem 2.6, the Clarke normal cone to $\mathrm{gr}\, F$ at the point $z_0 \in \{x_0\} \times \omega(x_0)$ is of the form

$$N_F(z_0) = -[T_{\mathrm{gr}\, F}^C(z_0)]^+ = -[\Gamma_F(z_0)]^+$$

$$= \left\{ z^* = \sum_{i=1}^{p} \lambda_i \nabla_y h_i(z_0) = 0, \ \lambda_i \geq 0, \ \lambda_i h_i(z_0) = 0, \ i = 1, \ldots, r \right\}.$$

We want to describe the set $K(z_0)$ from Theorem 4.9. In the case under study, we have

$$\begin{aligned}
K(z_0) &= \{x^* \mid (x^*, 0) \in \partial^0 f(z_0) + N_F(z_0)\} \\
&= \{x^* \mid x^* = \nabla_x f(z_0) + \sum_{i=1}^{p} \lambda_i \nabla_x h_i(z_0) \text{ subject to} \\
&\qquad \nabla_y f(z_0) + \sum_{i=1}^{p} \lambda_i \nabla_y h_i(z_0) = 0, \ \lambda_i \geq 0, \\
&\qquad \lambda_i h_i(z_0) = 0, \ i = 1, \ldots, r\} \\
&= \{x^* \mid x^* = \nabla_x L(z_0, \lambda), \ \lambda \in \Lambda(z_0)\}.
\end{aligned}$$

With regard to Lemma 6.18, $K(z_0) \neq \emptyset$ for all $y_0 \in \omega(x_0)$. Applying now Theorem 4.9, all conditions of which are satisfied, we get

$$\partial^0 \varphi(x_0) \subset \mathrm{cl}\, \mathrm{co} \bigcup_{y_0 \in \omega(x_0)} \bigcup_{\lambda \in \Lambda(z_0)} \nabla_x L(z_0, \lambda),$$

and, finally,

$$\partial^0 \varphi(x_0) \subset \mathrm{co} \bigcup_{y_0 \in \omega(x_0)} \bigcup_{\lambda \in \Lambda(z_0)} \nabla_x L(z_0, \lambda). \tag{6.16}$$

In this way, the following theorem established by Gauvin and Dubeau [69] has been proved.

THEOREM 6.33 *([69]) Let the problem (P_x) be (R)-regular at the point x_0. Then the function φ is Lipschitz continuous in a neighbourhood of the point x_0 and the estimate (6.16) is valid.*

2.5 Second-order Derivatives of Mappings

Let $z_0 \in \mathrm{gr}\, F$, $\bar{x}_1 \in X$, $\bar{y}_1 \in Y$, $\bar{z}_1 = (\bar{x}_1, \bar{y}_1)$ and $o(\varepsilon)$ be a m-vector function such that $o(\varepsilon)/\varepsilon \to 0$ for $\varepsilon \downarrow 0$.

We consider the *sets of second-order tangential directions* defined by

$$\hat{T}_{\mathrm{gr}F}^{L2}(z_0, \bar{z}_1) \triangleq \{\bar{z}_2 = (\bar{x}_2, \bar{y}_2) \mid \exists o(\varepsilon):$$
$$y_0 + \varepsilon \bar{y}_1 + \varepsilon^2 \bar{y}_2 + o(\varepsilon^2) \in F(x_0 + \varepsilon \bar{x}_1 + \varepsilon^2 \bar{x}_2), \ \varepsilon \geq 0\},$$

$$\hat{T}_{\mathrm{gr}F}^{U2}(z_0, \bar{z}_1) \triangleq \{\bar{z}_2 = (\bar{x}_2, \bar{y}_2) \mid \exists \varepsilon_k \downarrow 0, \exists o(\varepsilon):$$
$$y_0 + \varepsilon_k \bar{y}_1 + \varepsilon_k^2 \bar{y}_2 + o(\varepsilon_k^2) \in F(x_0 + \varepsilon_k \bar{x}_1 + \varepsilon_k^2 \bar{x}_2), \ k = 1, 2, \ldots\}.$$

Following Demyanov [55], [61], we introduce the *second-order lower* and *upper derivatives of the mapping* F at the point z_0 in the directions $\bar{z}_1$, $\bar{x}_2$ as

$$\hat{D}_L^2 F(z_0, \bar{z}_1; \bar{x}_2) \triangleq \{\bar{y}_2 \in Y \mid (\bar{x}_2, \bar{y}_2) \in \hat{T}_{\mathrm{gr}F}^{L2}(z_0, \bar{z}_1)\},$$

$$\hat{D}_U^2 F(z_0, \bar{z}_1; \bar{x}_2) \triangleq \{\bar{y}_2 \in Y \mid (\bar{x}_2, \bar{y}_2) \in \hat{T}_{\mathrm{gr}F}^{U2}(z_0, \bar{z}_1)\},$$

respectively. Moreover, we define

$$I^2(z_0, \bar{z}_1) \triangleq \{i \in I(z_0) \mid \langle \nabla h_i(z_0), \bar{z}_1 \rangle = 0\},$$

$$\Gamma_F^2(z_0, \bar{z}_1) \triangleq \Big\{ \bar{z}_2 = (\bar{x}_2, \bar{y}_2) \mid \langle \nabla h_i(z_0), \bar{z}_2 \rangle + \tfrac{1}{2}\langle \bar{z}_1, \nabla^2 h_i(z_0) \bar{z}_1 \rangle = 0, \ i \in I_0,$$
$$\langle \nabla h_i(z_0), \bar{z}_2 \rangle + \tfrac{1}{2}\langle \bar{z}_1, \nabla^2 h_i(z_0) \bar{z}_1 \rangle \leq 0, \ i \in I^2(z_0, \bar{z}_1) \Big\},$$

$$\Gamma_F^2(z_0, \bar{z}_1; \bar{x}_2) \triangleq \{\bar{y}_2 \in Y \mid (\bar{x}_2, \bar{y}_2) \in \Gamma_F^2(z_0, \bar{z}_1)\}.$$

LEMMA 6.34 *Let $z_0 \in \mathrm{gr}\, F$, $\bar{z}_1 \in \Gamma_F(z_0)$. Then the following inclusions hold:*

$$\hat{T}_{\mathrm{gr}F}^{L2}(z_0, \bar{z}_1) \subset \hat{T}_{\mathrm{gr}}^{U2}(z_0, \bar{z}_1) \subset \Gamma_F^2(z_0, \bar{z}_1).$$

Proof. The inclusion

$$\hat{T}_{\mathrm{gr}F}^{L2}(z_0, \bar{z}_1) \subset \hat{T}_{\mathrm{gr}F}^{U2}(z_0, \bar{z}_1)$$

follows directly from the definitions. Let us prove the second inclusion.

We take an arbitrary $\bar{z}_2 = (\bar{x}_2, \bar{y}_2) \in \hat{T}_{\mathrm{gr}F}^{U2}(z_0, \bar{z}_1)$. Then there exist a sequence $\varepsilon_k \downarrow 0$ and a m-vector function $o(\varepsilon)$ such that $o(\varepsilon)/\varepsilon \to 0$ for $\varepsilon_k \downarrow 0$ and

$$h_i(x_0 + \varepsilon_k \bar{x}_1 + \varepsilon_k^2 \bar{x}_2, y_0 + \varepsilon_k \bar{y}_1 + \varepsilon_k^2 \bar{y}_2 + o(\varepsilon_k^2)) = 0, \qquad i \in I_0, \quad (6.17)$$

$$h_i(x_0 + \varepsilon_k \bar{x}_1 + \varepsilon_k^2 \bar{x}_2, y_0 + \varepsilon_k \bar{y}_1 + \varepsilon_k^2 \bar{y}_2 + o(\varepsilon_k^2)) \leq 0, \qquad i \in I^2(z_0, \bar{z}_1) \quad (6.18)$$

We use the notation $x_k = x_0 + \varepsilon_k \bar{x}_1 + \varepsilon_k^2 \bar{x}_2$, $y_k = y_0 + \varepsilon_k \bar{y}_1 + \varepsilon_k^2 \bar{y}_2 + o(\varepsilon_k^2)$, $z_k = (x_k, y_k)$. Then

$$h_i(z_k) = \varepsilon_k^2 \left[\langle \nabla h_i(z_0), \bar{z}_2 \rangle + \frac{1}{2} \langle \bar{z}_1, \nabla^2 h_i(z_0) \bar{z}_1 \rangle \right] + o_i(\varepsilon_k^2).$$

But $h_i(z_k) = 0$ if $i \in I_0$ and $h_i(z_k) \leq 0$ if $i \in I^2(z_0, \bar{z}_1)$. Therefore, dividing relations (6.17) and (6.18) by ε_k and passing to the limit, we get

$$\langle \nabla h_i(z_0), \bar{z}_2 \rangle + \tfrac{1}{2} \langle \bar{z}_1, \nabla^2 h_i(z_0) \bar{z}_1 \rangle = 0, \; i \in I_0,$$

$$\langle \nabla h_i(z_0), \bar{z}_2 \rangle + \tfrac{1}{2} \langle \bar{z}_1, \nabla^2 h_i(z_0) \bar{z}_1 \rangle \leq 0, \; i \in I^2(z_0, \bar{z}_1),$$

i.e. $\bar{z}_2 \in \Gamma_F^2(z_0, \bar{z}_1)$. In this way, $\hat{T}_{\mathrm{gr} F}^{U2}(z_0, \bar{z}_1) \subset \Gamma_F^2(z_0, \bar{z}_1)$. ∎

LEMMA 6.35 *Let* $z_0 \in \mathrm{gr}\, F$, $\bar{z}_1 \in X \times Y$. *If* $\hat{D}_U^2 F(z_0, \bar{z}_1; \bar{x}_2) = \emptyset$, *then for any* $\bar{y}_2 \in Y$ *the relation*

$$\liminf_{\varepsilon \downarrow 0} \varepsilon^{-2} d_F(z_0 + \varepsilon \bar{z}_1 + \varepsilon^2 \bar{z}_2) = +\infty$$

is true.

Proof. Suppose the opposite, i.e., one can find an element $\bar{y}_2 \in Y$ and a sequence $\varepsilon_k \downarrow 0$ such that

$$\lim_{\varepsilon_k \downarrow 0} \varepsilon_k^{-2} d_F(z_0 + \varepsilon_k \bar{z}_1 + \varepsilon_k^2 \bar{z}_2) \leq \delta < +\infty.$$

Then there exists a number k_0 such that for $k \geq k_0$ we have

$$y_0 + \varepsilon_k \bar{y}_1 + \varepsilon_k^2 \bar{y}_2 \in F(x_0 + \varepsilon_k \bar{x}_1 + \varepsilon_k^2 \bar{x}_2) + 2\delta \varepsilon_k^2 B.$$

This means that one can indicate a vector $\xi_k \in 2\delta B$ such that

$$y_0 + \varepsilon_k \bar{y}_1 + \varepsilon_k^2 \bar{y}_2 \in F(x_0 + \varepsilon_k \bar{x}_1 + \varepsilon_k^2 \bar{x}_2) - \varepsilon_k^2 \xi_k$$

for $k \geq k_0$. Without loss of generality we can assume that $\xi_k \to \xi$. Then from the last inclusion it follows that

$$\bar{y}_2 + \xi \in \hat{D}_U^2 F(z_0, \bar{z}_1; \bar{x}_2).$$

The contradiction obtained proves the statement of the lemma. ∎

2.6 Directional Regularity

In the following we suppose that the functions f and h_i, $i = 1, \ldots, p$, are twice continuously differentiable. The point x_0 is assumed to belong to dom F but not necessarily in the interior of dom F. Our aim is to weaken the concept of (R)-regularity by considering this property for fixed direction $\bar{x}$.

DEFINITION 6.36 *We shall say that the multivalued mapping F is (R)-regular in the direction $\bar{x}$ or, equivalently, it satisfies the $(R_{\bar{x}})$-regularity condition at the point $z \in \operatorname{gr} F$ if for any $\bar{y} \in Y$ one can find quantities $\alpha > 0$, $\varepsilon_0 > 0$ and $o(\varepsilon)$ such that $o(\varepsilon)/\varepsilon \to 0$ if $\varepsilon \downarrow 0$ and for all $\varepsilon \in [0, \varepsilon_0]$ the inequality*

$$\rho(y + \varepsilon\bar{y} + o(\varepsilon), F(x + \varepsilon\bar{x}))$$
$$\leq \alpha \max\{0, h_i(z + \varepsilon\bar{z}), i \in I, |h_i(z + \varepsilon\bar{z})|, i \in I_0\},$$

holds, where $\bar{z} = (\bar{x}, \bar{y})$.

DEFINITION 6.37 *1) The multivalued mapping F is referred to as (Γ)-regular or, equivalently, to satisfy the Kuhn-Tucker regularity condition at the point $z \in \operatorname{gr} F$ if $\hat{T}^L_{\operatorname{gr} F}(z) = \Gamma_F(z)$.*

2) The multivalued mapping F is said to be (Γ)-regular in the direction $\bar{x}$ at the point z if $\hat{D}_L F(z; \bar{x}) = \Gamma_F(z; \bar{x}) \neq \emptyset$.

It is easy to see that (R)-regularity of the mapping F implies its (Γ)-regularity as well as (Γ)- and (R)-regularity in all directions $\bar{x}$.

THEOREM 6.38 *The following statements are equivalent:*
1. the mapping F is (R)-regular at the point $z \in \operatorname{gr} F$ in the direction $\bar{x}$;

2. the mapping F is (Γ)-regular at the point z in the direction $\bar{x}$.

Proof. 1. $\Rightarrow$ 2. Suppose that the $(R_{\bar{x}})$-regularity condition holds at the point z. We take an arbitrary vector $\bar{y} \in Y$ and repeat the proof of Lemma 6.28, denoting $\bar{z} = (\bar{x}, \bar{y})$. As a result we get the validity of statement 2.

2. $\Rightarrow$ 1. Let $\hat{D}_L F(z; \bar{x}) = \Gamma_F(z; \bar{x}) \neq \emptyset$. Taking into account Theorem 3.56 and the obvious inclusion $\hat{D}_L F(z; \bar{x}) \subset \Gamma_F(z; \bar{x})$, it immediately follows that the function $d_F(z)$ is differentiable in all directions $\bar{z} = (\bar{x}, \bar{y})$, where $\bar{y}$ is an arbitrary element from Y.

We fix $\varepsilon_0 > 0$ such that $x + \varepsilon\bar{x} \in \operatorname{dom} F$ for all $\varepsilon \in [0, \varepsilon_0]$. Then due to Theorem 3.56, for any $\bar{y} \in Y$ we get

$$\rho(y + \varepsilon\bar{y}, F(x + \varepsilon\bar{x})) = d_F(z + \varepsilon\bar{z}) - d_F(z)$$
$$= \varepsilon d'_F(z; \bar{z}) + o(\varepsilon) = \varepsilon\rho(\bar{y}, \Gamma_F(z; \bar{x})) + o(\varepsilon),$$

where $\varepsilon \in [0, \varepsilon_0]$, $o(\varepsilon)/\varepsilon \to 0$ for $\varepsilon \downarrow 0$.

Applying Lemma 6.16, where $\Gamma_F(z; \bar{x})$ is assumed to be fixed, we obtain

$$\rho(y + \varepsilon\bar{y}, F(x + \varepsilon\bar{x})) = \varepsilon\rho(\bar{y}, \Gamma_F(z; \bar{x})) + o(\varepsilon)$$
$$\leq \varepsilon\alpha \max\{0, \ \langle\nabla h_i(z), \bar{z}\rangle, \ i \in I(z), \ |\langle\nabla h_i(z), \bar{z}\rangle|, \ i \in I_0\} + o(\varepsilon)$$
$$= \alpha \max\{0, h_i(z) + \varepsilon\langle\nabla h_i(z), \bar{z}\rangle, \ i \in I(z), |h_i(z) + \varepsilon\langle\nabla h_i(z), \bar{z}\rangle|, \ i \in I_0\}$$
$$+ o(\varepsilon)$$
$$\leq \alpha \max\{0, h_i(z + \varepsilon\bar{z}), i \in I(z), \ |h_i(z + \varepsilon\bar{z})|, i \in I_0\} + \gamma(\varepsilon) + o(\varepsilon),$$

where

$$\gamma(\varepsilon) = \max_{i \in I(z) \cup I_0} |h_i(z + \varepsilon\bar{z}) - h_i(z) - \varepsilon\langle\nabla h_i(z), \bar{z}\rangle|.$$

It is easy to see that $\gamma(\varepsilon)/\varepsilon \to 0$ for $\varepsilon \downarrow 0$. Since $h_i(z) < 0$ for $i \in I \backslash I(z)$, then $h_i(z + \varepsilon\bar{z}) < 0$ for all $\varepsilon \in [0, \varepsilon_0]$, where $\varepsilon_0 > 0$ is sufficiently small. In this case we get

$$\rho(y + \varepsilon\bar{y}, F(x + \varepsilon\bar{x}))$$
$$\leq \alpha \max\{0, \ h_i(z + \varepsilon\bar{z}), \ i \in I, \ |h_i(z + \varepsilon\bar{z})|, \ i \in I_0\} + \tilde{o}(\varepsilon),$$

where $\tilde{o}(\varepsilon)/\varepsilon \to 0$ for $\varepsilon \downarrow 0$. The last relation is clearly equivalent to Definition 6.36. In this way, the $(R_{\bar{x}})$-regularity condition holds at the point z. ∎

DEFINITION 6.39 *We shall say that the multivalued mapping F satisfies the $(R_{\bar{x}_1\bar{x}_2})$-regularity condition at the point $z \in \text{gr}\, F$ if for any $\bar{y}_1 \in \Gamma_F(z; \bar{x}_1)$ and $\bar{y}_2 \in Y$ one can find quantities $\alpha > 0$, $\varepsilon_0 > 0$ and $o(\varepsilon)$ such that $o(\varepsilon)/\varepsilon \to 0$ if $\varepsilon \downarrow 0$ and for all $\varepsilon \in [0, \varepsilon_0]$ the inequality*

$$\rho(y + \varepsilon\bar{y}_1 + \varepsilon^2\bar{y}_2 + o(\varepsilon^2), F(x + \varepsilon\bar{x}_1 + \varepsilon^2\bar{x}_2))$$
$$\leq \alpha \max\{0, h_i(z_\varepsilon), i \in I, \ |h_i(z_\varepsilon)|, \ i \in I_0\},$$

holds, where $z_\varepsilon = (x + \varepsilon\bar{x}_1 + \varepsilon^2\bar{x}_2, y + \varepsilon\bar{y}_1 + \varepsilon^2\bar{y}_2)$.

LEMMA 6.40 *Let the $(R_{\bar{x}_1\bar{x}_2})$-regularity condition hold at the point $z_0 \in \text{gr}\, F$. Then $\hat{D}_L^2 F(z_0, \bar{z}_1; \bar{x}_2) = \Gamma_F^2(z_0, \bar{z}_1; \bar{x}_2) \neq \emptyset$.*

Proof. Let $\bar{z}_2 = (\bar{x}_2, \bar{y}_2) \in \Gamma_F^2(z_0, \bar{z}_1)$. Due to Definition 6.39, we have

$$d_F(x_\varepsilon, y_\varepsilon + o(\varepsilon^2)) \leq \alpha \max\{0, h_i(z_\varepsilon), i \in I(z_0), \ |h_i(z_\varepsilon)|, \ i \in I_0\},$$

where $x_\varepsilon = x_0 + \varepsilon\bar{x}_1 + \varepsilon^2\bar{x}_2$, $y_\varepsilon = y_0 + \varepsilon\bar{y}_1 + \varepsilon^2\bar{y}_2$, $z_\varepsilon = (x_\varepsilon, y_\varepsilon)$. In this case there exists $\varepsilon_0' > 0$ such that $0 < \varepsilon_0' < \varepsilon_0$. Furthermore, for

$\varepsilon \in [0, \varepsilon_0']$ we get

$$d_F(x_\varepsilon, y_\varepsilon + o(\varepsilon^2))$$

$$\leq \alpha \max \left\{ 0, \varepsilon \langle \nabla h_i(z_0), \bar{z}_1 \rangle + \varepsilon^2 \langle \nabla h_i(z_0), \bar{z}_2 \rangle + \frac{\varepsilon^2}{2} \langle \bar{z}_1, \nabla^2 h_i(z_0) \bar{z}_1 \rangle \right.$$

$$+ o_i(\varepsilon^2), \ i \in I(z_0), \ \left| \varepsilon \langle \nabla h_i(z_0), \bar{z}_1 \rangle + \varepsilon^2 \langle \nabla h_i(z_0), \bar{z}_2 \rangle \right.$$

$$\left. + \frac{\varepsilon^2}{2} \langle \bar{z}_1, \nabla^2 h_i(z_0) \bar{z}_1 \rangle + o_i(\varepsilon^2) \right|, \ i \in I_0 \right\},$$

$$\leq \alpha \varepsilon^2 \max \left\{ 0, \ \langle \nabla h_i(z_0), \bar{z}_2 \rangle + \frac{1}{2} \langle \bar{z}_1, \nabla^2 h_i(z_0) \bar{z}_1 \rangle, \ i \in I^2(z_0, \bar{z}_1), \right.$$

$$\left. |\langle \nabla h_i(z_0), \bar{z}_2 \rangle + \frac{1}{2} \langle \bar{z}_1, \nabla^2 h_i(z_0) \bar{z}_1 \rangle|, \ i \in I_0 \right\} + o(\varepsilon^2) = o(\varepsilon^2), (6.19)$$

where $o_i(\varepsilon^2) \triangleq \langle \bar{z}_1, \nabla^2 h_i(z_0 + \tau_\varepsilon^i(\varepsilon \bar{z}_1 + \varepsilon^2 \bar{z}_2)) \bar{z}_1 \rangle - \langle \bar{z}_1, \nabla^2 h_i(z_0) \bar{z}_1 \rangle$, $o(\varepsilon^2) \triangleq \max \{o_i(\varepsilon^2) \mid i \in I_0 \cup I^2(z_0, \bar{z}_1)\}$, $0 < \tau_\varepsilon^i < 1$, $i \in I_0 \cup I^2(z_0, \bar{z}_1)$.

With regard to inequality (6.18) and the definitions of the function d_F and the set $\hat{T}_{grF}^{L2}(z_0, \bar{z}_1)$, it immediately follows that $\bar{z}_2 \in \hat{T}_{grF}^{L2}(z_0, \bar{z}_1)$.

In this way, we get the inclusion $\Gamma_F^2(z_0, \bar{z}_1; \bar{x}_2) \subset \hat{D}_L^2 F(z_0, \bar{z}_1; \bar{x}_2)$. Lemma 6.35 yields the opposite inclusion. Thus, equality holds, which means $\hat{D}_L^2 F(z_0, \bar{z}_1; \bar{x}_2) = \Gamma_F^2(z_0, \bar{z}_1; \bar{x}_2)$. Now it remains to show that $\Gamma_F^2(z_0, \bar{z}_1; \bar{x}_2) \neq \emptyset$.

Due to Lemma 6.35, we have $\hat{D}_L^2 F(z_0, \bar{z}_1; \bar{x}_2) = \hat{D}_U^2 F(z_0, \bar{z}_1; \bar{x}_2) = \Gamma_F^2(z_0, \bar{z}_1; \bar{x}_2)$. On the other hand, (6.18) implies

$$\liminf_{\varepsilon \downarrow 0} \varepsilon^{-2} d_F(z_0 + \varepsilon \bar{z}_1 + \varepsilon^2 \bar{z}_2) < +\infty.$$

According to Lemma 6.35 this means that $\hat{D}_U^2 F(z_0, \bar{z}_1; \bar{x}_2)$ is nonempty. ∎

COROLLARY 6.41 *Let the* (R)-*regularity condition hold at the point* $z_0 \in$ gr F. *Then for all* $\bar{z}_1 \in \Gamma_F(z_0)$, $\bar{x}_2 \in X$, *one has*

$$\hat{D}_L^2 F(z_0, \bar{z}_1; \bar{x}_2) = \Gamma_F^2(z_0, \bar{z}_1; \bar{x}_2) \neq \emptyset.$$

LEMMA 6.42 *Let* $z_0 \in$ gr F. *Then for all* $\bar{z}_1, \bar{z}_2 \in X \times Y$ *the inequality*

$$\eta \triangleq \limsup_{\varepsilon \downarrow 0} \varepsilon^{-2} d_F(z_0 + \varepsilon \bar{z}_1 + \varepsilon^2 \bar{z}_2) \leq \rho(\bar{y}_2, \hat{D}_L^2 F(z_0, \bar{z}_1; \bar{x}_2))$$

is valid.

Proof. If $\hat{D}_L^2 F(z_0, \bar{z}_1; \bar{x}_2) = \emptyset$, then the statement of the lemma trivially holds since inf $\emptyset = +\infty$.

Now, suppose $\hat{D}_L^2 F(z_0, \bar{z}_1; \bar{x}_2) \neq \emptyset$, and let $\hat{y}_2$ be an arbitrary vector from this set. Then

$$d_F(z_0 + \varepsilon \bar{z}_1 + \varepsilon^2 \bar{z}_2) = \rho(y_0 + \varepsilon \bar{y}_1 + \varepsilon^2 \bar{y}_2, F(x_0 + \varepsilon \bar{x}_1 + \varepsilon^2 \bar{x}_2))$$
$$\leq |y_0 + \varepsilon \bar{y}_1 + \varepsilon^2 \bar{y}_2 - y_0 - \varepsilon \bar{y}_1 - \varepsilon^2 \hat{y}_2 - o(\varepsilon^2)| \leq \varepsilon^2 |\bar{y}_2 - \hat{y}_2| + |o(\varepsilon^2)|$$

and we get $\eta \leq |\bar{y}_2 - \hat{y}_2|$ for all $\hat{y}_2 \in \hat{D}_L^2 F(z_0, \bar{z}_1; \bar{x}_2)$. This is equivalent to the statement of the lemma. ∎

LEMMA 6.43 *Let $z_0 \in \mathrm{gr}\, F$, $\bar{y}_1 \in \Gamma_F(z_0; \bar{x}_1)$. If*

$$\hat{D}_L^2 F(z_0, \bar{z}_1; \bar{x}_2) = \Gamma_F^2(z_0, \bar{z}_1; \bar{x}_2) \neq \emptyset,$$

then the $(R_{\bar{x}_1 \bar{x}_2})$-condition holds at the point z_0.

Proof. Let us take $\varepsilon_0' > 0$ such that $x_0 + \varepsilon \bar{x}_1 + \varepsilon^2 \bar{x}_2 \in \mathrm{dom}\, F$ for $\varepsilon \in [0, \varepsilon_0']$. Then according to Lemma 6.42, for any $\bar{y}_2 \in Y$ we get

$$d_F(z_0 + \varepsilon \bar{z}_1 + \varepsilon^2 \bar{z}_2) \leq \varepsilon^2 \rho(\bar{y}_2, \hat{D}_L^2 F(z_0, \bar{z}_1; \bar{x}_2)) + o(\varepsilon^2)$$
$$= \varepsilon^2 \rho(\bar{y}_2, \Gamma_F^2(z_0, \bar{z}_1; \bar{x}_2)) + o(\varepsilon^2) \quad \text{for} \quad \varepsilon \in [0, \varepsilon_0'].$$

Applying Lemma 6.16, where the set $\Gamma_F^2(z_0, \bar{z}_1; \bar{x}_2)$ is assumed to be fixed, we obtain for $\varepsilon \in [0, \varepsilon_0]$, $0 < \varepsilon_0 < \varepsilon_0'$,

$$d_F(z_0 + \varepsilon \bar{z}_1 + \varepsilon^2 \bar{z}_2) \leq \varepsilon^2 \rho(\bar{y}_2, \Gamma_F^2(z_0, \bar{z}_1; \bar{x}_2)) + o(\varepsilon^2)$$

$$\leq \alpha \varepsilon^2 \max \Big\{ 0, \; \langle \nabla h_i(z_0), \bar{z}_2 \rangle + \frac{1}{2} \langle \bar{z}_1, \nabla^2 h_i(z_0) \bar{z}_1 \rangle, \; i \in I^2(z_0, \bar{z}_1),$$

$$|\langle \nabla h_i(z_0), \bar{z}_2 \rangle + \frac{1}{2} \langle \bar{z}_1, \nabla^2 h_i(z_0) \bar{z}_1 \rangle|, \; i \in I_0 \Big\} + o(\varepsilon^2)$$

$$= \alpha \max \Big\{ 0, \; h_i(z_0) + \varepsilon \langle \nabla h_i(z_0), \bar{z}_1 \rangle + \varepsilon^2 \langle \nabla h_i(z_0), \bar{z}_2 \rangle$$

$$+ \frac{1}{2} \varepsilon^2 \langle \bar{z}_1, \nabla^2 h_i(z_0) \bar{z}_1 \rangle, \; i \in I(z_0), \; |h_i(z_0) + \varepsilon \langle \nabla h_i(z_0), \bar{z}_1 \rangle$$

$$+ \varepsilon^2 \langle \nabla h_i(z_0), \bar{z}_2 \rangle + \frac{1}{2} \varepsilon^2 \langle \bar{z}_1, \nabla^2 h_i(z_0) \bar{z}_1 \rangle|, \; i \in I_0 \Big\} + o(\varepsilon^2)$$

$$= \alpha \max \{ 0, h_i(z_0 + \varepsilon \bar{z}_1 + \varepsilon^2 \bar{z}_2), i \in I, |h_i(z_0 + \varepsilon \bar{z}_1 + \varepsilon^2 \bar{z}_2)|, i \in I_0 \}$$
$$+ \tilde{o}(\varepsilon^2),$$

where $\tilde{o}(\varepsilon^2) = o(\varepsilon^2) + \max\limits_{i \in I \cup I_0} \gamma_{i\varepsilon}$ and $\gamma_{i\varepsilon} = h_i(z_0) + \varepsilon \langle \nabla h_i(z_0), \bar{z}_1 \rangle + \varepsilon^2 \langle \nabla h_i(z_0), \bar{z}_2 \rangle + \frac{1}{2} \varepsilon^2 \langle \bar{z}_1, \nabla^2 h_i(z_0) \bar{z}_1 \rangle - h_i(z_0 + \varepsilon \bar{z}_1 + \varepsilon^2 \bar{z}_2)$. Thus, for $\bar{y}_1 \in \Gamma_F(z_0; \bar{x}_1)$ Definition 6.39 is valid. ∎

Based on Lemmas 6.40 and 6.43, the following theorem can be proved.

THEOREM 6.44 *Let $z_0 \in \mathrm{gr}\, F$, $\bar{z}_1 \in \Gamma_F(z_0)$. Then the following statements are equivalent:*

1. *the multivalued mapping F satisfies the $(R_{\bar{x}_1\bar{x}_2})$-condition at the point z_0;*
2. $\hat{D}_L^2 F(z_0, \bar{z}_1; \bar{x}_2) = \Gamma_F^2(z_0, \bar{z}_1; \bar{x}_2) \neq \emptyset.$

Let us consider another concept of directional regularity introduced by Gollan in [74] and then studied by Auslender and Cominetti [9] as well as Gauvin and Janin [70].

DEFINITION 6.45 *Let $z \in \mathrm{gr}\, F$, $\bar{x} \in X$. We shall say that at the point z the Mangasarian-Fromowitz regularity condition in the direction $\bar{x}$ (or $(MF_{\bar{x}})$-regularity) holds if*
1. *the vectors $\nabla_y h_i(z)$, $i \in I_0$, are linearly independent;*
2. *one can find an element $\bar{y}_0 \in Y$ such that*

$$\langle \nabla h_i(z), \bar{z}_0 \rangle = 0, \quad i \in I_0,$$
$$\langle \nabla h_i(z_0), \bar{z}_0 \rangle < 0, \quad i \in I(z),$$

where $\bar{z}_0 = (\bar{x}, \bar{y}_0)$.

Note that the (MF_0)-condition coincides with the well-known (MF)-regularity condition.

Very often the $(MF_{\bar{x}})$-regularity condition is used in another form being equivalent to Definition 6.45.

LEMMA 6.46 *([9]) The following statements are equivalent:*
1. *the $(MF_{\bar{x}})$-regularity condition holds at the point z;*
2. $\sum_{i=1}^{p} \langle \lambda_i \nabla_x h_i(z), \bar{x} \rangle < 0$ *for all $\lambda \in \Lambda_0(z) \backslash \{0\}$.*

Proof. 1. $\Rightarrow$ 2. We choose an arbitrary vector $\lambda \in \Lambda_0(z) \backslash \{0\}$ and a vector $\bar{y}_0$ satisfying Definition 6.45. The $(MF_{\bar{x}})$-regularity implies $\lambda_i > 0$ for at least one $i \in I(z)$. Therefore $\sum_{i=1}^{p} \langle \lambda_i \nabla h_i(z), \bar{z}_0 \rangle < 0$ and we get the second statement of the lemma since $\sum_{i=1}^{p} \langle \lambda_i \nabla_y h_i(z), \bar{y}_0 \rangle = 0$ for $\lambda \in \Lambda_0(z)$.

2. $\Rightarrow$ 1. Suppose the vectors $\nabla_y h_i(z)$, $i \in I_0$, are linearly dependent. Then we can find a vector $\lambda \in \Lambda_0(z) \backslash \{0\}$ with λ_i equal to 0 for $i \in I$. But in this case $-\lambda \in \Lambda_0(z) \backslash \{0\}$ and, consequently, the strict inequality in the second statement is impossible. Thus the vectors $\nabla_y h_i(z)$, $i \in I_0$, are linearly independent.

Suppose now that the second condition in the definition of $(MF_{\bar{x}})$-regularity is not true. Then the convex set $\{a \in R^\nu \mid a_i = \langle \nabla h_i(z), \bar{z} \rangle,\ \bar{y} \in Y\}$ does not intersect with the set $\{a \in R^\nu \mid a_i < 0,\ i \in I(z),\ a_i = 0,\ i \in I_0\}$, where $\nu = |I(z) \cup I_0|$.

According to the Separation Theorem, there are a non-vanishing vector $\lambda \in R^\nu$ and a number α such that

$$\sum_{i=1}^{\nu} \langle \lambda_i \nabla h_i(z), \bar{z} \rangle \geq \alpha \geq \sum_{i \in I(z)} \lambda_i v_i$$

for all $\bar{y} \in Y$ and $v_i < 0$. These inequalities imply $\lambda_i \geq 0$ for $i \in I(z)$ and $\sum_{i \in I(z) \cup I_0} \lambda_i \nabla_y h_i(z) = 0$. Supplementing the vector λ to a vector in R^p by setting $\lambda_i = 0$ for $i \in I \backslash I(z)$, we get $\lambda \in \Lambda_0(z) \backslash \{0\}$. But this contradicts the second statement of the lemma since for $v_i \to 0$ the inequality $\sum_{i=1}^{p} \langle \lambda_i \nabla_x h_i(z), \bar{x} \rangle \geq 0$ holds. ∎

Let us denote $N = -[\Gamma_F(z)]^+$. Then due to Theorem 2.6

$$N = \left\{ z^* = \sum_{i=1}^{p} \lambda_i \nabla h_i(z) \;\middle|\; \lambda_i \geq 0 \;\; \text{and} \;\; \lambda_i h_i(z) = 0, \, i \in I \right\}.$$

We consider the set

$$N(0) \overset{\triangle}{=} \{ x^* \in R^n \,|\, (x^*, 0) \in N \}$$

$$= \left\{ x^* = \sum_{i=1}^{p} \lambda_i \nabla_x h_i(z) \,|\, \lambda_i \geq 0, \, \lambda_i h_i(z) = 0, \, i \in I, \, \sum_{i=1}^{p} \lambda_i \nabla_y h_i(z) = 0 \right\}$$

$$= \left\{ x^* = \sum_{i=1}^{p} \lambda_i \nabla_x h_i(z) \,|\, \lambda \in \Lambda_0(z) \right\}.$$

Let condition 2 of Lemma 6.46 hold. Due to Lemma 3.38 (cf. Section 1 concerning properties of polar cones and Theorem 19.3 from [154] on the projection of a polyhedral set), we get

$$\bar{x} \in -\text{int}\,[N(0)]^+ = \text{int cl dom}\,\Gamma_F(z; \cdot) = \text{int dom}\,\Gamma_F(z; \cdot).$$

REMARK. Due to what was said above, the $(MF_{\bar{x}})$-condition implies $\bar{x} \in \text{int dom}\,\Gamma_F(z; \cdot)$.

The following simple example shows that the opposite statement is not correct.

EXAMPLE 6.47 *Let us consider the set* $F(x) = \{ y \in R^2 \,|\, y_1 + y_2 - x \leq 0,$ $-y_1 - y_2 + x \leq 0, \, y_1 \geq 0, \, y_2 \geq 0 \}$, $x \in X = R$. *We choose the points* $y_0 = \left(\frac{1}{2}, \frac{1}{2} \right)$ *and* $x_0 = 1$. *It is not hard to determine the sets* $\Gamma_F(z_0; \bar{x}) = \{ \bar{y} \,|\, \bar{y}_1 + \bar{y}_2 = \bar{x} \}$. *Clearly* $dom\,\Gamma_F(z_0; \cdot) = X$. *Furthermore, it can be easily seen that* $\hat{T}^L_{gr\,F}(z_0) = \Gamma_F(z_0)$, *i. e. the Kuhn-Tucker regularity condition holds. Moreover, due to Lemma 6.16 the mapping* F *is* (R)-*regular at the point* z_0. *At the same time the* (MF)-*condition does not* hold at the point z_0. The same is true with respect to $(MF_{\bar{x}})$-regularity (which fails to be true for any $\bar{x}$) since

$$\Lambda_0(z_0) = \{\lambda \in R^4 \,|\, \lambda_1 = \alpha, \,\lambda_2 = \alpha, \,\lambda_3 = \lambda_4 = 0, \,\alpha \geq 0\} \neq \{0\},$$

$$\sum_{i=1}^{4} \lambda_i \nabla_x h_i \equiv 0 \quad \text{for} \quad \lambda \in \Lambda_0(z_0).$$

This example reveals a certain weakness of the regularity conditions (MF) and $(MF_{\bar{x}})$ compared with (R)-regularity and Kuhn-Tucker regularity.

In the same way as (MF)-regularity is a sufficient condition for (R)-regularity (see Theorem 6.15), $(MF_{\bar{x}})$-regularity implies Kuhn-Tucker regularity in the direction $\bar{x}$. This will be established in the following assertion.

LEMMA 6.48 *Let the mapping F satisfy the $(MF_{\bar{x}})$-regularity condition at the point $z \in \operatorname{gr} F$. Then the following relations hold at the point z:*

$$\hat{D}_L F(z; \bar{x}) = \Gamma_F(z; \bar{x}) \neq \emptyset,$$

$$\hat{D}_L^2 F(z, \bar{z}; \bar{x}_2) = \Gamma_F^2(z, \bar{z}; \bar{x}_2) \quad \text{for any} \quad \bar{z} \in \Gamma_F(z), \,\bar{x}_2 \in X.$$

Proof. Let $\bar{y}$ satisfy the second condition from the definition of $(MF_{\bar{x}})$-regularity. If $I_0 = \emptyset$, then we set $\xi(\varepsilon) = 0$. Otherwise, if $I_0 \neq \emptyset$, we consider the system of equations

$$h_i(x + \varepsilon\bar{x}, y + \varepsilon\bar{y} + \xi(\varepsilon)) = 0, \quad i \in I_0$$

assuming the components ξ_i to be equal to zero for those indices i which correspond to nonbasic rows of the matrix $[\nabla_y h_i(z), \,i \in I_0]$. In this case, due to the implicit function theorem, the system has a solution $\xi = \xi(\varepsilon)$ defined on some interval $[0, \varepsilon_1]$ and satisfying the conditions $\xi(0) = 0$ and $\xi(\varepsilon)/\varepsilon \to 0$ for $\varepsilon \downarrow 0$. Since $h_i(z) < 0, \,i \in I\backslash I(z)$, then in view of the continuity of the functions h_i there exists a scalar $\varepsilon_2 \leq \varepsilon_1$ such that

$$h_i(x + \varepsilon\bar{x}, y + \varepsilon\bar{y} + \xi(\varepsilon)) \leq 0, \,i \in I\backslash I(z), \,\varepsilon \in [0, \varepsilon_2].$$

We define $\bar{\xi} = (0, \xi(\varepsilon))$. Let $i \in I(z)$. Then $\langle \nabla h_i(z), \bar{z} \rangle \overset{\triangle}{=} -\delta_i < 0$ and

$$h_i(z + \varepsilon\bar{z} + \bar{\xi}(\varepsilon)) = -\varepsilon\delta_i + o_i(\varepsilon),$$

where $o_i(\varepsilon)/\varepsilon \to 0$ for $\varepsilon \downarrow 0$. We can find a number $\varepsilon_0 > 0, \,\varepsilon_0 \leq \varepsilon_2$, such that $-\delta_i + o_i(\varepsilon)/\varepsilon \leq 0$ for all $\varepsilon \in (0, \varepsilon_0]$. Thus $h_i(z + \varepsilon\bar{z} + \bar{\xi}(\varepsilon)) \leq 0$, $\varepsilon \in [0, \varepsilon_0], \,i \in I$, and therefore $\bar{y} \in \hat{D}_L F(z; \bar{x})$. In this way, we conclude

$$\emptyset \neq \operatorname{ri} \Gamma_F(z; \bar{x}) \subset \hat{D}_L F(z; \bar{x}).$$

Passing to the closure on both sides of this inclusion and taking into account the obvious inclusion $\hat{D}_L F(z; \bar{x}) \subset \Gamma_F(z; \bar{x})$, we obtain

$$\Gamma_F(z; \bar{x}) = \hat{D}_L F(z; \bar{x}) \neq \emptyset.$$

Let us now take arbitrary vectors $\bar{y} \in \Gamma_F(z; \bar{x})$ and $\bar{y}_2 \in G_F^2(z, \bar{z}; \bar{x}_2)$, where

$$G_F^2(z, \bar{z}; \bar{x}_2) \overset{\triangle}{=} \left\{ \bar{y}_2) \,|\, \langle \nabla h_i(z), \bar{z}_2 \rangle + \frac{1}{2} \langle \bar{z}, \nabla^2 h_i(z) \bar{z} \rangle = 0, \; i \in I_0, \right.$$

$$\left. \langle \nabla h_i(z), \bar{z}_2 \rangle + \frac{1}{2} \langle \bar{z}, \nabla^2 h_i(z) \bar{z} \rangle < 0, \; i \in I^2(z, \bar{z}) \right\},$$

and $\bar{z}_2 = (\bar{x}_2, \bar{y}_2)$. In this case, repeating the first part of the proof of the lemma with respect to $\bar{y}$ and $\bar{y}_2$, we may assert that there exists a function $\xi = \xi(\varepsilon)$ such that $\xi(\varepsilon) \to 0$ if $\varepsilon \downarrow 0$ and for $\varepsilon \in [0, \varepsilon_1]$ the relations

$$h_i(x + \varepsilon\bar{x} + \varepsilon^2\bar{x}_2, y + \varepsilon\bar{y} + \varepsilon^2\bar{y}_2 + \varepsilon^2\xi(\varepsilon)) = 0, \qquad i \in I_0,$$

$$h_i(x + \varepsilon\bar{x} + \varepsilon^2\bar{x}_2, y + \varepsilon\bar{y} + \varepsilon^2\bar{y}_2 + \varepsilon^2\xi(\varepsilon)) \leq 0, \qquad i \in I\backslash I(z)$$

are valid. For $i \in I(z)$ we have

$$h_i(z + \varepsilon\bar{z} + \varepsilon^2\bar{z}_2 + \varepsilon^2\bar{\xi}(\varepsilon)) = \varepsilon\langle \nabla h_i(z), \bar{z} \rangle$$

$$+ \varepsilon^2 \left\{ \langle \nabla h_i(z), \bar{z}_2 \rangle + \frac{1}{2}\langle \bar{z}, \nabla^2 h_i(z)\bar{z} \rangle \right\} + o_i(\varepsilon^2),$$

where $\bar{\xi} = (0, \xi)$. Therefore $h_i(z + \varepsilon\bar{z} + \varepsilon^2\bar{z}_2 + \varepsilon^2\bar{\xi}(\varepsilon)) < 0$, $i \in I(z)$, for sufficiently small $\varepsilon > 0$.

As a consequence we get $\bar{y}_2 \in \hat{D}_L^2 F(z, \bar{z}; \bar{x}_2)$. From this we conclude $G_F^2(z, \bar{z}; \bar{x}_2) \subset \hat{D}_L^2 F(z, \bar{z}; \bar{x}_2)$ for any $\bar{x}_2 \in X$ and $\bar{y} \in \Gamma_F(z; \bar{x})$. Taking the closure in this relation and making use of the obvious inclusion $\hat{D}_L^2 F(z, \bar{z}; \bar{x}_2) \subset \Gamma_F^2(z, \bar{z}; \bar{x}_2)$ (see Lemma 6.34), we obtain the required relation

$$\hat{D}_L^2 F(z, \bar{z}; \bar{x}_2) = \Gamma_F^2(z, \bar{z}; \bar{x}_2). \qquad \blacksquare$$

3.   First-Order Directional Derivatives of Optimal Value Functions and Sensitivity Analysis of Suboptimal Solutions

In this section we discuss first-order differentiability properties of the optimal value function φ. Together with the usual derivative $\varphi'(x; \bar{x})$

at the point x in the direction $\bar{x}$ we consider the upper and lower Dini derivatives defined as

$$D_+\varphi(x;\bar{x}) = \liminf_{\varepsilon\downarrow 0} \varepsilon^{-1}[\varphi(x+\varepsilon\bar{x}) - \varphi(x)],$$

$$D^+\varphi(x;\bar{x}) = \limsup_{\varepsilon\downarrow 0} \varepsilon^{-1}[\varphi(x+\varepsilon\bar{x}) - \varphi(x)],$$

respectively.

3.1 General Case

Let us return to the general mathematical programming problem under abstract constraints

$$(\bar{P}_x): \qquad \begin{cases} f(x,y) \to \inf_y \\ y \in F(x), \end{cases}$$

where the function $f : X \times Y \to R$ is continuously differentiable.

Together with (A1) we suppose the following assumptions to be fulfilled as well:

(A2) – the multivalued mapping F is closed, i. e., its graph

$$\operatorname{gr} F = \{(x,y) \mid x \in X, \ y \in F(x)\}$$

is a closed set in the space $X \times Y$;

(A3) – the function f is locally Lipschitz continuous on the set $X_0 \times [Y_0 + \varepsilon_0 B]$, where $\varepsilon_0 > 2\operatorname{diam} Y_0$, B is the open unit ball;

(A4) – the multivalued mapping F is differentiable in the direction $\bar{x} \in X$ at all points $z_0 = (x_0, y_0) \in \{x_0\} \times \omega(x_0)$ having the derivative $\hat{D}F(z_0;\bar{x})$ (see Definition 3.55);

(A5) – the function f is directionally differentiable at the points $z_0 \in \{x_0\} \times \omega(x_0)$.

THEOREM 6.49 *Let the assumptions (A1)–(A5) hold and let the set of optimal solutions be sequentially Lipschitz continuous at the point x_0 in the direction $\bar{x}$ (see Definition 6.6). Then the function φ is differentiable at the point x_0 in the direction $\bar{x}$ and*

$$\varphi'(x_0;\bar{x}) = \inf_{y_0\in\omega(x_0)} \inf_{\bar{y}\in\hat{D}F(z_0;\bar{x})} f'(z_0;\bar{z}), \qquad (6.20)$$

where $z_0 = (x_0, y_0)$, $\bar{z} = (\bar{x}, \bar{y})$.

Proof. 1. Let us take arbitrary elements $y_0 \in \omega(x_0)$ and $\bar{y} \in \hat{D}F(z_0;\bar{x})$. Then there exists a vector function o(t) such that $y_0 + t\bar{y} + \mathrm{o}(t) \in F(x_0 + t\bar{x})$ for $t \geq 0$ and

$$\varphi(x_0 + t\bar{x}) - \varphi(x_0) \leq f(x_0 + t\bar{x}, y_0 + t\bar{y} + \mathrm{o}(t)) - f(x_0, y_0).$$

Dividing this inequality by t and passing to the limit, we get

$$D^+\varphi(x_0; \bar{x}) \leq f'(z_0; \bar{z}).$$

Moreover

$$D^+\varphi(x_0; \bar{x}) \leq \inf_{y_0 \in \omega(x_0)} \inf_{\bar{y} \in \hat{D}F(z_0;\bar{x})} f'(z_0; \bar{z}). \qquad (6.21)$$

2. Let $D_+\varphi(x_0; \bar{x})$ be attained on the sequence $t_k \downarrow 0$. We choose an arbitrary sequence $y_k \in \omega(x_0 + t_k\bar{x})$, $k = 1, 2, \ldots$ Due to the assumptions of the theorem, without loss of generality we can assume that $y_k \to \tilde{y}_0$. Moreover, in view of (A1), we have $\tilde{y}_0 \in F(x_0)$. From Lemma 3.71 we deduce the upper semicontinuity of the function $\varphi(x_0 + t_k\bar{x})$. Now, passing to the limit in the equality

$$f(x_0 + t_k\bar{x}, y_k) - \varphi(x_0) = \varphi(x_0 + t_k\bar{x}) - \varphi(x_0),$$

we get $f(x_0, \tilde{y}_0) - \varphi(x_0) \leq 0$, i.e. $\tilde{y}_0 \in \omega(x_0)$. Therefore, from the sequential Lipschitz continuity it follows that

$$|y_k - \tilde{y}_0| \leq lt_k, \quad k = 1, 2, \ldots,$$

where $l = \text{const} > 0$. Without loss of generality we can assume that the relation $t_k^{-1}(y_k - \tilde{y}_0) \to \bar{y}_0$ holds for $k \to \infty$, where, due to the differentiability of F, one obviously has $\bar{y}_0 \in \hat{D}F(z_0; \bar{x}) = \hat{D}_L F(z_0; \bar{x}) = \hat{D}_U F(z_0; \bar{x})$. In this case, passing to the limit in the equality

$$t_k^{-1}[\varphi(x_0 + t_k\bar{x}) - \varphi(x_0)] = t_k^{-1}[f(x_0 + t_k\bar{x}, y_k) - f(x_0, \tilde{y}_0)],$$

we get

$$D_+\varphi(x_0; \bar{x}) = f'((x_0, \tilde{y}_0); (\bar{x}, \bar{y}_0)),$$

and, therefore

$$D_+\varphi(x_0; \bar{x}) \geq \inf_{y_0 \in \omega(x_0)} \inf_{\bar{y} \in \hat{D}F(z_0;\bar{x})} f'(z_0; \bar{z}).$$

Comparing the last relation with (6.21), we obtain the statement of the theorem and the formula (6.20). ∎

COROLLARY 6.50 *The assumption on sequential Lipschitz continuity of the optimal set $\omega(x)$ in the conditions of Theorem 6.49 can be replaced by the assumption about pseudo-Lipschitz continuity of the mapping F at the point (x_0, y_0) and the condition $\omega(x_0) = \{y_0\}$, where y_0 is a first-order local isolated minimizer of the problem $(\bar{P}_x)$.*

Proof. It follows directly from Theorem 6.3.

COROLLARY 6.51 *The assumption concerning sequential Lipschitz continuity of the optimal set $\omega(x)$ in the conditions of Theorem 6.49 can be replaced by the less strong condition of weak Lipschitz continuity of the set $\omega(x)$ at the point x_0 in the direction $\bar{x}$ (see Definition 6.7).*

Proof. The first part of the proof of Theorem 6.49 remains unchanged.

In the second part, in view of the weak Lipschitz continuity of $\omega(x)$ at the point x_0 in the direction $\bar{x}$, for the sequence $\{t_k\}$ one can find elements $y_0 \in \omega(x_0)$ and $y_k \in \omega(x_0 + t_k \bar{x})$ such that

$$|y_k - y_0| \leq lt_k, \quad k = 1, 2, \ldots$$

Therefore, without loss of generality, we can assume that $t_k^{-1}(y_k - y_0) \to \bar{y}_0$ if $k \to \infty$. For the remaining part we can exactly repeat the proof of Theorem 6.49. ∎

Now we fix the direction $\bar{x} \in X$. Together with the set of optimal solutions $\omega(x_0 + t\bar{x})$ we consider the set of ε-optimal (suboptimal) solutions

$$\omega_\varepsilon(x_0 + t\bar{x}) = \{y \in F(x_0 + t\bar{x}) \mid f(x_0 + t\bar{x}, y) \leq \varphi(x_0 + t\bar{x}) + \varepsilon\},$$

where $\varepsilon = \varepsilon(t) \geq 0$ for all t.

THEOREM 6.52 *Let the assumptions (A1)-(A5) hold. Then the following statements are equivalent:*

1. the derivative $\varphi'(x_0; \bar{x})$ exists and is defined by (6.20);

2. the set $\omega_\varepsilon(x)$ is weakly Lipschitz continuous at the point x_0 in the direction $\bar{x}$, i.e., there exist $M > 0$, $t_0 > 0$ and $\varepsilon(t) = o(t)$ such that

$$\rho(y_0, \omega_\varepsilon(x_0 + t\bar{x})) \leq Mt, \quad t \in [0, t_0].$$

Proof. 1. ⇒ 2. Let the solution of the programming problem at the right-hand side of (6.20) be attained at the points $y_0 \in \omega(x_0)$ and $\bar{y} \in \hat{D}F(z_0; \bar{x})$. Then there exists a vector function $o(t)$ such that $y_0 + t\bar{y} + o(t) \in F(x_0 + t\bar{x})$. We denote $y(t) = y_0 + t\bar{y} + o(t)$, $\varepsilon(t) = f(x_0 + t\bar{x}, y(t)) - \varphi(x_0 + t\bar{x})$. Then $\varepsilon(t) \geq 0$, $t^{-1}\varepsilon(t) \to 0$ for $t \downarrow 0$ and, consequently, $y(t) \in \omega_\varepsilon(x_0 + t\bar{x})$ with $\varepsilon = o(t)$. Moreover

$$|y(t) - y_0| \leq t[|\bar{y}| + |o(t)||t^{-1}|] \leq t(|\bar{y}| + 1).$$

Therefore

$$\rho(y_0, \omega_\varepsilon(x_0 + t\bar{x})) \leq Mt$$

for $t \in [0, t_0]$, where t_0 is a sufficiently small positive number, and $M = |\bar{y}| + 1$.

2. ⇒ 1. This implication is a consequence of Theorem 6.49 and Corollary 6.51. ∎

EXAMPLE 6.53 *Let $f \in C^1$, $F(x) \equiv C$, where C is a compact set in R^m and*

$$\varphi(x) = \min_{y \in C} f(x, y).$$

It is easy to see that the assumptions (A1)–(A3) and (A5) are fulfilled in this example. To satisfy (A4), we suppose that for all points $y_0 \in \omega(x_0)$ the condition $T_C^L(y_0) = T_C^U(y_0)$ holds, where $T_C^L(y_0)$ and $T_C^U(y_0)$ are the lower and the upper tangent cones to C at the point y_0, respectively.

In this case, for any element $\bar{x} \in X$ one can find a suboptimal solution $y(t) \in \omega_\varepsilon(x_0 + t\bar{x})$ such that

$$|y(t) - y_0| \leq Mt, \qquad t \in [0, t_0],$$

where $\varepsilon = o(t)$, $t_0 > 0$.

EXAMPLE 6.54 *Let $\varphi(x) = \inf\{f(x, y) \mid y \in F(x)\}$, where $F(x) = \{y \in Y \mid h_i(x, y) \leq 0, \ i = 1, \ldots, r, \ h_i(x, y) = 0, \ i = r + 1, \ldots, p\}$. Suppose f and h_i to be continuously differentiable. Under the assumption (A1) suboptimal Lipschitz continuous solutions exist for $\varepsilon = o(t)$ in the following cases which guarantee the existence of $\varphi'(x_0; \bar{x})$ and the validity of (6.20); see [123]:*

1) $h_i(x, y) = \langle a_i, y \rangle + b_i(x)$, $a_i \in R^m$, $b_i : R^n \to R$ are continuous functions for all $i = 1, \ldots, p$;
2) the (LI)-regularity condition holds at the points $y_0 \in \omega(x_0)$;
3) the mapping F is R-regular at all points $z_0 \in \{x_0\} \times \omega(x_0)$, the functions $f(x_0, y)$, $h_i(x_0, y)$, $i = 1, \ldots, r$ are convex in a neighbourhood of the set $\omega(x_0)$ and the functions $h_i(x_0, y)$, $i = r + 1, \ldots, p$, are affine.

Let us introduce the sets

$$\Omega(x) = \{y \in F(x) \mid \exists o(t) \text{ such that } f(x, y) \leq \varphi(x) + o(|x - x_0|)\},$$

$$\Gamma_\Omega(z; \bar{x}) = \{\bar{y} \in \hat{D}F(z; \bar{x}) \mid f'(z; \bar{z}) \leq \varphi'(x; \bar{x})\}.$$

It is obvious that

$$\Omega(x) = \bigcup_{\varepsilon = o(t)} \omega_\varepsilon(x).$$

LEMMA 6.55 *Let the assumptions (A1)–(A5) hold and let the relation (6.20) be valid. Then the mapping Ω is differentiable in the direction $\bar{x}$ at all points $z_0 = (x_0, y_0)$, where y_0 is the solution of the problem (6.20), and*

$$\hat{D}\Omega(z_0; \bar{x}) = \Gamma_\Omega(z_0; \bar{x}). \tag{6.22}$$

Proof. 1. Let $\bar{y} \in \hat{D}_L \Omega(z_0; \bar{x})$. Then we can find a function $o(t)$ such that $o(t)/t \to 0$ if $t \downarrow 0$ and $y_0 + t\bar{y} + o(t) \in \Omega(x_0 + t\bar{x})$, $t \geq 0$. The last relation means that $y_0 + t\bar{y} + o(t) \in F(x_0 + t\bar{x})$ and $f(x_0 + t\bar{x}, y_0 + t\bar{y} + o(t)) \leq \varphi(x_0 + t\bar{x}) + o(t)$, $t \geq 0$. Therefore $\bar{y} \in \hat{D}F(z_0; \bar{x})$ and $f'(z_0; \bar{z}) \leq \varphi'(x_0; \bar{x})$, i.e. $\bar{y} \in \Gamma_\Omega(z_0; \bar{x})$ and, consequently, $\hat{D}_L\Omega(z_0; \bar{x}) \subset \Gamma_\Omega(z_0; \bar{x})$.

2. We take $\bar{y} \in \Gamma_\Omega(z_0; \bar{x})$. Then $-\alpha(t) + f(z_0 + t\bar{z}) \leq \varphi(x_0 + t\bar{x})$ and $y_0 + t\bar{y} + o(t) \in F(x_0 + t\bar{x})$ for $t \geq 0$, where $o(t)/t \to 0$ and $\alpha(t)/t \to 0$ if $t \downarrow 0$. In view of the Lipschitz continuity of f we get $f(x_0 + t\bar{x}, y_0 + t\bar{y} + o(t)) \leq \varphi(x_0 + t\bar{x}) + o(t)$, which means $y_0 + t\bar{y} + o(t) \in \Omega(x_0 + t\bar{x})$, i.e. $\bar{y} \in \hat{D}_L\Omega(z_0; \bar{x})$. In this way $\Gamma_\Omega(z_0; \bar{x}) \subset \hat{D}_L\Omega(z_0; \bar{x})$. With regard to the opposite inclusion obtained in part 1 of the proof, we get (6.22).

It remains to notice that, according to the definitions of the derivatives, $\hat{D}_L\Omega(z_0; \bar{x}) \subset \hat{D}_U\Omega(z_0; \bar{x})$. Repeating the first part of this proof for $\hat{D}_U\Omega(z_0; \bar{x})$ and taking into account (A4), we get $\hat{D}_U\Omega(z_0; \bar{x}) \subset \Gamma_\Omega(z_0; \bar{x})$. Thus

$$\hat{D}_L\Omega(z_0; \bar{x}) = \hat{D}_U\Omega(z_0; \bar{x}) = \hat{D}\Omega(z_0; \bar{x}),$$

i.e., the mapping Ω is differentiable in the direction $\bar{x}$ at the point z_0. ∎

Let us denote by $\omega(x_0; \bar{x})$ the set of solutions belonging to problem (6.20), i.e.

$$\omega(x_0; \bar{x}) = \{(y_0, \bar{y}) \mid y_0 \in \omega(x_0),\ \bar{y} \in \hat{D}F(z_0; \bar{x}),\ f'(z_0; \bar{z}) \leq \varphi(x_0; \bar{x})\}.$$

THEOREM 6.56 *For any* $(y_0, \bar{y}) \in \omega(x_0; \bar{x})$ *one can find such quantities* $\varepsilon(t) = o(t)$ *and* $y(t) \in \omega_\varepsilon(x_0 + t\bar{x})$, $t \geq 0$, *that*

$$\lim_{t \downarrow 0} t^{-1}(y(t) - y_0) = \bar{y}.$$

Proof. Since $\bar{y} \in \Gamma_\Omega(z_0; \bar{x})$, then due to Lemma 6.55 there exists $o(t)$ such that $y_0 + t\bar{y} + o(t) \in \Omega(x_0 + t\bar{x})$ and $o(t)/t \to 0$ for $t \downarrow 0$. It follows that $y(t) = y_0 + t\bar{y} + o(t) \in F(x_0 + t\bar{x})$ and $f(x_0 + t\bar{x}, y(t)) \leq \varphi(x_0 + t\bar{x}) + o(t)$, i.e. $y(t) \in \omega_\varepsilon(x_0 + t\bar{x})$ for $\varepsilon(t) = o(t)$ and $t^{-1}(y(t) - y_0) \to \bar{y}$ if $t \downarrow 0$. ∎

The next statement will be required in the following.

LEMMA 6.57 (First-order optimality condition) *Assume the function* $f(x_0, y)$ *to be locally Lipschitz continuous and directionally differentiable at the point* $y_0 \in F(x_0)$. *Then for* $y_0 \in \omega(x_0)$, *we have*

$$f'_y(z_0; \bar{y}) \geq 0 \tag{6.23}$$

for all $\bar{y} \in \hat{D}F(z_0; 0)$, *where* $z_0 = (x_0, y_0)$.

Proof. Let $\bar{y} \in \hat{D}F(z_0; 0)$. Then there exists a function o(t) such that $y_0 + t\bar{y} + o(t) \in F(x_0)$ for $t \geq 0$. In this case, for $y_0 \in \omega(x_0)$ we get

$$\frac{1}{t}\left[f(x_0, y_0 + t\bar{y} + o(t)) - f(x_0, y_0)\right] \geq 0, \quad t > 0.$$

The passage to the limit yields (6.23). ∎

Let us denote by $D(z)$ the *cone of critical directions* of the problem $(\bar{P}_x)$ at the point $z = (x, y) \in \operatorname{gr} F$, i.e.

$$D(z) = \{\bar{y} \in \hat{D}F(z; 0) \mid f_y'(z; \bar{y}) \leq 0\}.$$

It is quite obvious that $D(z)$ is a convex closed cone.

If we deal with the classical nonlinear problem (P_x) with a feasible set which is defined by the multivalued mapping $F(x) = \{y \in Y \mid h_i(x, y) \leq 0, \ i = 1, \ldots, r, \ h_i(x, y) = 0, \ i = r+1, \ldots, p\}$, then the cone of critical directions can be described by

$$D(z) = \{\bar{y} \in Y \mid \langle \nabla_y f(z), \bar{y} \rangle \leq 0,$$

$$\langle \nabla_y h_i(z), \bar{y} \rangle \leq 0, \ i \in I(z), \ \langle \nabla_y h_i(z), \bar{y} \rangle = 0, \ i \in I_0\}.$$

Thus $D(z) = \Gamma_F(z; 0) \cap \{\bar{y} \mid \langle \nabla_y f(z), \bar{y} \rangle \leq 0\}$.

Based on the definition of $D(z)$ we can formulate the following optimality condition.

LEMMA 6.58 *Let $\bar{y} \in D(z_0)$, where $z_0 \in \{x_0\} \times \omega(x_0)$. Then $f_y'(z_0; \bar{y}) = 0$.*

Proof. It follows from the first-order optimality condition and the definition of $D(z_0)$. ∎

We will also need the following technical lemma (see Gauvin and Janin [70]).

LEMMA 6.59 *Let $t_k \downarrow 0$, $y_k \in \omega(x_0 + t_k\bar{x})$, $k = 1, 2, \ldots$, and $y_k \to y_0 \in \omega(x_0)$ if $k \to \infty$. If the multivalued mapping F is pseudo-Lipschitz continuous and differentiable at the point $z_0 = (x_0, y_0) \in \operatorname{gr} F$ in the direction $\bar{x}$ and if $t_k^{-1}|y_k - y_0| \to \infty$ for $k \to \infty$, then all limit points of the sequence $\{(y_k - y_0)|y_k - y_0|^{-1}\}$ belong to $D(z_0)$.*

Proof. Without loss of generality, we can assume that $(y_k - y_0)|y_k - y_0|^{-1} \to \bar{y}$. Let us denote $\beta_k = |y_k - y_0|$. Then

$$y_k = y_0 + \beta_k\bar{y} + o(\beta_k) \in F(x_0 + o(\beta_k))$$

and, in view of the pseudo-Lipschitz continuity of the mapping F, we get

$$y_k = y_0 + \beta_k\bar{y} + o(\beta_k) \in F(x_0), \quad k = 1, 2, \ldots$$

Due to the differentiability of F, this inclusion means that $\bar{y} \in \hat{D}F(z_0; 0)$. Since $f(x_0+t_k\bar{x}, y_k) - f(x_0, y_0) = \varphi(x_0+t_k\bar{x}) - \varphi(x_0)$, then, according to the Lipschitz continuity of φ (see Corollary 4.23), the inequality $f(x_0 + t_k\bar{x}, y_k) - f(x_0, y_0) \leq lt_k\bar{x}$ holds. Dividing it by $|y_k - y_0|$ and passing to the limit, we get

$$f'_y(x_0, y_0; \bar{y}) \leq 0,$$

Due to the definition of $D(z_0)$, we obtain the statement of the lemma. ∎

In Section 2 we considered the concept of first-order approximation (see Definition 3.59). In order to obtain more meaningful results concerning differential properties of the optimal value function, it is necessary to suppose a stronger differentiability assumption on the mapping F, namely the existence of Hölder approximations.

DEFINITION 6.60 *The multivalued mapping F is said to have a first-order Hölder approximation at the point $z_0 = (x_0, y_0) \in \mathrm{gr}\, F$ in the direction $\bar{x} \in X$ if for any sequences $\{t_k\}$ and $\{y_k\}$ such that $y_k \in F(x_0 + t_k\bar{x})$, $k = 1, 2, \ldots$, $t_k \downarrow 0$, $y_k \to y_0 \in F(x_0)$ and $t_k^{-\frac{1}{2}}(y_k - y_0) \to \bar{y} \in D(z_0)$ for $k \to \infty$ the representation*

$$y_k = y_0 + \sqrt{t_k}\bar{y} + t_k\hat{y}_k + o(t_k)$$

holds, where $\bar{y} \in D(z_0)$, $\hat{y}_k \in \hat{D}_L^2 F(z_0, (0, \bar{y}); \bar{x})$, $\sqrt{t_k}\hat{y}_k \to 0$ if $k \to \infty$.

We denote

$$\Phi(z_0, \bar{z}_1, \bar{z}_2) = \langle \nabla f(z_0), \bar{z}_2 \rangle + \frac{1}{2} \left\langle \bar{z}_1, \nabla^2 f(z_0)\bar{z}_1 \right\rangle.$$

THEOREM 6.61 *Let the assumptions (A1)–(A3) be satisfied and let, in addition,*
1) the set of optimal solution $\omega(x)$ be weakly Hölder continuous at the point x_0 in the direction $\bar{x}$;
2) the multivalued mapping F have a first-order Hölder approximation at any point $z_0 = (x_0, y_0) \in \{x_0\} \times \omega(x_0)$ in the direction $\bar{x}$;
3) the function f be twice continuously differentiable at the points $z_0 \in \{x_0\} \times \omega(x_0)$.

Then the function φ is differentiable at the point x_0 in the direction $\bar{x}$ and

$$\varphi'(x_0; \bar{x}) = \inf_{y_0 \in \omega(x_0)} \inf_{\bar{y}_1 \in D(z_0)} \inf_{\bar{y}_2 \in \hat{D}_L^2 F(z_0, (0, \bar{y}_1); \bar{x})} \Phi(z_0, (0, \bar{y}_1), (\bar{x}, \bar{y}_2)).$$

$$(6.24)$$

Proof. 1.   Let us take arbitrary elements $y_0 \in \omega(x_0)$ and $\bar{y}_1 \in D(z_0)$. We denote $\bar{z}_1 = (0, \bar{y}_1)$, $\bar{z}_2 = (\bar{x}, \bar{y}_2)$. Then, according to the definition of the lower second derivative of the mapping F, for any $\bar{y}_2 \in \hat{D}_L^2 F(z_0, z_1; \bar{x})$ one can find a function $o(t)$ such that

$$y_0 + \sqrt{t}\bar{y}_1 + t\bar{y}_2 + o(t) \in F(x_0 + t\bar{x}) \quad \text{for } t \geq 0.$$

Therefore

$$\varphi(x_0 + t\bar{x}) - \varphi(x_0) \leq f(x_0 + t\bar{x}, y_0 + \sqrt{t}\bar{y}_1 + t\bar{y}_2 + o(t)) - f(x_0, y_0)$$

$$= \sqrt{t}\, \langle \nabla_y f(z_0), \bar{y}_1 \rangle + t \left\{ \langle \nabla f(z_0), \bar{z}_2 \rangle + \frac{1}{2} \left\langle \bar{y}_1, \nabla_{yy}^2 f(z_0)\bar{y}_1 \right\rangle \right\} + \tilde{o}(t),$$

where $\tilde{o}(t)/t \to 0$ for $t \downarrow 0$.

Since due to Lemma 6.58 $\langle \nabla_y f(z_0), \bar{y}_1 \rangle = 0$, then dividing the latter relation by t and passing to the limit, we get

$$D^+\varphi(x_0; \bar{x}) \leq \Phi(z_0, \bar{z}_1, \bar{z}_2).$$

Therefore

$$D^+\varphi(x_0; \bar{x}) \leq \inf_{y_0 \in \omega(x_0)} \ \inf_{\bar{y}_1 \in D(z_0)} \ \inf_{\bar{y}_2 \in \hat{D}_L^2 F(z_0, z_1; \bar{x})} \Phi(z_0, \bar{z}_1, \bar{z}_2).$$

2.   Let $D_+\varphi(x_0; \bar{x})$ be attained on the sequence $t_k \downarrow 0$. Due to the weak Hölder continuity of $\omega(x)$, we can find a point $y_0 \in \omega(x_0)$ such that

$$\rho(y_0, \omega(x_0 + t_k\bar{x})) \leq l\sqrt{t_k}, \quad k = 1, 2, \ldots, \ l = \text{const}.$$

Thus there exists a sequence $\{y_k\}$ such that $y_k \in \omega(x_0 + t_k\bar{x})$ and the inequality $|y_k - y_0| \leq l\sqrt{t_k}$, $k = 1, 2, \ldots$ holds. In this case, without loss of generality, we can assume that $t_k^{-\frac{1}{2}}(y_k - y_0) \to \bar{y}_1$ for $k \to \infty$, where $\bar{y}_1 \in D(z_0)$ according to Lemma 6.59.

Using the existence of a first-order Hölder approximation of the mapping F, we can represent y_k as follows: $y_k = y_0 + \sqrt{t_k}\bar{y}_1 + t_k\bar{y}_k + o(t_k)$. Here $\sqrt{t_k}\bar{y}_k \to 0$ and $\bar{y}_k \in \hat{D}_L^2 F(z_0, (0, \bar{y}_1); \bar{x})$ for $k = 1, 2, \ldots$

In doing so, we obtain

$$\varphi(x_0 + t_k\bar{x}) - \varphi(x_0) = f(x_0 + t_k\bar{x}, y_0 + \sqrt{t_k}\bar{y}_1 + t_k\bar{y}_k + o(t_k)) - f(x_0, y_0)$$

$$= \sqrt{t_k}\, \langle \nabla_y f(z_0), \bar{y}_1 \rangle + t_k \langle \nabla f(z_0), (\bar{x}, \bar{y}_k) \rangle + \frac{t_k}{2} \left\langle \bar{y}_1, \nabla_{yy}^2 f(z_0)\bar{y}_1 \right\rangle + \tilde{o}(t_k),$$

where $\tilde{o}(t)/t \to 0$ for $t \downarrow 0$. Taking into account Lemma 6.58, from the latter relation we obtain

$$\varphi(x_0 + t_k\bar{x}) - \varphi(x_0) = t_k \langle \nabla f(z_0), (\bar{x}, \bar{y}_k) \rangle + \frac{t_k}{2} \left\langle \bar{y}_1, \nabla_{yy}^2 f(z_0)\bar{y}_1 \right\rangle + \tilde{o}(t_k)$$

$$\geq t_k \inf_{\bar{y}_2 \in \hat{D}_L^2 F(z_0, (0, \bar{y}_1); \bar{x})} \left\{ \langle \nabla f(z_0), (\bar{x}, \bar{y}_2) \rangle + \frac{1}{2} \left\langle \bar{y}_1, \nabla_{yy}^2 f(z_0)\bar{y}_1 \right\rangle \right\} + \tilde{o}(t_k).$$

Dividing both parts of this inequality by $t_k > 0$ and passing to the limit for $k \to \infty$, we get

$$D_+\varphi(x_0; \bar{x}) \geq$$

$$\inf_{\bar{y}_2 \in \hat{D}_L^2 F(z_0,(0,\bar{y}_1);\bar{x})} \left\{ \langle \nabla f(z_0), (\bar{x}, \bar{y}_2) \rangle + \tfrac{1}{2} \left\langle \bar{y}_1, \nabla_{yy}^2 f(z_0)\bar{y}_1 \right\rangle \right\} \geq$$

$$\inf_{y_0 \in \omega(x_0)} \inf_{\bar{y}_1 \in D(z_0)} \inf_{\bar{y}_2 \in \hat{D}_L^2 F(z_0,(0,\bar{y}_1);\bar{x})} \left\{ \langle \nabla f(z_0), (\bar{x}, \bar{y}_2) \rangle + \tfrac{1}{2} \left\langle \bar{y}_1, \nabla_{yy}^2 f(z_0)\bar{y}_1 \right\rangle \right\}.$$

Comparing this result with the estimate for $D^+\varphi(x_0; \bar{x})$ from the first part of the proof, we obtain the statement of the theorem. ∎

3.2 Directional Derivatives of Optimal Value Functions in Nonlinear Programming Problems

Let us now consider the nonlinear programming problem

$$(P_x): \quad \begin{cases} f(x, y) \to \inf_y \\ h_i(x, y) \leq 0, \quad i = 1, \ldots, r, \\ h_i(x, y) = 0, \quad i = r+1, \ldots, p, \end{cases}$$

supposing the functions $f : X \times Y \to R$ and $h_i : X \times Y \to R$ to be continuously differentiable.

Differential properties of the optimal value function φ associated with problem (P_x) were thoroughly investigated by many authors (see e. g. [32], [75], [84], [122], [126], [156]).

First of all for convex problems (see Gol'shtein [75]) and then under more general conditions concerning the problem (P_x), the formula

$$\varphi'(x_0; \bar{x}) = \inf_{y_0 \in \omega(x_0)} \min_{\bar{y} \in \Gamma(z_0;\bar{x})} \langle \nabla f(z_0), \bar{z} \rangle$$

$$= \inf_{y_0 \in \omega(x_0)} \max_{\lambda \in \Lambda(z_0)} \langle \nabla_x L(z_0, \lambda), \bar{x} \rangle \tag{6.25}$$

was proved. Gauvin and Dubeau [69] obtained relation (6.24) for (LI)-regular problems. Later on Auslender and Cominetti [9] proved its validity under $(MF_{\bar{x}})$-regularity and a weak second-order sufficient condition.

Let us consider the problem (P_x) under still more general assumptions.

THEOREM 6.62 *Let* $\bar{x} \in X$, *let the assumption (A1) hold and the problem* (P_x) *be* (R)-*regular at the point* x_0. *Then there exists a number*

$M > 0$ *such that*

$$D_+\varphi(x_0; \bar{x}) \geq \inf_{y_0 \in \omega(x_0)} \min_{\lambda \in \Lambda_M(z_0)} \langle \nabla_x L(z_0, \lambda), \bar{x} \rangle, \qquad (6.26)$$

$$D^+\varphi(x_0; \bar{x}) \leq \inf_{y_0 \in \omega(x_0)} \max_{\lambda \in \Lambda_M(z_0)} \langle \nabla_x L(z_0, \lambda), \bar{x} \rangle, \qquad (6.27)$$

where $\lambda_M(z) = \lambda(z) \cap \left\{ \lambda \in R^p \mid \sum_{i=1}^{p} |\lambda_i| \leq M \right\}$.

Proof. 1. Let $D_+\varphi(x_0; \bar{x})$ be attained on the sequence $\varepsilon_k \downarrow 0$. We take an arbitrary sequence $y_k \in \omega(x_0 + \varepsilon_k \bar{x})$, $k = 1, 2, \ldots$ In view of (A1), without loss of generality we can assume that $y_k \to y_0$. Moreover, due to Lemma 6.24, we have $y_0 \in \omega(x_0)$. Let us choose an arbitrary element $\bar{y} \in Y$ and denote $z_k = (x_0 + \varepsilon_k \bar{x}, y_k)$, $z_{0k} = (x_0, y_k - \varepsilon_k \bar{y})$. From the continuity of the functions h_i and the regularity of the mapping F at the point z_0 we get that for sufficiently large k the inequality $h_i(z_{0k}) < 0$ holds for all $i \in I \backslash I(z_0)$ and

$$\varphi(x_0) \leq f(z_{0k}) + \beta d_F(z_{0k})$$
$$\leq f(z_{0k}) + \alpha\beta \max\{0, h_i(z_{0k}), i \in I(z_0), |h_i(z_{0k})|, i \in I_0\}.$$

Consequently, denoting $M = \alpha\beta$, we get

$$\varphi(x_0 + \varepsilon_k \bar{x}) - \varphi(x_0)$$
$$\geq f(z_k) - f(z_{0k}) - M \max\{0, h_i(z_{0k}), i \in I(z_0), |h_i(z_{0k})|, i \in I_0\}. \tag{6.28}$$

Two cases are possible.

Case 1. There exists a number k_0 such that $h_i(z_{0k}) \leq 0$, $i \in I(z_0)$, and $h_i(z_{0k}) = 0$, $i \in I_0$, for all $k \geq k_0$. Then

$$\varphi(x_0 + \varepsilon_k \bar{x}) - \varphi(x_0) \geq f(z_k) - f(z_{0k}),$$

and we immediately get

$$D_+\varphi(x_0; \bar{x}) \geq \langle \nabla f(z_0), \bar{z} \rangle$$
$$\geq \langle \nabla f(z_0), \bar{z} \rangle + M \min\{0, \langle \nabla h_i(z_0), \bar{z} \rangle, i \in I(z_0), |\langle \nabla h_i(z_0), \bar{z} \rangle|, i \in I_0\}. \tag{6.29}$$

Case 2. There exists a subsequence of $\{z_{0k}\}$ (for simplicity we keep for it the same notation $\{z_{0k}\}$) such that for every k one can find an index $i \in I(z_0)$ with $h_i(z_{0k}) > 0$ or an index $i \in I_0$ with $|h_i(z_{0k})| > 0$.

In this case we get

$$\max\{0,\, h_i(z_{0k}),\, i \in I(z_0),\, |h_i(z_{0k})|,\, i \in I_0\}$$
$$= \max\{h_i(z_{0k}),\, i \in I(z_0),\, |h_i(z_{0k})|,\, i \in I_0\}$$
$$= \max\{h_i(z_k) - \varepsilon_k \langle \nabla h_i(z_0), \bar{z} \rangle + \varepsilon_k \gamma_{ik},\, i \in I(z_0),$$
$$|h_i(z_k) - \varepsilon_k \langle \nabla h_i(z_0), \bar{z} \rangle + \varepsilon_k \gamma_{ik}|,\, i \in I_0\}$$
$$\leq -\varepsilon_k \min\{\langle \nabla h_i(z_0), \bar{z} \rangle,\, i \in I(z_0), |\langle \nabla h_i(z_0), \bar{z} \rangle|,\, i \in I_0\}$$
$$+ \varepsilon_k M \gamma_k.$$

$$(6.30)$$

Here $\gamma_k = \max\{|\gamma_{ik}| \mid i \in I(z_0) \cup I_0\}$, where the quantities γ_{ik} are defined as follows: $\gamma_{ik} = \langle \nabla h_i(z_0) - \nabla h_i(x_0 + (1 - \Theta_{ik})\varepsilon_k \bar{x}, y_k - \Theta_{ik}\varepsilon_k \bar{y}), \bar{z} \rangle$ for $i \in I(z_0)$ with $0 < \Theta_{ik} < 1$, while for $i \in I_0$ we have $\gamma_{ik} = |\langle \nabla h_i(z_0) - \nabla h_i(x_0 + (1 - \Theta_{ik})\varepsilon_k \bar{x}, y_k - \Theta_{ik}\varepsilon_k \bar{y}), \bar{z} \rangle|$, $0 < \Theta_{ik} < 1$.

Taking into account (6.30), we recognize from (6.28) that

$$\varphi(x_0 + \varepsilon_k \bar{x}) - \varphi(x_0) \geq f(z_k) - f(z_{0k})$$
$$+ \varepsilon_k M \min\{\langle \nabla h_i(z_0), \bar{z} \rangle,\, i \in I(z_0),\, |\langle \nabla h_i(z_0), \bar{z} \rangle|,\, i \in I_0\} - \varepsilon_k M \gamma_k,$$

and, therefore

$$D_+\varphi(x_0; \bar{x}) \geq \langle \nabla f(z_0), \bar{z} \rangle + M \min\{\langle \nabla h_i(z_0), \bar{z} \rangle,\, i \in I(z_0),$$
$$|\langle \nabla h_i(z_0), \bar{z} \rangle|,\, i \in I_0\} \geq \langle \nabla f(z_0), \bar{z} \rangle$$
$$+ M \min\{0,\, \langle \nabla h_i(z_0), \bar{z} \rangle,\, i \in I(z_0),\, |\langle \nabla h_i(z_0), \bar{z} \rangle|,\, i \in I_0\}.$$

Comparing this inequality with (6.29), we get in both cases that

$$D_+\varphi(x_0; \bar{x}) \geq \sup_{\bar{y}} \Big\{ \langle \nabla f(z_0), \bar{z} \rangle$$
$$+ M \min\{0,\, \langle \nabla h_i(z_0), \bar{z} \rangle,\, i \in I(z_0),\, |\langle \nabla h_i(z_0), \bar{z} \rangle|,\, i \in I_0\} \Big\}.$$

Due to the Minimax Theorem, from the last inequality it follows that

$$D_+\varphi(x_0; \bar{x}) \geq \sup_{\bar{y}} \Big\{ \langle \nabla f(z_0), \bar{z} \rangle + \min_{\lambda \in \bar{\Lambda}_M(z_0)} \sum_{i=1}^{p} \lambda_i \langle \nabla h_i(z_0), \bar{z} \rangle \Big\}$$

$$= \sup_{\bar{y}} \min_{\lambda \in \bar{\Lambda}_M(z_0)} \Big\langle \nabla f(z_0) + \sum_{i=1}^{p} \lambda_i \nabla h_i(z_0), \bar{z} \Big\rangle$$

$$= \sup_{\bar{y}} \min_{\lambda \in \bar{\Lambda}_M(z_0)} \langle \nabla L(z_0, \lambda), \bar{z} \rangle = \inf_{\lambda \in \bar{\Lambda}_M(z_0)} \sup_{\bar{y}} \langle \nabla L(z_0, \lambda), \bar{z} \rangle$$

$$= \inf_{\lambda \in \Lambda_M(z_0)} \langle \nabla_x L(z_0, \lambda), \bar{x} \rangle = \min_{\lambda \in \Lambda_M(z_0)} \langle \nabla_x L(z_0, \lambda), \bar{x} \rangle.$$

From the last relation inequality (6.26) results.

2.   Let $D^+\varphi(x_0; \bar{x})$ be attained on the sequence $\varepsilon_k \downarrow 0$. We take arbitrary $y_0 \in \omega(x_0)$ and $\bar{y} \in Y$ and denote $\tilde{z}_k = (x_0 + \varepsilon_k \bar{x}, y_0 + \varepsilon_k \bar{y})$. Let us choose a number k_0 such that $x_0 + \varepsilon_k \bar{x} \in X_0$, $\varepsilon_k |\bar{y}| < \varepsilon_0/4$ for $k > k_0$, where $\varepsilon_0 > 2\,\mathrm{diam}\,Y_0$. Then due to Lemma 3.68, for all $k \geq k_0$ we get

$$\varphi(x_0 + \varepsilon_k \bar{x}) \leq f(\tilde{z}_k) + \beta d_F(\tilde{z}_k).$$

In view of the (R)-regularity of problem (P_x) as well as the continuity of h_i, $i = 1, \ldots, p$, there exist numbers $\tilde{k}_0 \geq k_0$ and $\alpha > 0$ such that for all $k \geq \tilde{k}_0$

$$\varphi(x_0 + \varepsilon_k \bar{x}) - \varphi(x_0) \leq f(\tilde{z}_k) - f(z_0) + \beta d_F(\tilde{z}_k)$$
$$\leq f(\tilde{z}_k) - f(z_0) + M \max\{0, h_i(\tilde{z}_k), i \in I(z_0), |h_i(\tilde{z}_k)|, i \in I_0\}$$
$$\leq \varepsilon_k \langle \nabla f(z_0 + \Theta_k \varepsilon_k \bar{z}), \bar{z}\rangle + \varepsilon_k M \max\{0, \langle \nabla h_i(z_0), \bar{z}\rangle, i \in I(z_0),$$
$$|\langle \nabla h_i(z_0), \bar{z}\rangle|, i \in I_0\} + \varepsilon_k M \gamma_k,$$

where $0 < \Theta_k < 1$ and γ_k is defined as above.

From this it follows that

$$D^+\varphi(x_0; \bar{x}) \leq \langle \nabla f(z_0), \bar{z}\rangle$$
$$+ M \max\{0, \langle \nabla h_i(z_0), \bar{z}\rangle, i \in I(z_0), |\langle \nabla h_i(z_0), \bar{z}\rangle|, i \in I_0\},$$

Using the Minimax Theorem, we get

$$D^+\varphi(x_0; \bar{x}) \leq \inf_{y_0 \in \omega(x_0)} \inf_{\bar{y}} \left[\langle \nabla f(z_0), \bar{z}\rangle + \max_{\lambda \in \tilde{\Lambda}_M(z_0)} \sum_{i=1}^{p} \lambda_i \langle \nabla h_i(z_0), \bar{z}\rangle \right]$$
$$= \inf_{y_0 \in \omega(x_0)} \sup_{\lambda \in \Lambda_M(z_0)} \inf_{\bar{y}} \langle \nabla L(z_0, \lambda), \bar{z}\rangle$$
$$= \inf_{y_0 \in \omega(x_0)} \sup_{\lambda \in \Lambda_M(z_0)} \langle \nabla_x L(z_0, \lambda), \bar{x}\rangle$$
$$= \inf_{y_0 \in \omega(x_0)} \max_{\lambda \in \Lambda_M(z_0)} \langle \nabla_x L(z_0, \lambda), \bar{x}\rangle.$$

In this way, relation (6.27) holds. ∎

REMARK. As an estimate M we can take any value greater than αl_0, where α is the constant from Definition 6.12, l_0 is the Lipschitz constant of the function f on the set $X_0 \times [Y_0 + \varepsilon_0 B]$, and $\varepsilon_0 > 2\,\mathrm{diam}\,Y_0$.

COROLLARY 6.63 *For any $\bar{x} \in X$ the upper and lower derivatives of the function φ at the point x_0 in the direction $\bar{x}$ are finite.*

COROLLARY 6.64 *If under the conditions of the theorem the set $\Lambda(z_0)$ consists of only a single point for all $y_0 \in \omega(x_0)$, then there exists the directional derivative $\varphi'(x_0; \bar{x})$ and*

$$\varphi'(x_0; \bar{x}) = \min_{y_0 \in \omega(x_0)} \langle \nabla_x L(z_0, \lambda), \bar{x}\rangle. \tag{6.31}$$

Note that for a fixed direction $\bar{x} \in X$ the derivative $\varphi'(x_0; \bar{x})$ exists and is defined by (6.31) if the function $\langle \nabla_x L(z_0, \lambda), \bar{x} \rangle$ is constant on the set $\Lambda(z_0)$.

The following example shows that the estimates (6.26) and (6.27) cannot be strengthened.

EXAMPLE 6.65 *Consider* $x \in R^3$, $y \in R^2$. *Let* $f(x, y) = -y_2$, $F(x) = \{y \in R^2 \mid y_2 + y_1^2 \leq x_1,\ y_2 - y_1^2 \leq x_2,\ -y_2 - 1 \leq x_3\}$.

Let $x_0 = 0$. *It is easy to see that the assumption (A1) holds. Moreover,* $\omega(x_0) = \{y_0\}$, *where* $y_0 = (0, 0)$, *and* (MF)-*regularity holds at the point* $z_0 = (x_0, y_0)$. *Therefore, the mapping* F *is* (R)-*regular at* z_0.

It is not hard to verify that

$$\Lambda(z_0) = \{\lambda \in R^3 \mid \lambda_1 + \lambda_2 = 1,\ \lambda_1 \geq 0,\ \lambda_2 \geq 0,\ \lambda_3 = 0\}.$$

Then, for $\bar{x} = (1/\sqrt{2}, 1/\sqrt{2}, 0)$ *we get* $\omega(x_0 + \varepsilon\bar{x}) = \{(0, \varepsilon\sqrt{2}\}$ *and* $\varphi(x_0 + \varepsilon\bar{x}) = -\varepsilon/\sqrt{2}$. *Thus, the derivative* $\varphi'(x_0; \bar{x})$ *exists and is equal to*

$$\varphi'(x_0; \bar{x}) = -\frac{1}{\sqrt{2}} = \min_{\lambda_1 + \lambda_2 = 1,\, \lambda_1 \geq 0,\, \lambda_2 \geq 0} -\frac{1}{\sqrt{2}}(\lambda_1 + \lambda_2)$$

$$= \min_{\lambda \in \Lambda(z_0)} \langle \nabla_x L(z_0, \lambda), \bar{x} \rangle.$$

On the other hand, for $\bar{x} = (0, 1, 0)$ *we have*

$$\varphi'(x_0; \bar{x}) = 0 = \max_{\lambda_1 + \lambda_2 = 1,\, \lambda_1 \geq 0,\, \lambda_2 \geq 0} -\lambda_2 = \max_{\lambda \in \Lambda(z_0)} \langle \nabla_x L(z_0, \lambda), \bar{x} \rangle.$$

In this way, the example under consideration shows that the estimates (6.26) and (6.27) may be exact. Moreover, it can be proved that for $\bar{x} = (1, 0, 0)$ *the relation*

$$\min_{\lambda \in \Lambda(z_0)} \langle \nabla_x L(z_0, \lambda), \bar{x} \rangle < \varphi'(x_0; \bar{x}) < \max_{\lambda \in \Lambda(z_0)} \langle \nabla_x L(z_0, \lambda), \bar{x} \rangle.$$

is true.

LEMMA 6.66 *Let* $\bar{x} \in X$, $y_k \in \omega(x_0 + \varepsilon_k \bar{x})$, $k = 1, 2, \ldots$, *and* $y_k \to y_0 \in \omega(x_0)$. *Then, beginning from some* $k \geq k_0$, *the equality*

$$\varphi(x_0 + \varepsilon_k \bar{x}) - \varphi(x_0) = \max_{\lambda \in \tilde{\Lambda}_M(z_0)} [L(z_k, \lambda) - L(z_0, \lambda)]$$

holds for any M, *where* $z_k = (x_0 + \varepsilon_k \bar{x}, y_k)$, $z_0 = (x_0, y_0)$, *and* $\tilde{\Lambda}_M(z_0) = \{\lambda \in R^p \mid \lambda_i \geq 0,\ \lambda_i h_i(z_0) = 0,\ i = 1, \ldots, r,\ |\lambda_1| + \ldots + |\lambda_p| \leq M\}$.

Proof. Due to the continuity of the functions h_i, $i = 1, \ldots, p$, for all $i \in I \backslash I(z_0)$ the inequality $h_i(z_k) < 0$ holds for k sufficiently large. Consequently

$$
\begin{aligned}
\varphi(x_0 + \varepsilon_k \bar{x}) - \varphi(x_0) &= f(z_k) - f(z_0) \\
&= f(z_k) - f(z_0) + M \max\{0, h_i(z_k), i \in I(z_0), |h_i(z_k)|, i \in I_0\} \\
&= f(z_k) - f(z_0) + M \max\{0, h_i(z_k) - h_i(z_0), i \in I(z_0), \\
&\quad |h_i(z_k) - h_i(z_0)|, i \in I_0\} \\
&= \max_{\lambda \in \tilde{\Lambda}_M(z_0)} [L(z_k, \lambda) - L(z_0, \lambda)]. \quad \blacksquare
\end{aligned}
$$

THEOREM 6.67 *Let the assumption (A1) be satisfied, the problem* (P_x) *be* (R)-*regular at the point* x_0, *the functions* $f(x_0, y)$, $h_i(x_0, y)$, $i = 1, \ldots, r$, *be convex in a neighbourhood of the set* $\omega(x_0)$, *and the functions* $h_i(x_0, y)$, $i = r + 1, \ldots, p$, *be affine.*

Then the function φ *is differentiable at the point* x_0 *in all directions* $\bar{x}$ *and*

$$
\varphi'(x_0; \bar{x}) = \inf_{y_0 \in \omega(x_0)} \max_{\lambda \in \Lambda(z_0)} \langle \nabla_x L(z_0, \lambda), \bar{x} \rangle. \tag{6.32}
$$

Proof. Let $D_+\varphi(x_0; \bar{x})$ be attained on the sequence $\varepsilon_k \downarrow 0$. We choose an arbitrary sequence $y_k \in \omega(x_0 + \varepsilon_k \bar{x})$, $k = 1, 2, \ldots$ In view of (A1), without loss of generality we can assume that $y_k \to y_0$, where $y_0 \in \omega(x_0)$ due to the (R)-regularity and Lemma 6.24.

We introduce the notation $x_k = x_0 + \varepsilon_k \bar{x}$, $z_k = (x_k, y_k)$, $\bar{y}_k = \varepsilon^{-1}(y_k - y_0)$, $\bar{z}_k = (\bar{x}, \bar{y}_k)$. Applying Lemma 6.66 for $M \geq \alpha\beta$ and having regard to the convexity of the Lagrangian function L with respect to y, we get

$$
\begin{aligned}
\varphi(x_0 + \varepsilon_k \bar{x}) - \varphi(x_0) &= \max_{\lambda \in \tilde{\Lambda}_M(z_0)} [L(z_k, \lambda) - L(z_0, \lambda)] \\
&\geq \varepsilon_k \max_{\lambda \in \tilde{\Lambda}_M(z_0)} [\langle \nabla_x L(x_0 + \varepsilon_k \tau_k \bar{x}, y_k, \lambda), \bar{x} \rangle + \langle \nabla_y L(x_0, y_0, \lambda), \bar{y}_k \rangle] \\
&\geq \varepsilon_k \inf_{\bar{y}} \max_{\lambda \in \tilde{\Lambda}_M(z_0)} [\langle \nabla_x L(x_0 + \varepsilon_k \tau_k \bar{x}, y_k, \lambda), \bar{x} \rangle + \langle \nabla_y L(x_0, y_0, \lambda), \bar{y} \rangle] \\
&\geq \varepsilon_k \sup_{\lambda \in \tilde{\Lambda}_M(z_0)} \inf_{\bar{y}} [\langle \nabla_x L(x_0 + \varepsilon_k \tau_k \bar{x}, y_k, \lambda), \bar{x} \rangle + \langle \nabla_y L(z_0, \lambda), \bar{y} \rangle] \\
&= \varepsilon_k \sup_{\lambda \in \Lambda_M(z_0)} \langle \nabla_x L(x_0 + \varepsilon_k \tau_k \bar{x}, y_k, \lambda), \bar{x} \rangle,
\end{aligned}
$$

where $0 < \tau_k < 1$.

Dividing the last inequality by ε_k and passing to the limit for $\varepsilon_k \downarrow 0$, for any $M > \alpha\beta$ we obtain in view of Lemma 6.24

$$
\varphi'(x_0; \bar{x}) \geq \sup_{\lambda \in \Lambda_M(z_0)} \langle \nabla_x L(z_0, \lambda), \bar{x} \rangle
$$

$$\geq \inf_{y_0 \in \omega(x_0)} \max_{\lambda \in \Lambda_M(z_0)} \langle \nabla_x L(z_0, \lambda), \bar{x} \rangle. \tag{6.33}$$

Comparing this result with (6.27) we get the statement of the theorem. ∎

REMARK. In equation (6.32) the set $\Lambda(z_0)$ can be replaced by $\Lambda_M(z_0)$.

REMARK. Let us note that for the derivation of the estimate (6.32) we used in fact not the (R)-regularity of problem (P_x) at the point x_0, but only the lower semicontinuity of the mapping $\omega(x_0 + \varepsilon \bar{x})$ at the point $\varepsilon = 0$ and the non-emptiness of the set $\Lambda_M(z_0)$ (or $\Lambda(z_0)$, resp.).

Let us reformulate Theorem 6.67. Under the conditions of this theorem the function $L(x_0, y, \lambda)$ is convex with respect to y and concave with respect to λ. We consider the set of its saddle points (y_0, λ_0), i.e. points satisfying the inequalities

$$L(x_0, y_0, \lambda) \leq L(x_0, y_0, \lambda_0) \leq L(x_0, y, \lambda_0) \tag{6.34}$$

for all $y \in Y$ and all λ such that $\lambda_i \geq 0$ for $i \in I_0$. It is not hard to see that all saddle points form some set $Y(x_0) \times \Lambda(x_0)$. Moreover, applying to (6.34) the necessary optimality conditions (in this case they are sufficient as well), we get that $Y(x_0) = \omega(x_0)$, $\Lambda(x_0) = \Lambda(x_0, y_0)$ for all $y_0 \in \omega(x_0)$. (Note that $\Lambda(x_0, y_0)$ does not depend on the choice of y_0 from $\omega(x_0)$.) In addition, $\omega(x_0)$ and $\Lambda(x_0)$ are convex sets and the function $\langle \nabla_x L(x_0, y, \lambda), \bar{x} \rangle$ is convex with respect to y on $\omega(x_0)$. Therefore, from the Minimax Theorem it follows that

$$\varphi'(x_0; \bar{x}) = \inf_{y_0 \in \omega(x_0)} \max_{\lambda \in \Lambda(x_0)} \langle \nabla_x L(x_0, y_0, \lambda), \bar{x} \rangle$$

$$= \sup_{\lambda \in \Lambda(x_0)} \min_{y_0 \in \omega(x_0)} \langle \nabla_x L(x_0, y_0, \lambda), \bar{x} \rangle. \tag{6.35}$$

In this way, the following theorem holds.

THEOREM 6.68 *Let the assumptions of Theorem 6.67 hold. Then the function φ is differentiable at the point x_0 in all directions $\bar{x}$ and its derivative is defined by (6.34).*

Based on Theorem 6.62 we can obtain a strengthened variant of Theorem 6.33 concerning estimates of the subdifferential $\partial^0 \varphi(x_0)$, i.e., the following theorem is valid.

THEOREM 6.69 *Let the assumption (A1) hold and the problem (P_x) be (R)-regular at the point x_0. Then the function φ is Lipschitz continuous in a neighbourhood of the point x_0 and there exists a constant $M > 0$ such that*

$$\partial^0 \varphi(x_0) \subset \text{co} \bigcup_{y_0 \in \omega(x_0)} \bigcup_{\lambda \in \Lambda_M(z_0)} \nabla_x L(z_0, \lambda). \tag{6.36}$$

Proof. Let us prove the Lipschitz continuity of φ. From the uniform boundedness of the mapping F, Lemma 6.18 as well as the relations (6.26) and (6.27), it follows that there exist numbers $l_0 > 0$ and $\delta_0 > 0$ such that

$$-l_0|\bar{x}| \leq D_+\varphi(x;\bar{x}) \leq D^+\varphi(x;\bar{x}) \leq l_0|\bar{x}| \qquad (6.37)$$

for all $x \in x_0 + \delta_0 B$.

Let us fix elements $x_1, x_2 \in x_0 + \delta_0 B$ and denote $\bar{x} = (x_2 - x_1)|x_2 - x_1|^{-1}$, $\bar{\varepsilon} = |x_2 - x_1|$.

Since relation (6.37) implies that $D_+\varphi(x_1+\varepsilon\bar{x};\bar{x})$ and $D^+\varphi(x_1+\varepsilon\bar{x};\bar{x})$ are bounded on $[0, \bar{\varepsilon}]$, then the function $\varphi(x_1+\varepsilon\bar{x})$ is Lipschitz continuous on $[0, \bar{\varepsilon}]$. Hence, there exists its derivative with respect to ε and

$$\varphi(x_2) - \varphi(x_1) = \varphi(x_1 + \bar{\varepsilon}\bar{x}) - \varphi(x_1) = \int_0^{\bar{\varepsilon}} \frac{d}{d\varepsilon}\varphi(x_1 + \varepsilon\bar{x})\, d\varepsilon.$$

But $\frac{d}{d\varepsilon}\varphi(x_1+\varepsilon\bar{x}) = D_+\varphi(x_1+\varepsilon\bar{x};\bar{x}) = D^+\varphi(x_1+\varepsilon\bar{x};\bar{x})$, and from (6.37) we get

$$|\varphi(x_2) - \varphi(x_1)| \leq l_0\bar{\varepsilon} = l_0|x_2 - x_1|,$$

i. e., the function φ is Lipschitz continuous.

Let the derivative

$$\varphi^0(x_0;\bar{x}) \stackrel{\triangle}{=} \limsup_{x \to x_0, \varepsilon\downarrow 0} \varepsilon^{-1}[\varphi(x + \varepsilon\bar{x}) - \varphi(x)]$$

be attained on sequences $x_k \to x_0$, $\varepsilon_k \downarrow 0$. We consider an arbitrary sequence $\{y_k\}$ such that $y_k \in \omega(x_k)$, $k = 1, 2, \ldots$ Due to (A1) without loss of generality we can assume the sequence $\{y_k\}$ to be convergent: $y_k \to y_0$. Moreover, according to Lemma 6.24, we have $y_0 \in \omega(x_0)$. We denote $z_k = (x_k, y_k)$ and $\bar{z} = (\bar{x}, \bar{y})$ (here $\bar{y}$ is an arbitrary element from Y), $\tilde{z}_k = z_k + \varepsilon_k\bar{z}$. Let us choose a number k_0 such that $x_k \in X_0$, $x_k + \varepsilon_k\bar{x} \in X_0$ and $\varepsilon_k|\bar{y}| < \varepsilon_0/4$ for $k \geq k_0$, where $\varepsilon_0 > 2\operatorname{diam} Y_0$. Then due to Lemma 3.68, for all $k \geq k_0$ we have $\varphi(x_k+\varepsilon_k\bar{x}) \leq f(\tilde{z}_k)+\beta d_F(\tilde{z}_k)$. Furthermore, repeating the proof of Lemma 6.25, we obtain

$$\varphi(x_k + \varepsilon_k\bar{x}) - \varphi(x_k) \leq f(\tilde{z}_k) + \beta d_F(\tilde{z}_k) - f(z_k) - \beta d_F(z_k)$$
$$\leq \varepsilon_k \langle \nabla f(z_k), \bar{z}\rangle + \varepsilon_k\alpha\beta \max\{0, \langle \nabla h_i(z_k), \bar{z}\rangle,\, i \in I(z_0),$$
$$|\langle \nabla h_i(z_k), \bar{z}\rangle|,\, i \in I_0\} + \varepsilon_k\gamma_k + \varepsilon_k \langle \nabla f(z_k+\tau_{0k}\varepsilon_k\bar{z}) - \nabla f(z_k), \bar{z}\rangle,$$

where $\gamma_k = \max_{i \in I(z_0)\cup I_0} \{|\langle \nabla h_i(z_k + \tau_{ik}\varepsilon_k\bar{z}) - \nabla h_i(z_k), \bar{z}\rangle|\}$, while $\tau_{ik} \in (0, 1)$, $i \in \{0\} \cup I(z_0) \cup I_0$.

Dividing this inequality by $\varepsilon_k > 0$ and passing to the limit, we get

$$\varphi^0(x_0; \bar{x}) \leq \langle \nabla f(z_0), \bar{z} \rangle + \alpha\beta \max\{0,$$

$$\langle \nabla h_i(z_0), \bar{z} \rangle, \, i \in I(z_0), \, |\langle \nabla h_i(z_0), \bar{z} \rangle|, \, i \in I_0\}.$$

Moreover, applying the Minimax Theorem and setting $M = \alpha\beta$, we deduce

$$\varphi^0(x_0; \bar{x}) \leq \sup_{y_0 \in \omega(x_0)} \inf_{\bar{y}} \left[\langle \nabla f(z_0), \bar{z} \rangle + \max_{\lambda \in \bar{\Lambda}_M(z_0)} \sum_{i=1}^{p} \lambda_i \langle \nabla h_i(z_0), \bar{z} \rangle \right]$$

$$= \sup_{y_0 \in \omega(x_0)} \sup_{\lambda \in \bar{\Lambda}_M(z_0)} \inf_{\bar{y}} \langle \nabla L(z_0, \lambda), \bar{z} \rangle = \sup_{y_0 \in \omega(x_0)} \sup_{\lambda \in \Lambda_M(z_0)} \langle \nabla_x L(z_0, \lambda), \bar{x} \rangle.$$

Since $\varphi^0(x_0; \bar{x}) = \delta^*(\bar{x} \,|\, \partial^0 \varphi(x_0))$, then in view of the properties of the support function (see Section 1) the inclusion (6.36) follows. ∎

The condition of (R)-regularity of the multivalued mapping F at the point $z_0 = (x_0, y_0)$ implicitly assumes that $x_0 \in \operatorname{int} \operatorname{dom} F$. To eliminate this constraining assumption, we introduce the concept of relative (R)-regularity.

Let us choose a set $U \subset X$ such that $x_0 \in U \subset \operatorname{dom} F$.

DEFINITION 6.70 *The multivalued mapping F is said to be R-regular at the point z_0 with respect to the set $U \subset X$ if there exist numbers $\alpha > 0$, $\delta_1 > 0$, and $\delta_2 > 0$ such that*

$$d_F(z) \leq \alpha \max\{0, \, h_i(z), \, i \in I, \, |h_i(z)|, \, i \in I_0\}$$

for all $x \in V_{\delta_1}(x_0) \cap U$, $y \in V_{\delta_2}(y_0)$, where $z = (x, y)$.

In particular, as such a set U we can take the point x_0 or the set $\operatorname{dom} F$. If $U = X$, then we get the standard definition of (R)-regularity (see Section 2).

The analysis of the main results from this and the above sections allows us to reformulate some statements by making use of the assumption of relative (R)-regularity.

LEMMA 6.71 *Let the mapping F be (R)-regular at the point $z_0 = (x_0, y_0)$ with respect to $\operatorname{dom} F$. Then*

1. for $y_0 \in \omega(x_0)$ we can find a number $M > 0$ such that $\Lambda_M(z_0) \neq \emptyset$;

2. the function φ and the multivalued mapping ω are upper semicontinuous at x_0 with respect to $\operatorname{dom} F$;

3. $\hat{D}_L F(z_0; \bar{x}) = \Gamma_F(z_0; \bar{x})$ for all $\bar{x} \in \gamma_{\operatorname{dom} F}(x_0)$.

LEMMA 6.72 *Let the mapping F be (R)-regular at all points $z_0 \in \{x_0\} \times \omega(x_0)$ with respect to $\operatorname{dom} F$. Then for any $\bar{x} \in \gamma_{\operatorname{dom} F}(x_0)$ the estimates (6.26) and (6.27) are valid.*

THEOREM 6.73 *Let the assumption (A1) hold and, in addition, the mapping F be (R)-regular at all points $z_0 \in \{x_0\} \times \omega(x_0)$ with respect to $\mathrm{dom}\, F$ and $\bar{x} \in \gamma_{\mathrm{dom}\, F}(x_0)$.*

Then the function φ is differentiable at the point x_0 in the direction $\bar{x}$ if one of the following conditions holds:

1. $\Lambda_M(z_0) = \{\lambda\}$;

2. the functions $f(x_0, y)$, $h_i(x_0, y)$, $i = 1, \ldots, r$, are convex and the functions $h_i(x_0, y)$, $i = r+1, \ldots, p$, are affine with respect to y.

Let us present another result based on the new regularity condition introduced in [84].

DEFINITION 6.74 *We shall say that at the point $z_0 \in \mathrm{gr}\, F$ the condition of constant rank (CR) is fulfilled if for any subset of indices $J \subset I(z_0) \cup I_0$ the matrix $[\nabla_y h_i(z_0), i \in J]$ has a constant rank in a certain neighbourhood of the point z_0.*

It is easy to see that the condition of constant rank (CR) generalizes the (LI)-regularity condition. The following examples shows that the (CR)-condition is not connected with (MF)-regularity.

EXAMPLE 6.75 *Let $X = R^2$, $Y = R^2$, $x_0 = (0,0)$, $y_0 = (0,0)$, $h_1(z) = y_1 + y_2^2 - x_1 \leq 0$, $h_2(z) = y_1 - x_2 \leq 0$. Then $\nabla_y h_1(z) = (1, 2y_2)$, $\nabla_y h_2(z) = (1,0)$ and, therefore, (MF)-regularity holds at the point z_0, but the (CR)-condition fails to be true.*

EXAMPLE 6.76 *Let $X = R^2$, $Y = R$, $x_0 = (0,0)$, $y_0 = 0$, $h_1(z) = y - x_1 \leq 0$, $h_2(z) = -y - x_2 \leq 0$. Here the (CR)-condition obviously holds, but (MF)-regularity is not valid.*

LEMMA 6.77 *(Janin [84]) Let the (CR)-condition hold at the point $z_0 = (x_0, y_0) \in \mathrm{gr}\, F$. Then there exist neighbourhoods $V(x_0)$, $V(y_0)$ and a number $M_0 > 0$ such that for any $x \in V(x_0)$ and $y' \in F(x) \cap v(y_0)$ one can find a point $y'' \in F(x_0)$ having the properties*

1. $h_i(x, y') \leq h_i(x_0, y'') \leq 0$, $i \in I(z_0)$;

2. $|y' - y''| \leq M_0 |x - x_0|$.

THEOREM 6.78 *Let the assumption (A1) hold, and let at every point $z_0 \in \{x_0\} \times \omega(x_0)$ the conditions of constant rank and (R)-regularity with respect to $\mathrm{dom}\, F$ be satisfied.*

Then the function φ is differentiable at the point x_0 in any direction $\bar{x} \in \gamma_{\mathrm{dom}\, F}(x_0)$, and

$$\varphi'(x_0; \bar{x}) = \inf_{y_0 \in \omega(x_0)} \max_{\lambda \in \Lambda_M(z_0)} \langle \nabla_x L(z_0, \lambda), \bar{x} \rangle, \qquad (6.38)$$

where M is some positive number.

Proof. From the conditions of the theorem it follows that for $\bar{x} \in \gamma_{\text{dom } F}(x_0)$ the estimate (6.27) holds. This is due to Lemma 6.72.

On the other hand, let $D_+\varphi(x_0; \bar{x})$ be attained on the sequence $\varepsilon_k \downarrow 0$. In view of the assumption (A1) and the fact that $\bar{x} \in \gamma_{\text{dom } F}(x_0)$, we get that $\omega(x_0 + \varepsilon_k\bar{x}) \neq \emptyset$ beginning from some $k = k_0$. Therefore, we can take a sequence $y_k \in \omega(x_0 + \varepsilon_k\bar{x})$, $k \geq k_0$. Due to (A1) without loss of generality we can assume that $y_k \to y_0$, where $y_0 \in \omega(x_0)$ according to Lemma 6.71. Let us denote $z_k = (x_0 + \varepsilon_k\bar{x}, y_k)$. In view of Lemma 6.77 there exist numbers k_1 and $M_0 > 0$ such that for every z_k, and $k > k_1$, one can find a vector $y_{0k} \in F(x_0)$ for which the inequalities

$$h_i(z_k) \leq h_i(z_{0k}) \leq 0 \quad \text{for all } i \in I(z_0), \tag{6.39}$$

$$|y_k - y_{0k}| \leq M_0\varepsilon_k|\bar{x}| \tag{6.40}$$

hold, where $z_{0k} = (x_0, y_{0k})$.

Due to (6.39) and the continuity of h_i, $i = 1, \ldots, p$, we obtain the following relations for sufficiently large k and for any $M > 0$:

$$
\begin{aligned}
\varphi(x_0 + \varepsilon_k\bar{x}) - \varphi(x_0) &= f(z_k) - f(z_0) \\
&= [f(z_k) - f(z_{0k})] + [f(z_{0k} - f(z_0)] \\
&\quad + M \max\{0, [h_i(z_k) - h_i(z_{0k})], i \in I(z_0), \\
&\quad |h_i(z_k) - h_i(z_{0k})|, i \in I_0\} \\
&\geq f(z_k) - f(z_{0k}) + M \max\{0, h_i(z_k) - h_i(z_{0k}), i \in I(z_0), \\
&\quad |h_i(z_k) - h_i(z_{0k})|, i \in I_0\} \\
&\geq \varepsilon_k \langle \nabla f(z_0), \bar{z}_k \rangle + M\varepsilon_k \max\{0, \langle \nabla h_i(z_0), \bar{z}_k \rangle, i \in I(z_0), \\
&\quad |\langle \nabla h_i(z_0), \bar{z}_k \rangle|, i \in I_0\} - \varepsilon_k\gamma_k,
\end{aligned}
\tag{6.41}
$$

where $\bar{z}_k = (\bar{x}, \varepsilon_k^{-1}(y_k - y_{0k}))$, while $\gamma_k = |\langle \nabla f(z_{0k} + \tau_{0k}(z_k - z_{0k})) - \nabla f(z_0), \bar{z}_k \rangle| + M \cdot \sum_{i \in I(z_0) \cup I_0} |\langle \nabla h_i(z_{0k} + \tau_{ik}(z_k - z_{0k})) - \nabla h_i(z_0), \bar{z}_k \rangle|$ with $0 < \tau_{0k}, \tau_{ik} < 1$, $i \in I(z_0) \cup I_0$.

Due to (6.40) and without loss of generality we can assume that the following sequences converge: $\varepsilon_k^{-1}(y_k - y_{0k}) \to \bar{y}_0$, $\bar{z}_k \to \bar{z}_0 = (\bar{x}, \bar{y}_0)$. Now, from (3.2) and the Minimax Theorem it follows that

$$
\begin{aligned}
D_+\varphi(x_0; \bar{x}) &\geq \langle \nabla f(z_0), \bar{z}_0 \rangle + M \max\{0, \langle \nabla h_i(z_0), \bar{z}_0 \rangle, i \in I(z_0), \\
&\quad |\langle \nabla h_i(z_0), \bar{z}_0 \rangle|, i \in I_0\} \\
&\geq \inf_{\bar{y}} \left\{ \langle \nabla f(z_0), \bar{z} \rangle + \max_{\lambda \in \tilde{\Lambda}_M(z_0)} \sum_{i=1}^{p} \lambda_i \langle \nabla h_i(z_0), \bar{z} \rangle \right\}
\end{aligned}
$$

$$= \inf_{\bar{y}} \max_{\lambda \in \bar{\Lambda}_M(z_0)} \langle \nabla L(z_0, \lambda), \bar{z} \rangle$$

$$= \sup_{\lambda \in \bar{\Lambda}_M(z_0)} \inf_{\bar{y}} \langle \nabla L(z_0, \lambda), \bar{z} \rangle = \sup_{\lambda \in \Lambda_M(z_0)} \langle \nabla_x L(z_0, \lambda), \bar{x} \rangle .$$

Therefore

$$D_+\varphi(x_0; \bar{x}) \geq \inf_{y_0 \in \omega(x_0)} \sup_{M>0} \sup_{\lambda \in \Lambda_M(z_0)} \langle \nabla_x L(z_0, \lambda), \bar{x} \rangle$$

$$\geq \inf_{y_0 \in \omega(x_0)} \sup_{\lambda \in \Lambda_M(z_0)} \langle \nabla_x L(z_0, \lambda), \bar{x} \rangle .$$

Comparing the inequality obtained with (6.27), we get (6.38). ∎

REMARK. In equation (6.38) the set $\Lambda_M(z_0)$ can be replaced by the set $\Lambda(z_0)$.

COROLLARY 6.79 *Let the assumption (A1) be fulfilled and let the functions h_i, $i = 1, \ldots, p$, be affine with respect to y, i. e.*

$$h_i(x, y) = \langle a_i, y \rangle + b_i(x), \quad a_i \in Y.$$

Then the derivative $\varphi'(x_0; \bar{x})$ exists for any direction $\bar{x} \in \gamma_{\text{dom} F}(x_0)$ and is finite.

Proof. Under the assumptions of the corollary all conditions of Theorem 6.78 are valid. Thus, we can apply this theorem by noting that (R)-regularity of the multivalued mapping F with respect to the set $\text{dom} F$ follows directly from Lemma 6.16 if we replace there the set X by the set $\text{dom} F$. ∎

3.3   Hölder Behaviour of Optimal Solutions and Directional Differentiability of Optimal Value Functions in (R)-regular Problems

To simplify further notation, we assume without loss of generality that the problem (P_x) contains only inequality constraints, i. e.

$$(P_x): \qquad \begin{cases} f(x, y) \to \inf_y \\ h_i(x, y) \leq 0, \quad i \in I = \{1, \ldots, p\}, \end{cases}$$

where $x \in X = R^n$, $y \in Y = R^m$. Thus, the set of feasible points amounts to $F(x) = \{y \in Y \mid h_i(x, y) \leq 0, \ i \in I\}$.

In order to unify notation we denote

$$h_0(x, y) \overset{\triangle}{=} f(x, y).$$

3.3.1 Higher-Order Derivatives of Multivalued Mappings

Here we want to continue the construction of derivatives of mappings (see Subsection 2.5 of this chapter) for an arbitrary order j.

Let $z_0 = (x_0, y_0) \in \text{gr } F$. For all $\bar{z}_1 = (\bar{x}_1, \bar{y}_1), \ldots, \bar{z}_{j-1} = (\bar{x}_{j-1}, \bar{y}_{j-1})$ from the set $X \times Y$, $j = 2, 3, \ldots$, and for any $\bar{x}_j \in X$ we construct the following sets:

$$\hat{D}_L^j F(z_0, \bar{z}_1, \ldots, \bar{z}_{j-1}; \bar{x}_j) \triangleq$$

$$\{\bar{y}_j \in Y \mid \exists \, o(t) : t^{-1} o(t) \to 0 \text{ for } t \downarrow 0, y_0 + t\bar{y}_1 + \ldots + t^j \bar{y}_j + o(t^j)$$

$$\in F(x_0 + t\bar{x}_1 + \ldots + t^j \bar{x}_j), \, t \geq 0\},$$

$$\hat{D}_U^j F(z_0, \bar{z}_1, \ldots, \bar{z}_{j-1}; \bar{x}_j) \triangleq$$

$$\{\bar{y}_j \in Y \mid \exists \, t_k \downarrow 0 \text{ and } o(t) : t_k^{-1} o(t_k) \to 0 \text{ for } k \to \infty \text{ and }$$

$$y_0 + t_k \bar{y}_1 + \ldots + t_k^j \bar{y}_j + o(t_k^j) \in F(x_0 + t_k \bar{x}_1 + \ldots + t_k^j \bar{x}_j), \, k = 1, 2, \ldots\},$$

which we shall call the *derivative* and the *contingent derivative of order j* for the multivalued mapping F at the point z_0 in the directions $\bar{z}_1, \ldots, \bar{z}_{j-1}, \bar{x}_j$ (see [10], [55], [123]).

Immediately from this definition it follows that

$$\hat{D}_L^j F(z_0, \bar{z}_1, \ldots, \bar{z}_{j-1}; \bar{x}_j) \subset \hat{D}_U^j F(z_0, \bar{z}_1, \ldots, \bar{z}_{j-1}; \bar{x}_j) \text{ for any } j.$$

Let $g : X \times Y \to R$. We introduce the following notation to simplify further computations:

$$\nabla_y^k g(z) y^k \triangleq \sum_{j_1=1}^{m} \cdots \sum_{j_k=1}^{m} \left[\frac{\partial^k g(z)}{\partial y_{j_1} \ldots \partial y_{j_k}} \right] y_{j_1} \cdots y_{j_k},$$

i. e. $\nabla_y g(z) \bar{y} = \langle \nabla_y g(z), \bar{y} \rangle$, $\nabla_{yy} g(z) \bar{y}^2 = \langle \bar{y}, \nabla_{yy}^2 g(z) \bar{y} \rangle$ etc.

To give a complete characterization of $\hat{D}_L^j F(z_0, \bar{z}_1, \ldots, \bar{z}_{j-1}; \bar{x}_j)$, we assume that the objective and constraints functions h_i, $i = 0, 1, \ldots, p$, are s times continuously differentiable. By $M_{ij} = M_{ij}(z_0, \bar{z}_1, \ldots, \bar{z}_j)$ we denote the coefficient of the term having the jth power of t in the expansion

$$G(t) = \sum_{l=1}^{j} \frac{t^l}{l!} \nabla^l h_i(z_0)(\bar{z}_1 + t\bar{z}_2 + \ldots + t^{s-1} \bar{z}_s)^l,$$

i. e. $M_{ij} = \dfrac{1}{j!} \cdot \dfrac{d^j}{dt^j} G(t)|_{t=0}$, where $\bar{z}_j = (\bar{x}_j, \bar{y}_j)$, $i = 0, \ldots, p$, $j = 1, \ldots, s$.
In this way

$$M_{i1} = \nabla h_i(z_0)\bar{z}_1, \quad M_{i2} = \nabla h_i(z_0)\bar{z}_2 + \frac{1}{2}\nabla^2 h_i(z_0)\bar{z}_1^2,$$

$$M_{i3} = \nabla h_i(z_0)\bar{z}_3 + \nabla^2 h_i(z_0)\bar{z}_1\bar{z}_2 + \frac{1}{6}\nabla^3 h_i(z_0)\bar{z}_1^3.$$

Furthermore, for $j = 1, \ldots, s$ we denote

$$I^1 = I(z_0), \quad I^2(z_0, \bar{z}_1) = \{i \in I^1(z_0) \mid M_{i1} = 0\}, \quad \ldots,$$
$$I^j(z_0, \bar{z}_1, \ldots, \bar{z}_{j-1}) = \{i \in I^{j-1}(z_0, \bar{z}_1, \ldots, \bar{z}_{j-2}) \mid M_{i,j-1} = 0\}$$

and introduce the following sets:

$$\Gamma_F^1(z_0) = \Gamma_F(z_0), \quad \Gamma_F^1(z_0; \bar{x}_1) = \Gamma_F(z_0; \bar{x}_1), \ldots,$$
$$\Gamma_F^j(z_0, \bar{z}_1, \ldots, \bar{z}_{j-1}) = \{\bar{z}_j \in X \times Y \mid M_{ij} \leq 0, \ i \in I^j(z_0, \bar{z}_1, \ldots, \bar{z}_{j-1})\},$$
$$\Gamma_F^j(z_0, \bar{z}_1, \ldots, \bar{z}_{j-1}; \bar{x}_j) = \{\bar{y}_j \in Y \mid (\bar{x}_j, \bar{y}_j) \in \Gamma_F^j(z_0, \bar{z}_1, \ldots, \bar{z}_{j-1})\}.$$

In particular, we get

$$\Gamma_F^2(z_0, \bar{z}_1; \bar{x}_2) = \{\bar{y}_2 \in Y \mid \nabla h_i(z_0)\bar{z}_2 + \frac{1}{2}\nabla^2 h_i(z_0)\bar{z}_1^2 \leq 0, \ i \in I^2(z_0, \bar{z}_1)\}$$

as has been used in Subsection 2.5 as well as

$$\Gamma_F^3(z_0, \bar{z}_1, \bar{z}_2; \bar{x}_3) = \{\bar{y}_3 \in Y \mid \nabla h_i(z_0)\bar{z}_3$$
$$+ \nabla^2 h_i(z_0)\bar{z}_1\bar{z}_2 + \frac{1}{6}\nabla^3 h_i(z_0)\bar{z}_1^3 \leq 0, \ i \in I^3(z_0, \bar{z}_1, \bar{z}_2)\}.$$

Now we can prove a generalization of Lemma 6.28.

LEMMA 6.80 *Let $h_i \in C^s$, $i = 1, \ldots, p$, and let the mapping F be (R)-regular at the point $z_0 = (x_0, y_0) \in \operatorname{gr} F$. Then for any $x_j \in X$ and for all $j = 1, \ldots, s$, we have*

$$\hat{D}_L^j F(z_0, \bar{z}_1, \ldots, \bar{z}_{j-1}; \bar{x}_j) = \hat{D}_U^j F(z_0, \bar{z}_1, \ldots, \bar{z}_{j-1}; \bar{x}_j)$$
$$= \Gamma_F^j(z_0, \bar{z}_1, \ldots, \bar{z}_{j-1}; \bar{x}_j) \neq \emptyset$$

for all $\bar{z}_1 \in \Gamma_F^1(z_0), \ldots, \bar{z}_{j-1} \in \Gamma_F^{j-1}(z_0, \bar{z}_1, \ldots, \bar{z}_{j-2})$.

Proof. Since due to Lemma 6.28 the assertion is true for $j = 1$, it suffices to consider the case $1 < j \leq s$.

1.   Let $\bar{z}_1 \in \Gamma^1_F(z_0), \ldots, \bar{z}_{j-1} \in \Gamma^{j-1}_F(z_0, \bar{z}_1, \ldots, \bar{z}_{j-2})$ and let $\bar{y}_j \in Y$. From the (R)-regularity of F at the point z_0 and the definition of $\bar{z}_1, \ldots, \bar{z}_{j-1}$ and M_{ij} we get

$$
\begin{aligned}
\rho(y_0 + t\bar{y}_1 &+ \ldots + t^j\bar{y}_j, F(x_0 + t\bar{x}_1 + \ldots + t^j\bar{x}_j)) \\
&\leq \alpha \max\{0, h_i(z_0 + t\bar{z}_1 + \ldots + t^j\bar{z}_j) \mid i \in I\} \\
&= \alpha \max\{0, h_i(z_0 + t\bar{z}_1 + \ldots + t^j\bar{z}_j) \mid i \in I(z_0)\} \\
&= \alpha \max\left\{0, \sum_{l=1}^{j} t^l M_{il} + o_i(t^j) \mid i \in I(z_0)\right\} \\
&= \alpha \max\{0, t^j M_{ij} + o_i(t^j) \mid i \in I^j(z_0, \bar{z}_1, \ldots, \bar{z}_{j-1})\} \quad (6.42)
\end{aligned}
$$

for all $t \in [0, t_0]$, where t_0 is a sufficiently small positive number.

Suppose $\bar{y}_j \in \Gamma^j_F(z_0, \bar{z}_1, \ldots, \bar{z}_{j-1}; \bar{x}_j)$, i. e. $M_{ij} = M_{ij}(z_0, \bar{z}_1, \ldots, \bar{z}_j) \leq 0$ for all $i \in I^j(z_0, \bar{z}_1, \ldots, \bar{z}_{j-1})$, where $\bar{z}_j = (\bar{x}_j, \bar{y}_j)$. In this case from (6.41) it follows that

$$
\rho(y_0 + t\bar{y}_1 + \ldots + t^j\bar{y}_j, F(x_0 + t\bar{x}_1 + \ldots + t^j\bar{x}_j)) \leq o(t^j),
$$

where $o(t)/t \to 0$ for $t \downarrow 0$. But this means that the inclusion $\bar{y}_j \in \hat{D}^j_L F(z_0, \bar{z}_1, \ldots, \bar{z}_{j-1}; \bar{x}_j)$ holds. Consequently,

$$
\Gamma^j_F(z_0, \bar{z}_1, \ldots, \bar{z}_{j-1}; \bar{x}_j) \subset \hat{D}^j_L F(z_0, \bar{z}_1, \ldots, \bar{z}_{j-1}; \bar{x}_j).
$$

2.   The proof of the inclusion

$$
\hat{D}^j_U F(z_0, \bar{z}_1, \ldots, \bar{z}_{j-1}; \bar{x}_j) \subset \Gamma^j_F(z_0, \bar{z}_1, \ldots, \bar{z}_{j-1}; \bar{x}_j)
$$

repeats the argument of Lemma 6.25.

3.   Now it will be shown that $\Gamma^j_F(z_0, \bar{z}_1, \ldots, \bar{z}_{j-1}; \bar{x}_j) \neq \emptyset$. From inequality (6.41) we get

$$
\liminf_{t \downarrow 0} t^{-j} d_F(z_0 + t\bar{z}_1 + \ldots + t^j\bar{z}_j) < +\infty
$$

for any $\bar{x}_j \in X$ and, hence, for any $\bar{z}_j = (\bar{x}_j, \bar{y}_j) \in X \times Y$. Therefore, there exist a sequence $t_k \downarrow 0$ and a number $\delta > 0$ such that

$$
t_k^{-j} d_F(z_0 + t_k\bar{z}_1 + \ldots + t_k^j\bar{z}_j) \leq \delta \quad \text{for all } k = 1, 2, \ldots
$$

which means

$$
y_0 + t_k\bar{y}_1 + \ldots + t_k^j\bar{y}_j \in F(x_0 + t_k\bar{x}_1 + \ldots + t_k^j\bar{x}_j) + \delta t_k^j B.
$$

Consequently, there exists a vector $\xi_k \in \delta B$ such that, for all $k = 1, 2, \ldots$,

$$y_0 + t_k \bar{y}_1 + \ldots + t_k^j(\bar{y}_j + \xi_k) \in F(x_0 + t_k \bar{x}_1 + \ldots + t_k^j \bar{x}_j).$$

Without loss of generality we can assume that $\xi_k \to \xi$. From the last inclusion we then get $\bar{y}_j + \xi \in \hat{D}_U^j F(z_0, \bar{z}_1, \ldots, \bar{z}_{j-1}; \bar{x}_j)$. In this way

$$\Gamma_F^j(z_0, \bar{z}_1, \ldots, \bar{z}_{j-1}; \bar{x}_j) = D_U^j F(z_0, \bar{z}_1, \ldots, \bar{z}_{j-1}; \bar{x}_j) \neq \emptyset. \quad \blacksquare$$

Let us introduce the notations

$$\tilde{I}^j(z_0, \bar{z}_1, \ldots, \bar{z}_{j-1}) = \{0\} \cup I^j(z_0, \bar{z}_1, \ldots, \bar{z}_{j-1}),$$

$$\tilde{I}^j(x_0, y_0, \bar{y}_1, \ldots, \bar{y}_{j-1}) = \tilde{I}^j(z_0, (0, \bar{y}_1), \ldots, (0, \bar{y}_{j-1})),$$

$$m_{ij} = m_{ij}(y_0, \bar{y}_1, \ldots, \bar{y}_j) = M_{ij}(z_0, (0, \bar{y}_1), \ldots, (0, \bar{y}_j)),$$

$$D^j(x_0, y_0, \bar{y}_1, \ldots, \bar{y}_{j-1})$$
$$= \{\bar{y}_j \in Y \mid m_{ij} \leq 0, \ i \in \tilde{I}^j(x_0, y_0, \bar{y}_1, \ldots, \bar{y}_{j-1})\}$$
$$= \Gamma_F^j(z_0, (0, \bar{y}_1), \ldots, (0, \bar{y}_{j-1}); 0) \cap \{\bar{y}_j \in Y \mid m_{0j} \leq 0\}$$

for $j = 1, 2, \ldots$ In particular, we have

$$D^1(z_0) = D(z_0) = \Gamma_F(z_0; 0) \cap \{\bar{y}_1 \in Y \mid \nabla_y h_0(z_0)\bar{y}_1 \leq 0\},$$

i. e., $D^1(z_0)$ coincides with the cone of critical directions (see Subsection 3.1). Moreover

$$D^2(z_0, \bar{y}_1) = \Gamma_F^2(z_0, (0, \bar{y}_1); 0) \cap \left\{ \bar{y}_2 \in Y \ \middle| \ \nabla_y h_0(z_0)\bar{y}_2 + \tfrac{1}{2}\nabla_y^2 h_0(z_0)\bar{y}_1^2 \leq 0 \right\}.$$

In addition, we shall use the following notations:

$$\Phi_j(z_0, \bar{z}_1, \ldots, \bar{z}_j) = M_{0j}(z_0, \bar{z}_1, \ldots, \bar{z}_j),$$

$$C_\nu(\lambda, z_0, \bar{z}_1, \ldots, \bar{z}_\nu) = \sum_{i=1}^p \sum_{j=1}^\nu (M_{0j} + \lambda_i M_{ij}),$$

$$\Lambda^1(z_0) = \Lambda(z_0),$$

$$\Lambda^j(z_0, \bar{z}_1, \ldots, \bar{z}_{j-1}) = \{\lambda \in \Lambda(z_0) \mid \lambda_i = 0 \text{ for } i \notin I^j(z_0, \bar{z}_1, \ldots, \bar{z}_{j-1})\}.$$

It is not hard to see that the following set inclusion is always true: $\Lambda^{j+1}(z_0, \bar{z}_1, \ldots, \bar{z}_j) \subset \Lambda^j(z_0, \bar{z}_1, \ldots, \bar{z}_{j-1})$.

The following lemma generalizing a corresponding result from Auslender and Cominetti [9] to arbitrary order j will be applied below.

LEMMA 6.81 *Let $h_i \in C^s$, $i = 0, \ldots, p$, and let the mapping F be (R)-regular at the point $z_0 = (x_0, y_0) \in \{x_0\} \times \omega(x_0)$. Then, for all $j = 1, \ldots, s$, the condition*

$$(P^j) \qquad \inf_{\bar{y}_j \in \Gamma_F^j(z_0, \bar{z}_1, \ldots, \bar{z}_{j-1}; \bar{x}_j)} \Phi_j(z_0, \bar{z}_1, \ldots, \bar{z}_j)$$

$$= \sup_{\lambda \in \Lambda^j(z_0, \bar{z}_1, \ldots, \bar{z}_{j-1})} C_j(\lambda, z_0, \bar{z}_1, \ldots, \bar{z}_j)$$

holds true for every $\bar{x}_1, \ldots, \bar{x}_j \in X$ *and any* $\bar{y}_1 \in \Gamma_F^1(z_0, ; \bar{x}_1), \ldots, \bar{y}_{j-1} \in$
$\Gamma_F^{j-1}(z_0, \bar{z}_1, \ldots, \bar{z}_{j-2}; \bar{x}_{j-1})$.

Moreover, if $\bar{y}_1, \ldots, \bar{y}_j$ *are such points for which the minimum in the conditions* $(P^1), \ldots, (P^j)$ *is achieved, then the extremum is attained on both sides of* (P^j) *and the sets*

$$\tilde{\Lambda}^\nu(z_0, \bar{z}_1, \ldots, \bar{z}_\nu) = \Big\{ \lambda \in \Lambda^\nu(z_0, \bar{z}_1, \ldots, \bar{z}_{\nu-1}) \, \Big|$$
$$C_\nu(\lambda, z_0, \bar{z}_1, \ldots, \bar{z}_\nu) = \max_{\lambda \in \Lambda^\nu(z_0, \bar{z}_1, \ldots, \bar{z}_{\nu-1})} C_\nu(\lambda, z_0, \bar{z}_1, \ldots, \bar{z}_\nu) \Big\}$$

coincide with the sets $\Lambda^{\nu+1}(z_0, \bar{z}_1, \ldots, \bar{z}_\nu)$, $\nu = 1, \ldots, j$.

Proof. Condition (P^j) immediately follows by duality arguments of linear programming if one observes that, in accordance with Lemma 6.80, $\Gamma_F^j(z_0, \bar{z}_1, \ldots, \bar{z}_{j-1}; \bar{x}_j) \neq \emptyset$. If $\bar{y}_1, \ldots, \bar{y}_j$ provide the minimum in the conditions $(P^1), \ldots, (P^j)$, then the maximum on the right-hand side is also attained on the sets $\tilde{\Lambda}^\nu(z_0, \bar{z}_1, \ldots, \bar{z}_\nu)$, $\nu = 1, \ldots, j$. On the other hand, the set $\tilde{\Lambda}^\nu(z_0, \bar{z}_1, \ldots, \bar{z}_\nu)$ can be equivalently represented as

$$\tilde{\Lambda}^\nu(z_0, \bar{z}_1, \ldots, \bar{z}_\nu)$$
$$= \Big\{ \lambda \in \Lambda^\nu(z_0, \bar{z}_1, \ldots, \bar{z}_{\nu-1}) \, \Big| \, \Phi_\nu(z_0, \bar{z}_1, \ldots, \bar{z}_\nu) = C_\nu(\lambda, z_0, \bar{z}_1, \ldots, \bar{z}_\nu) \Big\}$$
$$= \Big\{ \lambda \in \Lambda^\nu(z_0, \bar{z}_1, \ldots, \bar{z}_{\nu-1}) \, \Big| \, \sum_{i \in I^\nu(z_0, \bar{z}_1, \ldots, \bar{z}_{\nu-1})} \lambda_i M_{i\nu}(z_0, \bar{z}_1, \ldots, \bar{z}_\nu) = 0 \Big\}.$$

Since $M_{i\nu}(z_0, \bar{z}_1, \ldots, \bar{z}_\nu) \leq 0$, we get

$$\tilde{\Lambda}^\nu(z_0, \bar{z}_1, \ldots, \bar{z}_\nu)$$
$$= \{ \lambda \in \Lambda^\nu(z_0, \bar{z}_1, \ldots, \bar{z}_{\nu-1}) \, | \, \lambda_i = 0, \; i \notin I^{\nu+1}(z_0, \bar{z}_1, \ldots, \bar{z}_\nu) \}$$
$$= \Lambda^{\nu+1}(z_0, \bar{z}_1, \ldots, \bar{z}_\nu). \;\blacksquare$$

Finally, we denote

$$\bar{v}_1 = y_0; \quad \bar{v}_j = (y_0, \bar{y}_1, \ldots, \bar{y}_{j-1}), \; j = 2, 3, \ldots,$$
$$\Omega^j(x_0) = \{ \bar{v}_j \, | \, y_0 \in \omega(x_0), \; \bar{y}_1 \in D^1(x_0, y_0), \ldots,$$
$$\bar{y}_{j-1} \in D^{j-1}(x_0, y_0, \bar{y}_1, \ldots, \bar{y}_{j-2}) \},$$
$$\Gamma_F^j(x_0, \bar{v}_j; \bar{x}) = \Gamma_F^j(z_0, (0, \bar{y}_1), \ldots, (0, \bar{y}_{j-1}); \bar{x}),$$
$$\Phi_j(x_0, \bar{v}_j, \bar{x}, \bar{y}) = \Phi_j(z_0, (0, \bar{y}_1), \ldots, (0, \bar{y}_{j-1}); (\bar{x}, \bar{y})),$$
$$C_j(\lambda, x_0, \bar{v}_j, \bar{x}, \bar{y}) = C_j(\lambda, z_0, (0, \bar{y}_1), \ldots, (0, \bar{y}_{j-1}); (\bar{x}, \bar{y})),$$
$$\Lambda^j(x_0, \bar{v}_j) = \Lambda^j(z_0, (0, \bar{y}_1), \ldots, (0, \bar{y}_{j-1})).$$

Of course, if $\lambda \in \Lambda(z_0)$, then the functions $C_j(\lambda, x_0, \bar{v}_j, \bar{x}, \bar{y})$ are independent of $\bar{y}$. This is the reason why in this case we shall write $C_j(\lambda, x_0, \bar{v}_j, \bar{x})$ in the following.

We also note that, due to Lemma 6.81, we have

$$\Lambda^2(x_0, y_0, \bar{y}_1) = \Lambda(x_0, y_0), \quad \Lambda^3(x_0, y_0, \bar{y}_1, \bar{y}_2) = \Lambda^3(x_0, y_0, \bar{y}_1)$$

if $y_0 \in \omega(x_0)$, $\bar{y}_1 \in D(z_0)$, $\bar{y}_2 \in D^2(z_0, \bar{y}_1)$, $\lambda \in \Lambda(z_0)$.

LEMMA 6.82 *Let* $h_i \in C^s$, $i = 0, \ldots, p$, *and let the mapping* F *be* (R)-*regular at the point* $z_0 = (x_0, y_0) \in \{x_0\} \times \omega(x_0)$. *Then for all* $j = 1, \ldots, s$ *the condition*

$$(P_0^j) \qquad \inf_{\bar{y}_j \in \Gamma_F^j(z_0, \bar{v}_j; 0)} \Phi_j(x_0, \bar{v}_j, 0, \bar{y}_j) = \sup_{\lambda \in \Lambda^j(x_0, \bar{v}_j)} C_j(\lambda, x_0, \bar{v}_j, 0) \geq 0$$

is valid for any $\bar{v}_j \in \Omega^j(x_0)$.

Proof. Let $\bar{y}_j \in \Gamma_F^j(x_0, \bar{v}_j; 0)$. Then in view of Lemma 6.80 we get

$$h_0(x_0, y_0 + t\bar{y}_1 + \ldots + t^j \bar{y}_j + o(t^j)) - h_0(x_0, y_0) \geq 0$$

and, consequently,

$$t\Phi_1(z_0, (0, \bar{y}_1)) + t^2 \Phi_2(z_0, (0, \bar{y}_1), (0, \bar{y}_2)) + \ldots$$
$$+ t^j \Phi_j(z_0, (0, \bar{y}_1), \ldots, (0, \bar{y}_j)) + o(t^j) \geq 0.$$

This means that
$$t^j \Phi_j(x_0, \bar{v}_j, 0, \bar{y}_j) + o(t^j) \geq 0.$$

Hence, we obtain $\Phi_j(x_0, \bar{v}_j, 0, \bar{y}_j) \geq 0$ for every $\bar{y}_j \in \Gamma_F^j(x_0, \bar{v}_j; 0)$. Now applying Lemma 6.81, we obtain condition (P_0^j). ∎

REMARK. From Lemma 6.81 it follows that the minimum in the condition (P_0^j) is attained on the set $\Omega^j(x_0)$.

3.3.2 Directional Derivatives of the Optimal Value Function

In the sequel we shall need the following assumptions:

(A2') – problem (P_x) is (R)-regular at the point x_0;

(A3') – for the direction $\bar{x} \in X$ and for any sequences $t_k \downarrow 0$ and $y_k \to y_0 \in \omega(x_0)$ such that $y_k \in \omega(x_0 + t_k \bar{x})$, $k = 1, 2, \ldots$, the inequality

$$\limsup_{k \to \infty} t_k^{-1} |y_k - y_0|^2 < +\infty,$$

holds, i.e., the set of optimal solutions $\omega(x)$ of the problem (P_x) is sequentially Hölder continuous at the point x_0 in the direction $\bar{x}$.

THEOREM 6.83 *Let $h_i \in C^2$, $i = 0, \ldots, p$, and let the assumptions (A1), (A2') and (A3') be fulfilled. Then the function φ is differentiable at the point x_0 in the direction $\bar{x}$ and*

$$\varphi'(x_0; \bar{x}) = \inf_{y_0 \in \omega(x_0)} \inf_{\bar{y}_1 \in D(z_0)} \inf_{\bar{y}_2 \in \Gamma_F^2(z_0, \bar{z}_1; \bar{x})} \Phi_2(\bar{y}_1, \bar{z}_2) \qquad (6.43)$$

$$= \inf_{y_0 \in \omega(x_0)} \inf_{\bar{y}_1 \in D(z_0)} \max_{\lambda \in \Lambda(z_0)} \left\{ \nabla_x L(z_0, \lambda)\bar{x} + \frac{1}{2}\nabla_{yy}^2 L(z_0, \lambda)\bar{y}_1^2 \right\},$$

where $\bar{z}_1 = (0, \bar{y}_1)$, $\bar{z}_2 = (\bar{x}, \bar{y}_2)$, $\Phi_2(\bar{y}_1, \bar{z}_2) = \nabla h_0(z_0)\bar{z}_2 + \frac{1}{2}\nabla_{yy}^2 h_0(z_0)\bar{y}_1^2$.

Proof. 1. Let us take arbitrary elements $y_0 \in \omega(x_0)$, $\bar{y}_1 \in D(z_0)$ and set $\bar{x}_1 = 0$, $\bar{x}_2 = \bar{x}$, $\bar{z}_1 = (0, \bar{y}_1)$, $\bar{z}_2 = (\bar{x}_2, \bar{y}_2)$.

For any $\bar{y}_2 \in \Gamma_F^2(z_0, \bar{z}_1; \bar{x})$ there exists a function $o(t)$, $o(t)/t \to 0$ for $t \downarrow 0$ such that

$$y_0 + t^{1/2}\bar{y}_1 + t\bar{y}_2 + o(t) \in F(x_0 + t\bar{x}).$$

Consequently

$$\varphi(x_0 + t\bar{x}) - \varphi(x_0) \leq h_0(x_0 + t\bar{x}, y_0 + t^{1/2}\bar{y}_1 + t\bar{y}_2 + o(t)) - h_0(x_0, y_0)$$
$$= t^{1/2}\nabla_y h_0(z_0)\bar{y}_1 + t\Phi_2(\bar{y}_1, \bar{z}_2) + \tilde{o}(t) \leq t\Phi_2(\bar{y}_1, \bar{z}_2) + \tilde{o}(t),$$
$$(6.44)$$

where $\tilde{o}(t)/t \to 0$ if $t \downarrow 0$. Hence, for all $\bar{y}_2 \in \Gamma_F^2(z_0, \bar{z}_1; \bar{x})$, $\bar{y}_1 \in D(z_0)$, $y_0 \in \omega(x_0)$ the inequality $D^+\varphi(x_0; \bar{x}) \leq \Phi_2(\bar{y}_1, \bar{z}_2)$ is satisfied. Therefore

$$D^+\varphi(x_0; \bar{x}) \leq \inf_{y_0 \in \omega(x_0)} \inf_{\bar{y}_1 \in D(z_0)} \inf_{\bar{y}_2 \in \Gamma_F^2(z_0, \bar{z}_1; \bar{x})} \Phi_2(\bar{y}_1, \bar{z}_2).$$

Due to Lemmas 6.81 and 6.82, we have $\Lambda^2(z_0, 0) = \Lambda(z_0)$ and, consequently,

$$D^+\varphi(x_0; \bar{x}) \leq$$
$$\inf_{y_0 \in \omega(x_0)} \inf_{\bar{y}_1 \in D(z_0)} \max_{\lambda \in \Lambda(z_0)} \left\{ \nabla_x L(z_0, \lambda)\bar{x} + \frac{1}{2}\nabla_{yy}^2 L(z_0, \lambda)\bar{y}_1^2 \right\}. \qquad (6.45)$$

2. Let $\{t_k\}$, $t_k \downarrow 0$, be a sequence on which the limit

$$D_+\varphi(x_0; \bar{x}) = \liminf_{k \to \infty} t_k^{-1}(\varphi(x_0 + t_k\bar{x}) - \varphi(x_0))$$

is attained. We choose an arbitrary sequence $y_k \in \omega(x_0 + t_k\bar{x})$, $k = 1, 2, \ldots$ In view of assumption (A1), without loss of generality we can assume that $y_k \to y_0$. Furthermore, due to Lemma 6.24, $y_0 \in \omega(x_0)$. Denote $x_k = x_0 + t_k\bar{x}$, $z_k = (x_k, y_k)$. For any $\lambda \in \Lambda(z_0)$ we then get

$$\varphi(x_k) - \varphi(x_0) \geq L(x_k, y_k, \lambda) - L(x_0, y_0, \lambda)$$

$$= t_k \nabla_x L(z_0, \lambda)\bar{x} + \nabla_y L(z_0, \lambda)(y_k - y_0) + \nabla_{yy}^2 L(z_0, \lambda)(y_k - y_0)^2 + \eta_k$$

$$= t_k \nabla_x L(z_0, \lambda)\bar{x} + \frac{1}{2}\nabla_{yy}^2 L(z_0, \lambda)(y_k - y_0)^2 + \eta_k, \tag{6.46}$$

where $\eta_k = t_k(\nabla_x L(x_0 + \tau_k t_k\bar{x}, y_k, \lambda)\bar{x} - \nabla_x L(z_0, \lambda)\bar{x}) + \frac{1}{2}[\nabla_{yy}^2 L(x_0, y_0 + \bar{\tau}_k(y_k - y_0), \lambda) - \nabla_{yy}^2 L(z_0, \lambda)](y_k - y_0)^2$ and $0 < \tau_k, \bar{\tau}_k < 1$.

From the assumption (A3') we conclude $|y_k - y_0| \leq Ct_k^{1/2}$, where C is some constant. Thus, without loss of generality we can assume $(y_k - y_0)t_k^{-1/2} \to \bar{y}_1$. From (6.45) the inequality

$$D_+\varphi(x_0; \bar{x}) \geq \nabla_x L(z_0, \lambda)\bar{x} + \frac{1}{2}\nabla_{yy}^2 L(z_0, \lambda)\bar{y}_1^2 \tag{6.47}$$

results for all $\lambda \in \Lambda(z_0)$ and some $\bar{y}_1$. Owing to Lemma 6.24, for $k = 1, 2, \ldots$ we have

$$h_0(x_k, y_k) - h_0(x_0, y_0) \leq lt_k,$$

$$h_i(x_k, y_k) - h_i(x_0, y_0) \leq 0, \quad i \in I(z_0),$$

where $l \geq 0$. From these inequalities we derive

$$\nabla_y h_i(z_0)\bar{y}_1 \leq 0, \quad i \in \{0\} \cup I(z_0),$$

i.e. $\bar{y}_1 \in D(z_0)$. Therefore,

$$D_+\varphi(x_0; \bar{x}) \geq \inf_{y_0 \in \omega(x_0)} \inf_{\bar{y}_1 \in D(z_0)} \max_{\lambda \in \Lambda(z_0)} \left\{\nabla_x L(z_0, \lambda)\bar{x} + \frac{1}{2}\nabla_{yy}^2 L(z_0, \lambda)\bar{y}_1^2\right\}.$$

Comparing this result with relations (6.45) and (6.47), we get the equality (6.43). ∎

In order to describe a sufficient condition for the validity of assumption (A3'), we consider the following condition (cf. [70]).

DEFINITION 6.84 *We shall say that the weak second-order sufficient condition (SOSC) holds at the point* $z_0 \in \mathrm{gr}\, F$ *if*

$$\sup_{\lambda \in \Lambda(z_0)} \nabla_{yy}^2 L(z_0, \lambda)\bar{y}^2 > 0$$

for any $\bar{y} \in D(z_0)$, $\bar{y} \neq 0$.

LEMMA 6.85 *Let $D^+\varphi(x_0; \bar{x}) < +\infty$. Then the condition (SOSC) with respect to the point $z_0 = (x_0, y_0)$ implies sequential Hölder continuity of the set of optimal solutions, i. e.*

$$\limsup_{k \to \infty} t_k^{-1} |y_k - y_0|^2 < +\infty,$$

where $\varepsilon_k \downarrow 0$, $y_k \in \omega(x_0 + t_k \bar{x})$, $k = 1, 2, \ldots$, $y_k \to y_0 \in \omega(x_0)$.

Proof. Suppose the opposite. Then we can find a subsequence of the sequence $\{y_k\}$ (for simplicity we denote it also by $\{y_k\}$) such that $t_k^{-1}|y_k - y_0|^2 \to \infty$. In this case without loss of generality we can assume that

$$\hat{y}_k \stackrel{\triangle}{=} (y_k - y_0)|y_k - y_0|^{-1} \to \hat{y},$$

where $\hat{y} \in D(z_0)$ according to Lemma 6.59.

Furthermore, we denote $\beta_k = |y_k - y_0|$, $x_k = x_0 + t_k \bar{x}$, $z_k = (x_k, y_k)$. Then $y_k = y_0 + \beta_k \hat{y} + o(\beta_k)$, $x_k = x_0 + o(\beta_k^2)$. Applying Lemma 6.66, we get for any $\lambda \in \Lambda(z_0)$ that

$$\varphi(x_k) - \varphi(x_0) \geq \frac{1}{2}\beta_k^2 \nabla_{yy}^2 L(z_0, \lambda)\hat{y}^2 + o(\beta_k^2)$$

and, therefore, $0 \geq \nabla_{yy}^2 L(z_0, \lambda)\hat{y}^2$. But this contradicts (SOSC). ∎

In this way, assumption (A3') in Theorem 6.83 can be replaced by the condition (SOSC).

Now, referring to the well-known example of Gauvin and Tolle (cf. [73]), we want to demonstrate the efficiency of Theorem 6.83.

EXAMPLE 6.86 *Let $x = (x_1, x_2, x_3) \in R^3$, $y = (y_1, y_2) \in R^2$.*

Let us consider the problem

$$\begin{cases} -y_2 & \to & \min \\ y_2 + y_1^2 & \leq & x_1, \\ y_2 - y_1^2 & \leq & x_2, \\ -y_2 - 1 & \leq & x_3, \end{cases}$$

i. e. $h_0(y) = -y_2$ and $F(x) = \{y \mid y_2 + y_1^2 \leq x_1, y_2 - y_1^2 \leq x_2, -y_2 - 1 \leq x_3\}$. Fix the point $x_0 = (0, 0, 0)$ and the direction $\bar{x} = (1, 0, 0)$. It is not hard to see that condition (A1) is satisfied and $\omega(x_0) = y_0 = (0, 0)$. Moreover, at the point $z_0 = (x_0, y_0)$ the Mangasarian-Fromowitz regularity condition holds. Consequently, the mapping F is (R)-regular at z_0, i. e. (A2') holds at x_0. Furthermore,

$$D(z_0) = \{\bar{y} = (\bar{y}_1, 0) \mid \bar{y}_1 \in R\},$$
$$\Lambda(z_0) = \{\lambda \mid \lambda_1 + \lambda_2 = 1, \lambda_1 \geq 0, \lambda_2 \geq 0, \lambda_3 = 0\}.$$

In this way, condition (SOSC) is equivalent to the inequality

$$\sup_{\lambda \in \Lambda(z_0)} (\lambda_1 - \lambda_2)\bar{y}_1^2 > 0 \ \ \text{for all } \bar{y}_1^2 \neq 0.$$

Consequently, (A3') is satisfied for every $\bar{x} \in R^3$. Applying Theorem 6.83, we recognize that the directional derivative $\varphi'(x_0; \bar{x})$ exists and

$$\varphi'(x_0; \bar{x}) = \min_{\bar{y}_1 \in R} \max_{\lambda \in \Lambda(z_0)} \{-\lambda_1 + (\lambda_1 - \lambda_2)\bar{y}_1^2\} = -\frac{1}{2}.$$

We emphasize that this example also demonstrates that the well-known minimax formula for $\varphi'(x_0; \bar{x})$ (see Gol'shtein [75]) cannot be applied. Indeed, in the example considered we get

$$\min_{y_0 \in \omega(x_0)} \max_{\lambda \in \Lambda(z_0)} \nabla_x L(z_0, \lambda)\bar{x} = 0$$

so that the formula

$$\varphi'(x_0; \bar{x}) = \min_{y_0 \in \omega(x_0)} \max_{\lambda \in \Lambda(z_0)} \nabla_x L(z_0, \lambda)\bar{x} \tag{6.48}$$

is not applicable.

However, it can be shown (cf. [156]) that the two formulas (6.43) and (6.48) yield the same result if in addition to the assumptions (A1), (A2') and (A3') we assume that $\Lambda(z_0)$ coincides with the set

$$\Lambda^2(z_0) = \{\lambda \in \Lambda(z_0) \,|\, \nabla_{yy}^2 L(z_0, \lambda)\bar{y}^2 \geq 0 \text{ for any } \bar{y} \in D(z_0)\}.$$

The last condition holds, in particular, if $L(x, y, \lambda)$ is convex with respect to y or if $\Lambda(z_0) = \{\lambda\}$.

Let us now formulate still another assumption:

(A4') – for the direction $\bar{x} \in X$ and for any sequences $t_k \downarrow 0$ and $y_k \to y_0 \in \omega(x_0)$ such that $y_k \in \omega(x_0 + t_k\bar{x})$, $k = 1, 2, \ldots$, the following inequality holds:

$$\limsup_{k \to \infty} t_k^{-1} |y_k - y_0|^3 < +\infty.$$

We denote

$$\tilde{D}^2(z_0, \bar{y}_1) = \Gamma_F^2(z_0, (0, \bar{y}_1); 0) \cap \{\bar{y}_2 \in Y \,|\, \Phi_2(z_0, (0, \bar{y}_1), (0, \bar{y}_2)) \leq \Phi_2^0(\bar{y}_1)\},$$

where $\Phi_2^0(\bar{y}_1) = \min_{\bar{y}_2 \in \Gamma_F^2(z_0, (0, \bar{y}_1); 0)} \Phi_2(z_0, (0, \bar{y}_1), (0, \bar{y}_2)).$

LEMMA 6.87 *Let assumptions (A1), (A2') and (A4') hold and let $y_k \in \omega(x_0 + t_k\bar{x})$, $k = 1, 2, \ldots$, where $y_k \to y_0$. Then there exist bounded* sequences $\{\bar{y}_{1k}\}$ and $\{\bar{y}_{2k}\}$ such that $\{\bar{y}_{1k}\} \in D(z_0)$, $\{\bar{y}_{2k}\} \in \tilde{D}^2(z_0, \bar{y}_{1k})$ and

$$y_k = y_0 + t_k^{1/3}\bar{y}_{1k} + t_k^{2/3}\bar{y}_{2k} + t_k^{2/3}\eta_k, \tag{6.49}$$

where $|\eta_k| \le M t_k^{1/3}$, $k = 1, 2, \ldots$, $M = \text{const} > 0$.

Proof. We denote $\beta_k = t_k^{1/3}$, $\hat{y}_k = \beta_k^{-1}(y_k - y_0)$, $x_k = x_0 + t_k\bar{x}$, $z_k = (x_k, y_k)$. Since the function φ is Lipschitz continuous at the point x_0 (see Theorem 6.69) and $y_0 \in \omega(x_0)$ (due to Lemma 6.24), then we get

$$h_0(x_k, y_k) - h_0(x_0, y_0) = \varphi(x_k) - \varphi(x_0) \le l\beta_k^3, \quad l = \text{const} > 0,$$
$$h_i(x_k, y_k) - h_i(x_0, y_0) \le 0, \quad i \in I(z_0).$$

From these inequalities it follows that

$$\nabla_y h_i(z_0)\hat{y}_k + \frac{\beta_k}{2}\nabla_{yy}^2 h_i(z_0)\hat{y}_k^2 \le M_0\beta_k^2, \tag{6.50}$$

where $M_0 = \text{const} > 0$, $i \in \{0\} \cup I(z_0)$, $k = 1, 2, \ldots$

Since $D(z_0) \ne \emptyset$, we can apply Lemma 6.16 to this system of inequalities. We obtain that there exist some bounded sequences $\{\bar{y}_{1k}\}$ and $\{\xi_k\}$ such that

$$\hat{y}_k = \bar{y}_{1k} + \beta_k\xi_k, \quad \bar{y}_{1k} \in D(z_0), \quad k = 1, 2, \ldots$$

Putting this expansion in (6.50), we get

$$\nabla_y h_i(z_0)\bar{y}_{1k} + \beta_k(\nabla_y h_i(z_0)\xi_k + \frac{1}{2}\nabla_{yy}^2 h_i(z_0)\bar{y}_{1k}^2) \le M_1\beta_k^2,$$

where $M_1 = \text{const} > 0$, $k = 1, 2, \ldots$ Then, for $i \in \tilde{I}^2(z_0, \bar{y}_{1k})$, the inequality

$$\nabla_y h_i(z_0)\xi_k + \frac{1}{2}\nabla_{yy}^2 h_i(z_0)\bar{y}_{1k}^2 \le M_2\beta_k \tag{6.51}$$

holds with $M_2 = \text{const} > 0$, $k = 1, 2, \ldots$ According to Lemma 6.82, we have

$$\Phi_2^0(\bar{y}_1) = \sup_{\lambda \in \Lambda(z_0)} \frac{1}{2}\nabla_{yy}^2 L(z_0, \lambda)\bar{y}_1^2 \ge 0$$

so that we can apply Lemma 6.16 to inequality (6.51). We obtain that there exists a bounded sequence $\{\bar{y}_{2k}\}$ such that $\bar{y}_{2k} \in \tilde{D}^2(z_0, \bar{y}_{1k})$, $\xi_k = \bar{y}_{2k} + \eta_k$, where $|\eta_k| \le M_2\beta_k$, $k = 1, 2, \ldots$ In this way,

$$\hat{y}_k = \bar{y}_{1k} + \beta_k\bar{y}_{2k} + \beta_k\eta_k, \quad k = 1, 2, \ldots,$$

and we get the representation (6.49). ∎

THEOREM 6.88 *Let* $h_i \in C^3$, $i = 0, \ldots, p$, *and let the assumptions (A1), (A2') and (A4') be satisfied. Then the function* φ *is differentiable at the point* x_0 *in the direction* $\bar{x}$ *and*

$$\varphi'(x_0; \bar{x}) = \inf_{y_0 \in \omega(x_0)} \inf_{\bar{y}_1 \in D(z_0)} \inf_{\bar{y}_2 \in \tilde{D}^2(z_0, \bar{y}_1)} \sup_{\lambda \in \Lambda(z_0, \bar{y}_1)} C_3(\lambda, z_0, \bar{z}_1, \bar{z}_2, \bar{z})$$

$$= \inf_{y_0 \in \omega(x_0)} \inf_{\bar{y}_1 \in D(z_0)} \inf_{\bar{y}_2 \in \tilde{D}^2(z_0, \bar{y}_1)} \inf_{\bar{y} \in \Gamma_F^3(z_0, \bar{z}_1, \bar{z}_2; \bar{x})} \Phi_3(z_0, \bar{z}_1, \bar{z}_2, \bar{z}), (6.52)$$

where

$\bar{z}_1 = (0, \bar{y}_1), \bar{z}_2 = (0, \bar{y}_2), \bar{z} = (\bar{x}, \bar{y}),$

$\Phi_3(z_0, \bar{z}_1, \bar{z}_2, \bar{z}) = \nabla h_0(z_0)\bar{z} + \nabla_{yy}^2 h_0(z_0)\bar{y}_1\bar{y}_2 + \frac{1}{2}\nabla_{yyy}^3 h_0(z_0)\bar{y}_1^3,$

$C_3(\lambda, z_0, \bar{z}_1, \bar{z}_2, \bar{z}) = \nabla_x L(z_0, \lambda)\bar{x} + \nabla_{yy}^2 L(z_0, \lambda)\bar{y}_1\bar{y}_2 + \frac{1}{6}\nabla_{yyy}^3 L(z_0, \lambda)\bar{y}_1^3,$

$\Lambda(z_0, \bar{y}_1) = \Lambda^3(z_0, (0, \bar{y}_1))$

$$= \{\lambda \in \Lambda(z_0) \mid \nabla_{yy}^2 L(z_0, \lambda)\bar{y}_1^2 = \max_{\lambda \in \Lambda(z_0)} \nabla_{yy}^2 L(z_0, \lambda)\bar{y}_1^2\}.$$

Proof. 1. Let us take arbitrary $y_0 \in \omega(x_0)$, $\bar{y}_1 \in D(z_0)$, $\bar{y}_2 \in \tilde{D}^2(z_0, \bar{y}_1)$, $\bar{y} \in \Gamma_F^3(z_0, \bar{z}_1, \bar{z}_2; \bar{x})$. Due to Lemma 6.80, there exists a function o(t), $o(t)/t \to 0$ for $t \downarrow 0$ such that

$$y_0 + \beta\bar{y}_1 + \beta^2\bar{y}_2 + \beta^3\bar{y} + o(\beta^3) \in F(x_0 + \beta^3\bar{x}),$$

where $\beta = t^{1/3}$, $t \geq 0$. Then

$$\varphi(x_0 + t\bar{x}) - \varphi(x_0)$$
$$\leq h_0(x_0 + \beta^3\bar{x}, y_0 + \beta\bar{y}_1 + \beta^2\bar{y}_2 + \beta^3\bar{y} + o(\beta^3)) - h_0(x_0, y_0)$$
$$= \beta\nabla_y h_0(z_0)\bar{y}_1 + \beta^2\Phi_2(z_0, \bar{z}_1, \bar{z}_2) + \beta^3\Phi_3(z_0, \bar{z}_1, \bar{z}_2, \bar{z}) + \tilde{o}(t),$$

where $\bar{z}_1 = (0, \bar{y}_1)$, $\bar{z}_2 = (0, \bar{y}_2)$, $\bar{z} = (\bar{x}, \bar{y})$, $\tilde{o}(t)/t \to 0$ if $t \downarrow 0$. From this estimate it follows that

$$D^+\varphi(x_0; \bar{x}) \leq \Phi_3(z_0, \bar{z}_1, \bar{z}_2, \bar{z})$$

for all $y_0 \in \omega(x_0)$, $\bar{y}_1 \in D(z_0)$, $\bar{y}_2 \in \tilde{D}^2(z_0, \bar{y}_1)$, $\bar{y} \in \Gamma_F^3(z_0, \bar{z}_1, \bar{z}_2; \bar{x})$. Consequently, we get

$$D^+\varphi(x_0; \bar{x})$$
$$\leq \inf_{y_0 \in \omega(x_0)} \inf_{\bar{y}_1 \in D(z_0)} \inf_{\bar{y}_2 \in \tilde{D}^2(z_0, \bar{y}_1)} \inf_{\bar{y} \in \Gamma_F^3(z_0, \bar{z}_1, \bar{z}_2; \bar{x})} \Phi_3(z_0, \bar{z}_1, \bar{z}_2, \bar{z}).$$
$$(6.53)$$

2. Let $\{t_k\}$, $t_k \downarrow 0$, be a sequence on which the limit

$$D_+\varphi(x_0; \bar{x}) = \liminf_{k \to \infty} t_k^{-1}(\varphi(x_0 + t_k\bar{x}) - \varphi(x_0))$$

is attained. We take any sequence $y_k \in \omega(x_0 + t_k\bar{x})$, $k = 1, 2, \ldots$ Without loss of generality we can assume that $y_k \to y_0$. Due to Lemma 6.24 the inclusion $y_0 \in \omega(x_0)$ holds. We denote $x_k = x_0 + t_k\bar{x}$. Then for any $\lambda \in \Lambda(z_0)$ we get

$$\varphi(x_k) - \varphi(x_0) = h_0(x_k, y_k) - h_0(x_0, y_0)$$

$$\geq L(x_k, y_k, \lambda) - L(x_0, y_0, \lambda) \tag{6.54}$$

$$= t_k \nabla_x L(z_0, \lambda)\bar{x} + \sum_{l=1}^{3} \frac{\nabla_y^l L(z_0, \lambda)}{l!} \cdot (y_k - y_0)^l + o(t_k).$$

Denote $\hat{y}_k = \beta_k^{-3}(y_k - y_0)$, where $\beta_k^3 = t_k$. Since from the assumption (A4') we have

$$|y_k - y_0| \leq C t_k^{1/3}, \quad C = \text{const},$$

then without loss of generality we can assume $\hat{y}_k \to \bar{y}$. With the help of Lemma 6.87, all assumptions of which are satisfied for $\{\hat{y}_k\}$, we obtain

$$\hat{y}_k = \bar{y}_{1k} + \beta_k \bar{y}_{2k} + \beta_k \eta_k,$$

where $\{\bar{y}_{1k}\} \in D(z_0)$, $\{\bar{y}_{2k}\} \in \tilde{D}^2(z_0, \bar{y}_{1k})$ and $|\eta_k| \leq M\beta_k$. Substituting this expression for $\hat{y}_k$ in (6.53), we get

$$t_k^{-1}(\varphi(x_k) - \varphi(x_0)) \geq \frac{1}{2}\beta_k^{-1}\nabla_{yy}^2 L(z_0, \lambda)\bar{y}_{1k}^2 + C_3(\lambda, z_0, \bar{z}_{1k}, \bar{z}_{2k}, \bar{z}) + o(t_k)t_k^{-1},$$

where $\bar{z}_{1k} = (0, \bar{y}_{1k})$, $\bar{z}_{2k} = (0, \bar{y}_{2k})$, $\bar{z} = (\bar{x}, \bar{y})$.

Applying Lemma 6.82, we observe that

$$t_k^{-1}(\varphi(x_k) - \varphi(x_0)) \geq C_3(\lambda, z_0, \bar{z}_{1k}, \bar{z}_{2k}, \bar{z}) + o(t_k)t_k^{-1}$$

for all $\lambda \in \Lambda(z_0, \bar{y}_{1k})$. Therefore

$$t_k^{-1}(\varphi(x_k) - \varphi(x_0)) \geq o(t_k)t_k^{-1} + \sup_{\lambda \in \Lambda(z_0, \bar{y}_{1k})} C_3(\lambda, z_0, \bar{z}_{1k}, \bar{z}_{2k}, \bar{z})$$

$$\geq o(t_k)t_k^{-1} + \inf_{y_0 \in \omega(x_0)} \inf_{\bar{y}_1 \in D(z_0)} \inf_{\bar{y}_2 \in \tilde{D}^2(z_0, \bar{y}_1)} \sup_{\lambda \in \Lambda(z_0, \bar{y}_1)} C_3(\lambda, z_0, \bar{z}_1, \bar{z}_2, \bar{z}).$$

From this inequality it follows that

$$D_+\varphi(x_0; \bar{x}) \geq \inf_{y_0 \in \omega(x_0)} \inf_{\bar{y}_1 \in D(z_0)} \inf_{\bar{y}_2 \in \tilde{D}^2(z_0, \bar{y}_1)} \sup_{\lambda \in \Lambda(z_0; \bar{y}_1)} C_3(\lambda, z_0, \bar{z}_1, \bar{z}_2, \bar{z}).$$

Comparing the last inequality with (6.53), we get the desired result (6.51). ∎

3.3.3   Hölder Directional Continuity of Optimal Solutions

Let us now try to find a sufficient condition ensuring the validity of assumption (A4'). Before we need the following technical lemma.

LEMMA 6.89 *Let the assumptions of Lemma 6.59 hold and let*

$$\limsup_{k \to \infty} t_k^{-1} |y_k - y_0|^2 = +\infty.$$

Then any accumulation point $\hat{y}$ of the sequence $\{(y_k - y_0)|y_k - y_0|^{-1}\}$ satisfies the condition

$$\sup_{\lambda \in \Lambda(z_0)} \nabla_{yy}^2 L(z_0, \lambda) \hat{y}_1^2 = 0. \tag{6.55}$$

Proof. Due to Lemma 6.59, we have $\hat{y} \in D(z_0)$. Consequently, according to Lemma 6.82, the condition

$$\sup_{\lambda \in \Lambda(z_0)} \nabla_{yy}^2 L(z_0, \lambda) \hat{y}_1^2 \geq 0$$

is true. We denote $\beta_k = |y_k - y_0|$, $\hat{y}_k = \beta_k^{-1}(y_k - y_0)$, $x_k = x_0 + t_k \bar{x}$, $z_k = (x_k, y_k)$. Since $t_k = o(\beta_k^2)$, for any $\lambda \in \Lambda(z_0)$ the relation

$$\varphi(x_k) - \varphi(x_0) \geq L(z_k, \lambda) - L(z_0, \lambda) = \frac{1}{2} \beta_k^2 \nabla_{yy}^2 L(z_0, \lambda) \hat{y}_k^2 + o(\beta_k^2).$$

holds. From this inequality it follows that $0 \geq \nabla_{yy}^2 L(z_0, \lambda) \hat{y}_k^2$ for all $\lambda \in \Lambda(z_0)$. Thus, the equality (6.55) is valid. ∎

Denote

$$D^*(z_0) = \left\{ \bar{y} \in D(z_0) \,\middle|\, \bar{y} \neq 0 \text{ and } \sup_{\lambda \in \Lambda(z_0)} \nabla_{yy}^2 L(z_0, \lambda) \bar{y}^2 = 0 \right\}.$$

Then the equality $D^*(z_0) = \emptyset$ for all $z_0 = (x_0, y_0) \in \{x_0\} \times \omega(x_0)$ means that condition (SOSC) holds at the point z_0.

Consider the set of second-order critical directions at the point z_0 for some $\hat{y} \in D(z_0)$, i.e. the set

$$D^2(z_0, \hat{y}) = \left\{ \bar{y} \in \Gamma_F^2(z_0, (0, \hat{y}); 0) \,\middle|\, \nabla_y h_0(z_0)\bar{y} + \frac{1}{2} \nabla_{yy}^2 h_0(z_0)\hat{y}^2 \leq 0 \right\}$$

as well as the set of second-order Lagrange multipliers

$$\Lambda^2(z_0) = \{ \lambda \in \Lambda(z_0) \,|\, \nabla_{yy}^2 L(z_0, \lambda)\bar{y}^2 \leq 0 \text{ for any } \bar{y} \in D(z_0) \}.$$

DEFINITION 6.90 *We shall say that condition (TOSC) holds at the point* $z_0 \in \{x_0\} \times \omega(x_0)$ *if*

$$\sup_{\lambda \in \Lambda^2(z_0)} \{6\nabla^2_{yy}L(z_0,\lambda)\bar{y}_1\bar{y}_2 + \nabla^3_{yyy}L(z_0,\lambda)\bar{y}_1^3\} > 0$$

for all $\bar{y}_1 \in D^*(z_0)$ *and all* $\bar{y}_2 \in D^2(z_0,\bar{y}_1)$.

THEOREM 6.91 *Let* $h_i \in C^3$, $i = 0,\ldots,p$, *and let the assumptions of Lemma 6.59 be satisfied. Then the condition (TOSC) at the point* $z_0 = (x_0,y_0) \in \{x_0\} \times \omega(x_0)$ *implies that*

$$\limsup_{k\to\infty} t_k^{-1}|y_k - y_0|^3 < +\infty.$$

Proof. Suppose the opposite, i. e.

$$\limsup_{k\to\infty} t_k^{-1}|y_k - y_0|^3 = +\infty.$$

Then there exists a subsequence from $\{y_k\}$ (to simplify notation, this subsequence will also be denoted by $\{y_k\}$) such that $t_k^{-1}\beta_k^3 \to \infty$, where $\beta_k = |y_k - y_0|$.

We denote $\hat{y}_k = (y_k - y_0)\beta_k^{-1}$, $x_k = x_0 + t_k\bar{x}$, $z_k = (x_k, y_k)$. Then without loss of generality we can assume that $\hat{y}_k \to \hat{y}$, where due to Lemma 6.89 the point $\hat{y}$ satisfies the condition (6.55). Since $t_k = o(\beta_k^3)$, we obtain for any $\lambda \in \Lambda(z_0)$ that

$$\varphi(x_k) - \varphi(x_0) \geq L(z_k,\lambda) - L(z_0,\lambda)$$
$$= \frac{1}{2}\beta_k^2\nabla^2_{yy}L(z_0,\lambda)\hat{y}_k^2 + \frac{1}{6}\beta_k^3\nabla^3_{yyy}L(z_0,\lambda)\hat{y}_k^3 + o(\beta_k^3).$$

From this inequality we conclude that

$$\beta_k^{-1}\nabla^2_{yy}L(z_0,\lambda)\hat{y}_k^2 + \frac{1}{3}\nabla^3_{yyy}L(z_0,\lambda)\hat{y}_k^3 \leq \eta_k, \tag{6.56}$$

where $\eta_k \to 0$ if $k \to \infty$. On the other hand, from the inequalities

$$h_0(x_k,y_k) - h_0(x_0,y_0) = \varphi(x_k) - \varphi(x_0),$$
$$h_i(x_k,y_k) - h_i(x_0,y_0) \leq 0, \quad i \in I(z_0),$$

it follows that there exist numbers $k_0 > 0$ and $M > 0$ such that

$$\nabla_y h_i(z_0)\hat{y}_k + (m_i - M\beta_k)\beta_k \leq 0$$

for all $k \geq k_0$ and $i \in \tilde{I}(z_0) = \{0\} \cup I(z_0)$, where $m_i = \frac{1}{2}\nabla^2_{yy}h_i(z_0)\hat{y}_k^2$.

Let us denote

$$g_i(y, \beta) = \nabla_y h_i(z_0)y + (m_i - M\beta)\beta, \quad G(\beta) = \{y \mid g_i(y, \beta) \le 0, \ i \in \tilde{I}(z_0)\}.$$

Then $\hat{y}_k \in G(\beta_k)$, $\hat{y} \in G(0)$. Due to Lemma 6.77, one can indicate an element $\bar{y}_k \in G(0)$ such that:

a) $g_i(\hat{y}_k, \beta_k) \le g_i(\bar{y}_k, 0) \le 0$ $\quad \forall i \in \tilde{I}(z_0, \hat{y}) = \{i \in \tilde{I}(z_0) \mid \nabla_y h_i(z_0)\hat{y} = 0\}$,

b) $|\hat{y}_k - \bar{y}_k| \le K\beta_k$, where $K = \text{const} > 0$.

It follows from what was said above that

$$g_i(\hat{y}_k, \beta_k) - g_i(\bar{y}_k, 0) \le 0,$$

i. e.

$$\nabla_y h_i(z_0)(\hat{y}_k - \bar{y}_k) + (m_i - M\beta_k)\beta_k \le 0$$

for every $i \in \tilde{I}(z_0, \hat{y})$.

Without loss of generality we may assume that $\beta_k^{-1}(\hat{y}_k - \bar{y}_k) \to \xi$, which means that $\hat{y}_k = \bar{y}_k + \beta_k \xi + o(\beta_k)$, where $\bar{y}_k \in G(0) = D(z_0)$. In this case, $\nabla_y h_i(z_0)\xi + m_i \le 0$ for all $i \in \tilde{I}(z_0, \hat{y})$ and thus $\xi \in D^2(z_0, \hat{y})$. Consequently, inequality (6.56) is equivalent to

$$\beta_k^{-1}\nabla_{yy}^2 L(z_0, \lambda)\bar{y}_k^2 + 2\nabla_{yy}^2 L(z_0, \lambda)\xi\bar{y}_k + \frac{1}{3}\nabla_{yyy}^3 L(z_0, \lambda)\bar{y}_k^3 \le \bar{\eta}_k,$$

where $\bar{\eta}_k \to 0$ for $k \to \infty$, $\bar{y}_k \in D(z_0)$, $\xi \in D^2(z_0, \hat{y})$. In this way, we have

$$2\nabla_{yy}^2 L(z_0, \lambda)\xi\bar{y}_k + \frac{1}{3}\nabla_{yyy}^3 L(z_0, \lambda)\bar{y}_k^3 \le \bar{\eta}_k$$

for any $\lambda \in \Lambda^2(z_0)$.

Since $\bar{y}_k \to \hat{y}$ for $k \to \infty$, from the last inequality we obtain that

$$2\nabla_{yy}^2 L(z_0, \lambda)\xi\hat{y} + \frac{1}{3}\nabla_{yyy}^3 L(z_0, \lambda)\hat{y}^3 \le 0,$$

where $\hat{y} \in D(z_0)$, $\hat{y} \ne 0$ and $\hat{y}$ satisfies (6.55), $\xi \in D^2(z_0, \hat{y})$. But this contradicts the condition (6.56). ∎

EXAMPLE 6.92 *Let $x = (x_1, x_2, x_3) \in R^3$, $y = (y_1, y_2) \in R^2$ and let us consider the problem*

$$\begin{cases} -y_2 & \to & \min \\ y_2 + y_1^3 & \le & x_1, \\ y_2 - y_1^3 & \le & x_2, \\ -y_2 - 1 & \le & x_3, \end{cases}$$

i. e. $h_0(y) = -y_2$, $F(x) = \{y \mid y_2 + y_1^3 \le x_1, \ y_2 - y_1^3 \le x_2, \ -y_2 - 1 \le x_3\}$.

We fix the point $x_0 = (0,0,0)$. It is not hard to see that $\omega(x_0) = y_0 = (0,0)$ and that conditions (A1), (A2') hold at the point $z_0 = (x_0, y_0)$.

Furthermore, we obtain

$$
\begin{aligned}
D(z_0) &= \{\bar{y} = (\bar{y}_1, 0) \,|\, \bar{y}_1 \in R\}, \\
\Lambda(z_0) &= \{\lambda \,|\, \lambda_1 + \lambda_2 = 1,\ \lambda_1 \geq 0,\ \lambda_2 \geq 0,\ \lambda_3 = 0\}.
\end{aligned}
$$

Condition (SOSC) is not fulfilled in this example, whereas condition (TOSC) is equivalent to

$$
\sup_{\lambda \in \Lambda(z_0)} (\lambda_1 - \lambda_2)\bar{y}_1^3 > 0 \quad \text{for all } \bar{y}_1 \neq 0.
$$

Consequently, (TOSC) holds true. Due to Theorem 6.91, this means that assumption (A4') is satisfied and Theorem 6.88 is valid for any $\bar{x} \in X$. Thus, the optimal value function φ is differentiable at the point x_0 in any direction $\bar{x}$.

3.4 Problems with Vertical Perturbations

DEFINITION 6.93 *Problem (P_x) is said to have vertical perturbations if*

$$
h_i(x, y) = g_i(y) + x_i, \quad i = 1, \dots, p,
$$

where x_i are the components of the vector $x \in X = R^n$ and $p = n$.

In the following we shall study problem (P_x) with vertical perturbations under the assumption (A1) of uniform boundedness in a neighbourhood of the point $x_0 = 0$ and under the condition that $F(0) \neq \emptyset$.

THEOREM 6.94 *Let $z_0 = (x_0, y_0) \in \operatorname{gr} F$. Then for the problem (P_x) with vertical perturbations the following statements are equivalent:*

1. *the mapping F is (MF)-regular at the point z_0;*
2. *the mapping F is R-regular at z_0;*
3. *$\hat{D}_L F(z_0; \bar{x}) = \Gamma_F(z_0; \bar{x}) \neq \emptyset$ for any $\bar{x} \in X$.*

Proof. 1. $\Rightarrow$ 2. This implication follows directly from Theorem 6.15.

2. $\Rightarrow$ 3. This statement results from Lemma 6.28.

3. $\Rightarrow$ 1. Since

$$
N(z_0) \triangleq -[\Gamma_F(z_0)]^+
$$

$$
= \{z^* = \sum_{i=1}^{p} \lambda_i \nabla h_i(z_0) \,|\, \lambda_i h_i(z_0) = 0,\ \lambda_i \geq 0,\ i = 1, \dots, r\},
$$

then from the condition $\operatorname{dom} \Gamma_F(z_0; \cdot) = X$ and from Lemma 3.38 we get

$$
N(z_0; 0) \triangleq \left\{ x^* = \sum_{i=1}^{p} \lambda_i \nabla_x h_i(z_0) \,\Big|\, \lambda \in \Lambda_0(z_0) \right\} = \{0\}.
$$

But in the case of vertical perturbations the relation $\sum_{i=1}^{p} \lambda_i \nabla_x h_i(z_0) = \lambda$ holds, i.e., we get $\Lambda_0(z_0) = \{0\}$, which is equivalent to (MF)-regularity. ∎

Let us consider the problem

$$(P_0): \quad \begin{cases} f(y) \to \inf \\ h_i(y) \le 0, \quad i = 1, \ldots, r, \\ h_i(y) = 0, \quad i = r+1, \ldots, p, \\ y \in C, \end{cases}$$

where f and h_i, $i = 1, \ldots, p$, are continuously differentiable functions, and C is a closed set.

Together with problem (P_0) we investigate the problem (P_x) with vertical perturbations

$$\begin{cases} f(y) \to \inf \\ h_i(y) + x_i \le 0, \quad i = 1, \ldots, r, \\ h_i(y) + x_i = 0, \quad i = r+1, \ldots, p, \\ y \in C, \end{cases}$$

where $x = (x_1, \ldots, x_n)$, i.e. $p = n$.

Let φ be the optimal value function of problem (P_x) involving vertical perturbations.

DEFINITION 6.95 *The problem (P_0) is said to be stable (in the sense of Clarke) if*

$$\liminf_{x \to 0} \frac{\varphi(x) - \varphi(0)}{|x|} > -\infty. \tag{6.57}$$

Stable problems have been introduced by Clarke in [42]. He showed that stability of the problem (P_0) implies non-emptiness of the set of Lagrange multipliers $\Lambda(z_0)$ at all points $z_0 \in \{x_0\} \times \omega(x_0)$. Moreover, in [42] it was proved that (MF)-regular problems are stable. In this way, Theorem 6.94 gives another sufficient condition for the problem (P_0) to be stable.

The following concept was introduced in [42].

DEFINITION 6.96 *The problem (P_0) is said to have the uniform penalty property (UP) if there exist numbers $M > 0$ and $\delta_0 > 0$ such that any element $y_0 \in \omega(0)$ is a solution of the minimum problem $\inf\{f(y) + M \max\{0, h_i(y), i = 1, \ldots, r, |h_i(y)|, i = r+1, \ldots, p\} \mid y \in C \cap (\omega(0) + \delta_0 B)\}$.*

THEOREM 6.97 *The following statements are equivalent:*
1. *the problem* (P_0) *is stable;*
2. *the problem* (P_0) *has the (UP)-property.*

Proof. 1. $\Rightarrow$ 2. We take an arbitrary point $y_0 \in \omega(0)$ and show that we can find numbers $M > 0$ and $\delta_0 > 0$ such that y_0 is a solution of the problem

$$f(y) + M \max\{0, \, h_i(y), \, i=1,\ldots,r, \, |h_i(y)|, \, i=r+1,\ldots,p\} \to \inf$$
$$y \in C \cap (y_0 + \delta_0 B).$$
$$(6.58)$$

Suppose the opposite. Then for any k there exists a point $y_k \in C \cap (y_0 + \frac{1}{k}B)$ such that

$$f(y_k) + k \max\{0, \, h_i(y_k), \, i=1,\ldots,r, \, |h_i(y_k)|, \, i=r+1,\ldots,p\} < f(y_0).$$

We denote

$$x_{ik} = \begin{cases} \max\{0, h_i(y_k)\}, & i=1,\ldots,r, \\ |h_i(y_k)|, & i=r+1,\ldots,p. \end{cases}$$

Then $x_k = (x_{1k},\ldots,x_{pk}) \to 0$ and we get $(f(y_k) - f(y_0))/|x_k| < -k$, which contradicts (6.57).

In this way, for any $y_0 \in \omega(0)$ there exist numbers $M = M(y_0) > 0$ and $\delta_0 = \delta(y_0) > 0$ such that condition (6.58) holds. Since $\omega(0)$ is a compact set, then from the system of neighbourhoods $\{y_0 + \delta_0(y_0)B | y_0 \in \omega(0)\}$ we can select a finite subsystem of neighbourhoods covering $\omega(0)$.

2. $\Rightarrow$ 1. Suppose the opposite, i. e., assume that inequality (6.57) fails to be true. This means that for any number k we can find a sequence $x_k \to 0$ such that $\varphi(x_k) - \varphi(0) < -k|x_k|$.

Let us take a sequence $y_k \in \omega(x_k)$, $k = 1, 2,\ldots$ (Such a sequence always exists, because otherwise $\varphi(x_k) = +\infty$.) Then

$$f(y_k) < \varphi(0) - k|x_k| < \varphi(0). \qquad (6.59)$$

In view of (A1), without loss of generality we can assume that $y_k \to \tilde{y}_0$, where $\tilde{y}_0 \in F(0)$. Thus, from (6.59) it follows that $f(\tilde{y}_0) \leq \varphi(0)$, $\tilde{y}_0 \in F(0)$, i. e. $\tilde{y}_0 \in \omega(0)$.

Therefore, for k large enough we have $y_k \in C \cap (\tilde{y}_0 + \delta_0 B)$. Due to the (UP)-property, we obtain

$$\varphi(0) \leq f(y_k) + M \max\{0, \, h_i(y_k), \, i=1,\ldots,r,$$
$$|h_i(y_k)|, \, i=r+1,\ldots,p\} \leq f(y_k) + M|x_k|,$$

which contradicts (6.59). $\blacksquare$

It is quite obvious that the problem with vertical perturbations is a particular case of the general problem (P_x) and, therefore, the conditions of directional differentiability of the optimal value function obtained in the previous subsections are valid for this problem, too. At the same time, the problem with vertical perturbations plays an important role among perturbed nonlinear programming problems. In particular, the optimal value function φ associated with problem (P_x) is also the optimal value function of the auxiliary problem

$$\varphi(u) = \inf_z \{ f(z) \mid h_i(z) \leq 0, \ i=1,\ldots,r, \ h_i(z)=0, \ i=r+1,\ldots,p,$$
$$-x_i + x_{i0} + u_i = 0, \ i = 1,\ldots,n \},$$

$$(6.60)$$

where $u = (u_1,\ldots,u_n) \in R^n$ is a vector of parameters, $z = (x,y)$, and $x_0 = (x_{10},\ldots,x_{n0})$.

In this way, problem (P_x) can be transformed into problem (6.60). This result obtained by Rockafellar [155] is of great importance. The key idea is that instead of problem (P_x) we can consider another problem simpler with respect to its structure, although of greater dimension.

As we emphasized in the previous subsections, regularity conditions play an important role in the investigation of differential properties of optimal value functions. Thus, there appears the natural question, whether regularity of problem (P_x) remains valid after turning over to the auxiliary problem (6.60). Our next aim is to show that this conjecture is true.

Suppose that the condition

$$\hat{T}^L_{\mathrm{gr}\, F}(z_0) = \Gamma_F(z_0) \tag{6.61}$$

holds at the points $z_0 \in \{x_0\} \times \omega(x_0)$ in problem (P_x).

We denote the feasible set of the problem (6.60) by $\tilde{F} : R^n \to 2^{X \times Y}$. Let $\tilde{T}(z_0, u_0) = \hat{T}^L_{\mathrm{gr}\, \tilde{F}}(z_0, u_0)$, and $\tilde{\Gamma}(z_0, u_0) = \Gamma_{\tilde{F}}(z_0, u_0)$, where $u_0 = 0$. Then

$$\tilde{\Gamma}(z_0, u_0) = \{ (\bar{z}, \bar{u}) \mid \langle \nabla h_i(z_0), \bar{z} \rangle \leq 0, \ i \in I(z_0),$$
$$\langle \nabla h_i(z_0), \bar{z} \rangle = 0, \ i \in I_0, \ \bar{x} = \bar{u} \} .$$

Suppose $(\bar{z}, \bar{u}) \in \tilde{\Gamma}(z_0, u_0)$. Then obviously $\bar{z} \in \Gamma_F(z_0)$, $\bar{u} = \bar{x}$. But for $\varepsilon \geq 0$ and any $\bar{u} \in R^n$ we have $u_0 + \varepsilon \bar{u} = x_0 + \varepsilon \bar{x} - x_0$. Owing to (6.61), there exists a function $o(\varepsilon)$ with $o(\varepsilon)/\varepsilon \to 0$ for $\varepsilon \downarrow 0$ such that

$$h_i(x_0 + \varepsilon \bar{x}, y_0 + \varepsilon \bar{y} + o(\varepsilon)) \leq 0, \ i \in I,$$
$$h_i(x_0 + \varepsilon \bar{x}, y_0 + \varepsilon \bar{y} + o(\varepsilon)) = 0, \ i \in I_0.$$

This means that $z_0 + \varepsilon\bar{z} + (0, o(\varepsilon)) \in \tilde{F}(u_0 + \varepsilon\bar{u})$, i.e. $(\bar{z}, \bar{u}) \in \tilde{T}(z_0, u_0)$. Hence $\tilde{\Gamma}(z_0, u_0) \subset \tilde{T}(z_0, u_0)$. Since the converse inclusion $\tilde{T}(z_0, u_0) \subset \tilde{\Gamma}(z_0, u_0)$ is always true (see Lemma 6.25), we get $\tilde{T}(z_0, u_0) = \tilde{\Gamma}(z_0, u_0)$. Moreover, from the condition $\mathrm{dom}\,\Gamma_F(z_0; \cdot) = X$ it follows that the equality $\mathrm{dom}\,\tilde{\Gamma}(z_0, u_0; \cdot) = U$ with $U = R^n$ is always true. In this way, if the condition

$$\hat{D}_L F(z_0; \bar{x}) = \Gamma_F(z_0; \bar{x}) \neq \emptyset \quad \text{for any } \bar{x} \in X$$

holds for problem (P_x), then it holds for problem (6.60) as well.

This fact, however, does not mean that in the studied case we can apply Theorem 6.94 and draw conclusions concerning (MF)- and (R)-regularity of the problem (6.57). The reason is that problem (6.60) is, strictly speaking, an incomplete problem with vertical perturbations (compare (6.60) with Definition 6.93)

It can be proved that the (MF)-condition remains valid when passing to problem (6.60)

Suppose now that the mapping F is (R)-regular at the point $z_0 \in \mathrm{gr}\,F$ (see Definition 6.12). We want to estimate the value $\rho(z, \tilde{F}(u))$. It is obvious that

$$\tilde{F}(u) = \mathrm{gr}\,F \cap \{x = u + x_0, y \in Y\} = \{u + x_0\} \times F(u + x_0).$$

Therefore, in a neighbourhood $V(x_0) \times V(y_0)$ of the point z_0 and in a neighbourhood $V(0)$ of the point $u_0 = 0$ we get

$$\rho((x, y), \tilde{F}(u)) \leq |x - x_0 - u| + \rho(y, F(u + x_0))$$

$$\leq \alpha_1 \max_{i=1,\ldots,n} |x_i - x_{0i} - u_i| + \alpha h_0(u + x_0, y)$$

$$\leq \alpha_2 \max \{0, h_i(u + x_0, y), i \in I, |h_i(u + x_0, y)|, i \in I_0,$$

$$|x_i - x_{0i} - u_i|, i = 1, \ldots, n\},$$

i.e. the mapping $\tilde{F}$ is also (R)-regular at the point (u_0, z_0).

3.5 Quasidifferentiable Programming Problems

Here we consider the constrained mathematical programming problem

$$f(x, y) \to \inf, \quad g(x, y) \leq 0, \tag{6.62}$$

where $f : R^n \times R^m \to R$ and $g : R^n \times R^m \to R^k$ are quasidifferentiable functions which are in general nonconvex and nondifferentiable. The components of the vector x can be seen as parameters or perturbations, but we act as follows. The vectors x and y are regarded as two groups of variables, and we are interested in decomposing problem (6.62) by fixing the vector x, thus getting the optimal value function

$$\varphi(x) = \inf_y \{f(x,y) \,|\, g(x,y) \le 0\}, \qquad (6.63)$$

where $\varphi(x)$ is assumed to be equal to $+\infty$ if there does not exist an y such that $g(x,y) \le 0$. The lower-level problem, i. e. the problem on the right-hand side of (6.63) will be denoted by (P_x). As above, the set of optimal solutions to (P_x) is

$$\omega(x) = \{y \,|\, g(x,y) \le 0,\, f(x,y) = \varphi(x)\}\,.$$

We are concerned with estimates for the upper Dini derivative of the optimal value function ω to (6.62), which plays an important role in deriving algorithms for minimizing ω and checking whether a current point x is inf-stationary (cf. Definition 2.34). By $f_y(x)$ we denote the function $x \to f(x,y)$ for fixed y, similarly $f_x(y)$, $g_{ix}(y)$, $g_{iy}(x)$. Let $\hat{x} \in \mathrm{dom}\varphi$. Then the Lagrangian associated with problem (6.62) is defined as

$$L(x,y,u) = f(x,y) + \sum_{i=1}^k u_i g_i(x,y)\,.$$

In this subsection, the functions f and g_i are assumed to be continuous and quasidifferentiable at the points (x^*, y) with $y \in \omega(x^*)$. This implies the quasidifferentiability of f_x, f_y, g_{ix}, g_{iy}. As usual, the quasidifferentials are described by $Df(x^*, y) = [\underline{\partial} f(x^*, y), \overline{\partial} f(x^*, y)]$, $Dg_i(x^*, y) = [\underline{\partial} g_i(x^*, y), \overline{\partial} g_i(x^*, y)]$, while fixed elements of the superdifferentials are denoted by $(v_2, v_2) \in \overline{\partial} f(x^*, y) \subset R^n \times R^m$ and $(w_{i2}, w_{i2}) \in \overline{\partial} g_i(x^*, y) \subset R^n \times R^m$, $i = 1, \dots, k$. Furthermore, we introduce the vector $w_2 = (w_{12}, \dots, w_{k2})$. The index set of active constraints will be denoted by $I(x^*, y) = \{i \,|\, g_i(x^*, y) = 0\}$, and let $\bar{x}$ and $\bar{y}$ be directions in R^n and R^m, respectively. We suppose y^* to be an arbitrary but fixed element of the set $\omega(x^*)$.

LEMMA 6.98 *For fixed elements* $v_2 \in \overline{\partial} f_{x^*}(y^*)$ *and* $w_{i2} \in \overline{\partial} g_{ix^*}(y^*)$, $i = 1, \dots, k$, *and the given directions* $\bar{x} \in R^n$, $\bar{y} \in R^m$, *the problems*

$(P_{v_2, w_2}) \quad f'_{x^*, v_2}(y^*; \bar{y}) \to \inf_{\bar{y}, \chi}$,

$$g'_{ix^*, w_{i2}}(y^*; \bar{y}) + g'_{iy^*}(x^*; \bar{x}) + \chi g_i(x^*, y^*) \le 0, \quad i = 1, \dots, k,$$

and

$$(D_{v_2,w_2}) \quad \sum_{i=1}^{k} u_i g'_{iy^*}(x^*;\bar{x}) \to \sup_{u},$$

$$u \in K(x^*,y^*,v_2,w_2) = \Big\{ u \mid u_i \geq 0, \ u_i g_i(x^*,y^*) = 0, \ i=1,\dots,k,$$

$$L'_{x^* u v_2 w_2}(y^*;\bar{y}) = f'_{x^* v_2}(y^*;\bar{y}) + \sum_{i=1}^{k} u_i g'_{ix^* w_{i2}}(y^*;\bar{y}) \geq 0 \ \forall \bar{x} \in R^n \Big\}$$

are dual to each other in the sense of Lagrange duality, where the functions $f_{x^* v_2}$ and $g_{ix^* w_{i2}}$ are defined according to (2.17) and related to the point (x^*,y^*).

Proof. The proof results immediately from Lemma 2.32 by noting that $f'_{x^* v_2}(y^*;\bar{y})$ and $g'_{ix^* w_2}(y^*;\bar{y})$ are homogeneous (and even convex) functions in $\bar{y}$. ∎

LEMMA 6.99 *Let $\bar{x} \in R^n$ be a fixed direction. Then, under the regularity condition*

$$(RC1) \ \exists \hat{y} \in R^m: \ \max_{z \in \underline{\partial} g_{ix^*}(y^*)} \langle z, \hat{y} \rangle + \max_{z \in \overline{\partial} g_{ix^*}(y^*)} \langle z, \hat{y} \rangle < 0 \ \forall i \in I(x^*,y^*),$$

the optimal values of the problems (P_{v_2,w_2}) and (D_{v_2,w_2}) coincide for any $v_2 \in \overline{\partial} f_{x^*}(y^*)$, $w_2 = (w_{12},\dots,w_{k2})$ with $w_{i2} \in \overline{\partial} f_{x^*}(y^*)$, $i = 1,\dots,k$.

Proof. Lemma 2.39 applied to problem (P_{x^*}) and hence related to the set $K(x^*,y^*,v_2,w_2)$ from Lemma 6.98 states that, under (RC1), the set $K(x^*,y^*,v_2,w_2)$ is non-empty and compact for each v_2, w_2. Consequently, there is a feasible element in (D_{v_2,w_2}), the optimal value of this problem is finite and provides a lower bound for the value of (P_{v_2,w_2}). Moreover, one can indicate a feasible solution $(\bar{y}_0, \chi_0)$ of (P_{v_2,w_2}) fulfilling the constraints as strict inequalities by choosing χ_0 large enough, $\bar{y}_0 = \tau \hat{y}$ with $\hat{y}$ from (RC1) and τ sufficiently large. Thus, by Lemma 6.98 and the strong duality theorem, $\inf (P_{v_2,w_2}) = \max (D_{v_2,w_2})$. ∎

For given v_2, w_2, $\bar{x} \in R^n$ and $\beta > 0$, we denote by $y(\bar{x},\beta,v_2,w_2)$ a β-optimal solution to (P_{v_2,w_2}). According to Lemma 6.98, we then have

$$f'_{x^* v_2}(y^*;y(\bar{x},\beta,v_2,w_2)) \leq \max_{u \in K(x^*,y^*,v_2,w_2)} \sum_{i=1}^{k} u_i g'_{iy^*}(x^*;\bar{x}) + \beta. \quad (6.64)$$

Now we still introduce the sets

$$V(\bar{x},x^*,y^*) =$$

$$\Big\{ v_2 \in \overline{\partial} f_{x^*}(y^*) \mid \exists v_1 : (v_1,v_2) \in \overline{\partial} f(x^*,y^*), \langle v_1,\bar{x} \rangle = \min_{z \in \overline{\partial} f_{y^*}(x^*)} \langle z, \bar{x} \rangle \Big\},$$

$$W_i(\bar{x}, x^*, y^*) = \left\{ w_{i2} \in \overline{\partial} g_{ix^*}(y^*) \mid \exists \, w_{i1} : (w_{i1}, w_{i2}) \in \overline{\partial} g_i(x^*, y^*), \right.$$

$$\left. \langle w_{i1}, \bar{x} \rangle = \min_{z \in \overline{\partial} g_{iy^*}(x^*)} \langle z, \bar{x} \rangle \right\}, \quad i \in I(x^*, y^*).$$

Because of the quasidifferentiability assumptions made above, we have $V(\bar{x}, x^*, y^*) \neq \emptyset$, $W_i(\bar{x}, x^*, y^*) \neq \emptyset \;\; \forall \bar{x} \in R^n$.

LEMMA 6.100 *Given $\bar{x} \in R^n$ and $x^* \in \mathrm{dom}\,\varphi$, we assume that (RC1) holds for $y^* \in \omega(x^*)$. Then, for any $v_2 \in V(\bar{x}, x^*, y^*)$, $w_{i2} \in W_i(\bar{x}, x^*, y^*)$, $i = 1, \ldots, k$, and $\beta > 0$, there exists a $\bar{t} > 0$ such that $\zeta(t, \bar{x}) = \zeta(t, \bar{x}, \beta, v_2, w_2) = y^* + \beta t \hat{y} + t y(\bar{x}, \beta, v_2, w_2)$ is feasible in $(P_{x^* + t\bar{x}})$ for $t \in (0, \bar{t})$, where $\hat{y}$ occurs in (RC1).*

Proof. In the case $i \notin I(x^*, y^*)$ the inequality $g_i(x^* + t\bar{x}, \zeta(t, \bar{x})) < 0$ results, for small t, from the continuity of g_i. Now, let $i \in I(x^*, y^*)$, i.e. $g_i(x^*, y^*) = 0$, choose $(v_2, w_2) = (v_2, w_{12}, \ldots, w_{k2}) \in V \times W_1 \times \ldots \times W_k$, and let w_{i2} be an element occurring in the set $W_i(\bar{x}, x^*, y^*)$, which especially means $\langle w_{i1}, \bar{x} \rangle = \min\{\langle z, \bar{x} \rangle \mid z \in \overline{\partial} g_{iy^*}(x^*)\}$. Due to the quasidifferentiability of g_i as well as the definition and properties of $g_{i(w_{i2}, w_{i2})}$, $g_{ix^* w_{i2}}$ and $g_{iy^* w_{i2}}$ (cf. relation (2.17); in particular, these functions are subdifferentiable, hence their directional derivatives are convex and homogeneous), we get, for t small enough,

$$g_i(x^* + t\bar{x}, y^* + \beta t \hat{y} + t y(\bar{x}, \beta, v_2, w_2))$$

$$= g_i(x^*, y^*) + g_i'((x^*, y^*); (t\bar{x}, t[\beta \hat{y} + y(\bar{x}, \beta, v_2, w_2)])) + o(t)$$

$$\leq g_{i(w_{i2}, w_{i2})}'((x^*, y^*); (t\bar{x}, t[\beta \hat{y} + y(\bar{x}, \beta, v_2, w_2)])) + o(t)$$

$$\leq g_{ix^* w_{i2}}'(x^*; t\bar{x}) + g_{iy^* w_{i2}}'(y^*; t[\beta \hat{y} + y(\bar{x}, \beta, v_2, w_2)])) + o(t)$$

$$= g_{ix^* w_{i2}}'(x^*; t\bar{x}) + g_{iy^*}'(y^*; t[\beta \hat{y} + y(\bar{x}, \beta, v_2, w_2)])) + o(t)$$

$$\leq t g_{iy^*}'(x^*; \bar{x}) + t\beta g_{ix^* w_{i2}}'(y^*; \hat{y}) + t g_{ix^* w_{i2}}'(y^*; y(\bar{x}, \beta, v_2, w_2)) + o(t)$$

$$\leq t\beta g_{ix^* w_{i2}}'(y^*; \hat{y}) + o(t) < 0.$$

The last two inequalities result from the definition of $y(\bar{x}, \beta, v_2, w_2)$ as a β-optimal (and thus feasible) solution to problem (P_{v_2, w_2}) and the regularity condition (RC1). ∎

LEMMA 6.101 *Under the conditions of Lemma 6.100,*

$$\varphi^+(x^*; \bar{x}) \leq \max_{u \in K(x^*, y^*, v_2, w_2)} L_{y^* u}'(x^*; \bar{x}).$$

Proof. Using the notation introduced above, Lemma 6.100 yields

$$\varphi(x^* + t\bar{x}) \le f(x^* + t\bar{x}, \zeta(t, \bar{x})).$$

Let v_1 be an element related to v_2 via the definition of $V(\bar{x}, x^*, y^*)$, which in particular means $\langle v_1, \bar{x}\rangle = \min\{\langle z, \bar{x}\rangle \mid z \in \overline{\partial} f_{y^*}(x^*)\}$. Then, in view of the quasidifferentiability of f, the definition of $f_{(v_1, v_2)}$ and the subdifferentiability of $f_{x^*v_2}$, by Lemma 6.100 and (6.64), we obtain

$$\varphi^+(x^*; \bar{x}) \le \lim_{t\downarrow 0} t^{-1}[f(x^* + t\bar{x}, y^* + \beta t\hat{y} + ty(\bar{x}, \beta, v_2, w_2) - f(x^*, y^*)]$$

$$= f'((x^*, y^*); (\bar{x}, \beta\hat{y} + y(\bar{x}, \beta, v_2, w_2))$$

$$\le f'_{(v_1, v_2)}((x^*, y^*); (\bar{x}, \beta\hat{y} + y(\bar{x}, \beta, v_2, w_2))$$

$$\le f'_{x^*v_2}(y^*; \beta\hat{y} + y(\bar{x}, \beta, v_2, w_2)) + f'_{y^*v_1}(x^*; \bar{x})$$

$$\le \beta f'_{x^*v_2}(y^*; \hat{y}) + f'_{x^*v_2}(y^*; y(\bar{x}, \beta, v_2, w_2)) + f'_{y^*}(x^*; \bar{x})$$

$$\le \beta f'_{x^*v_2}(y^*; \hat{y}) + \max_{u \in K(x^*, y^*, v_2, w_2)} \sum_{i=1}^{k} u_i g'_{iy^*}(x^*; \bar{x}) + \beta + f'_{y^*}(x^*; \bar{x})$$

$$= \beta[f'_{x^*v_2}(y^*; \hat{y}) + 1] + \max_{u \in K(x^*, y^*, v_2, w_2)} L'_{y^*u}(x^*; \bar{x}).$$

The claim follows by letting $\beta \downarrow 0$. ∎

THEOREM 6.102 *In problem (6.62), suppose the functions f and g_i, $i = 1, \ldots, k$, to be continuous and quasidifferentiable at the points (x^*, y), where $x^* \in \operatorname{dom}\varphi$, $y \in \omega(x^*)$. Moreover, let the direction $\bar{x} \in R^n$ be given, and assume the condition (RC1) to be fulfilled for every $y \in \omega(x^*)$. Then*

$$\varphi^+(x^*; \bar{x}) \le \inf_{y \in \omega(x^*)} \inf_{\substack{v_2 \in V(\bar{x}, x^*, y) \\ w_{i2} \in W_i(\bar{x}, x^*, y) \\ i \in I(x^*, y)}} \max_{u \in K(x^*, y, v_2, w_2)} L'_{yu}(x^*; \bar{x}). \quad (6.65)$$

Proof. The proof results immediately from Lemma 6.101 and the arbitrariness of $y \in \omega(x^*)$, $v_2 \in V(\bar{x}, x^*, y)$, $w_{i2} \in W_i(\bar{x}, x^*, y)$, $i \in I(x^*, y)$. ∎

Let us single out some special cases of problem (6.62).

1. An important case for practical aims is the additive problem

$$f(x, y) = f_1(x) + f_2(y) \to \inf, \quad g(x, y) = g_1(x) + g_2(y) \le 0, \quad (6.66)$$

where the functions $f_1 : R^n \to R$, $g_1 : R^n \to R^k$, $f_2 : R^m \to R$, $g_2 : R^m \to R^k$ are quasidifferentiable. In this case, the sets $V(\bar{x}, x^*, y)$

and $W_i(\bar{x}, x^*, y)$ in (6.65) turn into $\overline{\partial}f_2(y)$ and $\overline{\partial}g_i(y)$, $i \in I(x^*, y)$, respectively, which are independent of $\bar{x}$.

2. Let the functions f and g involved in (6.62) be subdifferentiable at (x^*, y), i.e. $Df(x^*, y) = [\partial f(x^*, y), \{0\}]$; similarly $Dg_i(x^*, y)$. This class of functions is closely related to quasidifferentiable functions in the sense of Pshenichnyi [148], regular locally convex functions (Ioffe and Tikhomirov [83]) and Clarke regular functions (Rockafellar [155]). Obviously, this class contains convex as well as differentiable functions. In this case, in estimate (6.65) the operation of taking the infimum over v_2 and w_{i2} may be omitted and the set $K(x^*, y, v_2, w_2) = K(x^*, y)$ no longer depends on v_2 and w_2.

3. In the case of continuously differentiable functions, we arrive at the well-known results of Gauvin and Dubeau [69].

In the following, we briefly want to discuss the situation concerning lower bounds of the (potential) directional derivative. In the light of so-called primal decomposition, i.e. the minimization of the optimal value function φ, we are mainly interested in getting estimates of the upper Dini derivative in order to determine directions of descent for φ. On the other hand, to exclude some vector as a direction of descent it is desirable to have bounds for the lower Dini directional derivative, too. In some cases (see e.g. [69], [155]) such estimates have been obtained. In these papers even the existence of the directional derivative have been shown under somewhat stronger conditions. In the case of a quasidifferentiable function and continuously differentiable constraints it is also possible to derive a lower bound for the potential directional derivative (see Luderer [102]).

Even in the differentiable case of problem (6.62) there are examples, where the bounds obtained are sharp and the directional derivative exists, but neither bound is attained (see [73]). Naturally, this statement applies all the more to the quasidifferentiable case considered here.

Finally, the function φ defined via (6.63) need not be quasidifferentiable or directionally differentiable in general, although the original problem is so. The following example demonstrates this phenomenon.

EXAMPLE 6.103 *Let* $f(x, y) = y$, $g(x, y) = \min\{(x-1)^2 + y^2; (x+1)^2 + y^2\} - 1$, $x^* = 0$, $\bar{x} = 1$. *Clearly,* $y^* = 0$ *is the only feasible point for* $x^* = 0$ *and* g *is quasidifferentiable at* $(0,0)$. *The optimal value function can be calculated explicitly and amounts to*

$$\varphi(x) = \begin{cases} -\sqrt{1 - (x-1)^2}, & 0 \leq x \leq 2, \\ -\sqrt{1 - (x+1)^2}, & -2 \leq x < 0, \\ \infty, & |x| > 2. \end{cases}$$

The directional derivative $\varphi'(x^*; \bar{x})$ is obviously $-\infty$ and thus fails to exist in the true sense. Therefore, φ is not quasidifferentiable at $x^* = 0$.

In Luderer [103] the quasidifferentiability of some function associated with the right-hand side of (6.65) and approximating the optimal value function (6.63) has been shown and a description of the corresponding quasidifferential has been given. These results are based on the application of Theorem 2.31.

4. Second-Order Analysis of the Optimal Value Function and Differentiability of Optimal Solutions

The purpose of this section is to describe sufficient conditions allowing us to guarantee the existence and to give a characterization of the second-order directional derivative of the optimal value function

$$\varphi(x) = \inf\{f(x, y) \mid y \in F(x)\},$$

where

$$F(x) = \{y \in Y \mid h_i(x, y) \leq 0, \ i=1,\ldots,r, \ h_i(x, y)=0, \ i=r+1,\ldots,p\}.$$

The most natural way to define a second-order directional derivative $\varphi''(x_0; \bar{x})$ is the approach for which the second-order Taylor expansion remains valid. In this way we arrive at the following definition (see Demyanov [55], Seeger [162]).

DEFINITION 6.104 *Let $\bar{x} \in X$, and let the derivative $\varphi'(x_0; \bar{x})$ exist. If the limits*

$$D^{2+}\varphi(x_0; \bar{x}) = \limsup_{t \downarrow 0} \frac{2}{t^2} \left[\varphi(x_0 + t\bar{x}) - \varphi(x_0) - t\varphi'(x_0; \bar{x})\right]$$

and

$$D^2_+\varphi(x_0; \bar{x}) = \liminf_{t \downarrow 0} \frac{2}{t^2} \left[\varphi(x_0 + t\bar{x}) - \varphi(x_0) - t\varphi'(x_0; \bar{x})\right]$$

are finite and equal to each other, then the function φ is called twice directionally differentiable at the point x_0 in the direction $\bar{x}$ and the value

$$\varphi''(x_0; \bar{x}) = \lim_{t \downarrow 0} \frac{2}{t^2} \left[\varphi(x_0 + t\bar{x}) - \varphi(x_0) - t\varphi'(x_0; \bar{x})\right] \tag{6.67}$$

is called the second-order directional derivative of φ at x_0 in the direction $\bar{x}$.

Based on the method of first-order approximation, Demyanov proved in [55] the existence of the derivative $\varphi''(x_0; \bar{x})$ in nonlinear programming problems with constraints convex with respect to y.

In a different context the function $\varphi''(x_0; \bar{x})$ has been considered by other authors either (see, e.g., Auslender [6], Levitin [97]). Although the definition of $\varphi''(x; \bar{x})$ was introduced in a very natural way, it is preferable to consider another second-order directional derivative better adapted to the special form of the optimal value function φ. A generalization of Definition (6.67) was proposed by Ben-Tal and Zowe (cf. [17]). This new type of second-order directional derivative contains more information about the behaviour of φ in a neighbourhood of x_0 than $\varphi''(x_0; \bar{x})$, since it requires the knowledge of φ not only over the half-lines $\{x_0 + t\bar{x} \,|\, t \in R_+\}$, but also over the sets of the form $\{x_0 + t\bar{x}_1 + t^2\bar{x}_2 \,|\, t \in R_+\}$.

DEFINITION 6.105 *Let $\bar{x}_1, \bar{x}_2 \in X$. If the limits*

$$D^{2+}\varphi(x_0; \bar{x}_1, \bar{x}_2) = \limsup_{t\downarrow 0} \frac{2}{t^2}\left[\varphi(x_0 + t\bar{x}_1 + t^2\bar{x}_2) - \varphi(x_0) - t\varphi'(x_0; \bar{x}_1)\right]$$

and

$$D^2_+\varphi(x_0; \bar{x}_1, \bar{x}_2) = \liminf_{t\downarrow 0} \frac{2}{t^2}\left[\varphi(x_0 + t\bar{x}_1 + t^2\bar{x}_2) - \varphi(x_0) - t\varphi'(x_0; \bar{x}_1)\right]$$

are finite and equal to each other, then the function φ is called twice directionally differentiable at the point x_0 in the sense of Ben-Tal and Zowe in the directions $\bar{x}_1$, $\bar{x}_2$, and the value

$$\varphi''(x_0; \bar{x}_1, \bar{x}_2) = \lim_{t\downarrow 0} \frac{2}{t^2}\left[\varphi(x_0 + t\bar{x}_1 + t^2\bar{x}_2) - \varphi(x_0) - t\varphi'(x_0; \bar{x}_1)\right] \tag{6.68}$$

is called the second-order directional derivative of φ at x_0 in the sense of Ben-Tal and Zowe in the directions $\bar{x}_1$, $\bar{x}_2$.

Thus, the existence of $\varphi''(x_0; \bar{x}_1, \bar{x}_2)$ is stronger in comparison with $\varphi''(x_0; \bar{x})$. The function $\varphi''(x_0; \bar{x}_1, \bar{x}_2)$ allows us to recover $\varphi'(x_0; \bar{x})$ and $\varphi''(x_0; \bar{x})$ simply by choosing $\bar{x}_1 = 0$ or $\bar{x}_2 = 0$, respectively.

The following properties of $\varphi''(x_0; \bar{x}_1, \bar{x}_2)$ are valid (see Seeger [162]).

LEMMA 6.106 *For any $\bar{x}_1, \bar{x}_2 \in X$ the following inequality holds:*

$$D^{2+}\varphi(x_0; \bar{x}_1, \bar{x}_2)$$

$$\leq D^{2+}\varphi(x_0; \bar{x}_1) + \limsup_{t\downarrow 0} \frac{2}{t^2}\left[\varphi(x_0 + t\bar{x}_1 + t^2\bar{x}_2) - \varphi(x_0 + t\bar{x}_1)\right].$$

Moreover, if the two limits $\varphi''(x_0; \bar{x}_1)$ *and*

$$A(x_0; \bar{x}_1, \bar{x}_2) = \lim_{t \downarrow 0} \frac{2}{t^2} \left[\varphi(x_0 + t\bar{x}_1 + t^2\bar{x}_2) - \varphi(x_0 + t\bar{x}_1) \right]$$

exist and are finite, then $\varphi''(x_0; \bar{x}_1, \bar{x}_2)$ *exists and is given by*

$$\varphi''(x_0; \bar{x}_1, \bar{x}_2) = A(x_0; \bar{x}_1, \bar{x}_2) + \varphi''(x_0; \bar{x}_1).$$

LEMMA 6.107 *Let* $\varphi(x) = h(k(x))$, *where* $k(\cdot)$ *and* $h(\cdot)$ *are twice directionally differentiable in the sense of Ben-Tal and Zowe. Then*

$$\varphi''(x_0; \bar{x}_1, \bar{x}_2) = h''(k(x_0); k'(x_0; \bar{x}_1), k''(x_0; \bar{x}_1, \bar{x}_2)).$$

LEMMA 6.108 *Let the functions* $\varphi_i(x)$ *be twice directionally differentiable at the point* x_0 *in the sense of Ben-Tal and Zowe in the direction* $\bar{x}_1, \bar{x}_2$. *Then for the function* $\varphi(x) = \max\{\varphi_i(x) \mid 1 \leq i \leq n\}$ *the derivative* $\varphi''(x_0; \bar{x}_1, \bar{x}_2)$ *exists and*

$$\varphi''(x_0; \bar{x}_1, \bar{x}_2) = \max_{i \in I(x_0; \bar{x}_1)} \varphi_i''(x_0; \bar{x}_1, \bar{x}_2),$$

where $I(x_0; \bar{x}_1) = \{i \in I(x_0) \mid \varphi'(x_0; \bar{x}_1) = \varphi_i'(x_0; \bar{x}_1)\}$, $I(x_0) = \{1 \leq i \leq n \mid \varphi(x_0) = \varphi_i(x_0)\}$.

The basic tool for estimating the second-order derivative in the sense of Ben-Tal and Zowe of the optimal value function is presented in the next lemma.

LEMMA 6.109 *Let the mapping* F *satisfy the condition*

$$\hat{D}_L^2 F(z_0, \bar{z}_1; \bar{x}_2) = \Gamma_F^2(z_0, \bar{z}_1; \bar{x}_2) \neq \emptyset \tag{6.69}$$

at the points $z_0 = (x_0, y_0) \in \operatorname{gr} F$ *for all* $\bar{z}_1 \in \Gamma_F(z_0)$, $\bar{x}_2 \in X$. *Then for any* $\bar{y}_2 \in \Gamma_F^2(z_0, \bar{z}_1; \bar{x}_2)$ *the inequality*

$$\varphi(x_0 + t\bar{x}_1 + t^2\bar{x}_2) \leq \varphi(x_0) + t\langle \nabla f(z_0), \bar{z}_1 \rangle + t^2 \Phi(z_0, \bar{z}_1, \bar{z}_2) + o(t^2)$$

holds, where $t^{-1}o(t) \to 0$ *if* $t \downarrow 0$, $\bar{z}_1 = (\bar{x}_1, \bar{y}_1)$, $\bar{z}_2 = (\bar{x}_2, \bar{y}_2)$ *and* $\Phi(z_0, \bar{z}_1, \bar{z}_2) = \langle \nabla f(z_0), \bar{z}_2 \rangle + \frac{1}{2}\langle \bar{z}_1, \nabla^2 f(z_0)\bar{z}_1 \rangle$.

Proof. From the condition (6.69) we get that for any $\bar{y}_2 \in \Gamma_F^2(z_0, \bar{z}_1; \bar{x}_2)$,

$$\varphi(x_0 + t\bar{x}_1 + t^2\bar{x}_2) - \varphi(x_0)$$
$$\leq f(x_0 + t\bar{x}_1 + t^2\bar{x}_2, y_0 + t\bar{y}_1 + t^2\bar{y}_2 + o(t^2)) - f(x_0, y_0)$$
$$= t\langle \nabla f(z_0), \bar{z}_1 \rangle + t^2 \Phi(z_0, \bar{z}_1, \bar{z}_2) + o(t^2). \quad \blacksquare$$

With the help of this lemma we can prove the next result about directional Lipschitz behaviour of the set of optimal solutions.

As we have seen in Subsection 3.3 (Definition 6.84) the condition (SOSC) appears to be a sufficient condition for Hölder continuity. In order to strengthen this result, we will make use of a directional form of condition (SOSC) (see Shapiro [163]).

Let us denote

$$\Lambda^2(z_0, \bar{x}) = \{\lambda \in \Lambda(z_0) \,|\, \langle \nabla_x L(z_0, \lambda), \bar{x} \rangle = \max_{\lambda \in \Lambda(z_0)} \langle \nabla_x L(z_0, \lambda), \bar{x} \rangle\}.$$

DEFINITION 6.110 *We shall say that the strong second-order sufficient condition (SOSC$_{\bar{x}}$) holds at the point $z_0 \in \mathrm{gr}\, F$ if $\Lambda(z_0) \neq \emptyset$ and*

$$\sup_{\lambda \in \Lambda^2(z_0, \bar{x})} \left\langle \bar{y}, \nabla^2_{yy} L(z_0, \lambda) \bar{y} \right\rangle > 0$$

for any $\bar{y} \in D(z_0)$, $\bar{y} \neq 0$.

REMARK. Using the representation

$$\Lambda^2(z_0, \bar{x}) = \{\lambda \in \Lambda(z_0) \,|\, \lambda_i \langle \nabla h_i(z_0), \bar{z} \rangle = 0, \, i \in I(z_0)\},$$

we get another kind of (SOSC$_{\bar{x}}$) (cf. [32], [86]):

$$\sup_{\lambda \in \Lambda(z_0)} \left\langle \bar{y}, \nabla^2_{yy} L(z_0, \lambda) \bar{y} \right\rangle > 0$$

for every $\bar{y} \in D(z_0) \cap \{\langle \nabla h_i(z_0), \bar{z} \rangle = 0, \, i \in I_0 \cup I^*(z_0)\}$, where $I^*(z_0) = \{i \in I(z_0) \,|\, \lambda_i > 0\}$.

Note that the condition (SOSC$_{\bar{x}}$) reduces to (SOSC) in the following particular cases (see Shapiro [163]):

1) the constraint functions h_i, $i = 1, \ldots, p$, are independent of x, i.e. $h_i(x, y) = h_i(y)$, hence the feasible set does not depend on x: $F(x) = F_0$;

2) the constraint functions $h_i(x, y)$, $i = 1, \ldots, p$, are linear with respect to y so that $\nabla^2_{yy} L(z_0, \lambda) = \nabla^2_{yy} f(z_0)$ for any λ;

3) the set of Lagrange multipliers is a singleton, i.e. $\Lambda(z_0) = \{\lambda_0\}$.

LEMMA 6.111 *Let the conditions (SOSC$_{\bar{x}}$) and (6.69) hold at the point $z_0 = (x_0, y_0)$, $y_0 \in \omega(x_0)$. Then for any sequences $t_k \downarrow 0$, $y_k \in \omega(x_0 + t_k \bar{x}_1 + t_k^2 \bar{x}_2)$, $k = 1, 2, \ldots$, such that $y_k \to y_0 \in \omega(x_0)$ the following inequality holds:*

$$\limsup_{t_k \downarrow 0} |y_k - y_0| t_k^{-1} < \infty.$$

Proof. Suppose the opposite. Then, without loss of generality and taking into account Lemmas 6.59 and 6.109, we can assume that

$$(y_k - y_0)|y_k - y_0|^{-1} \to \hat{y} \in D(z_0).$$

Let us take some $\lambda \in \Lambda^2(z_0, \bar{x}_1)$ such that

$$\left\langle \hat{y}, \nabla^2_{yy} L(z_0, \lambda) \hat{y} \right\rangle > 0.$$

We denote $x_k = x_0 + t_k \bar{x}_1 + t_k^2 \bar{x}_2$, $z_k = (x_k, y_k)$. Applying Lemmas 6.66, 6.82 and 6.109, for sufficiently large k we get

$$L(z_k, \lambda) - L(z_0, \lambda) \le \varphi(x_k) - \varphi(x_0) \le t_k \langle \nabla_x L(z_0, \lambda), \bar{x}_1 \rangle + \mu t_k^2,$$

where $\mu = $ const. Therefore, due to the fact that $\nabla_y L(z_0, \lambda) = 0$, we obtain

$$\frac{L(z_k, \lambda) - L(z_0, \lambda) - t_k \langle \nabla_x L(z_0, \lambda), \bar{x}_1 \rangle}{|y_k - y_0|^2} \le \mu \left(\frac{t_k}{|y_k - y_0|} \right)^2$$

and the passage to the limit yields $\left\langle \hat{y}, \nabla^2_{yy} L(z_0, \lambda) \hat{y} \right\rangle \le 0$. This contradicts the choice of λ. ∎

The main result of this section is stated in the following theorem.

THEOREM 6.112 *Let the problem* (P_x) *satisfy the* (R)-*regularity condition at all points* $z_0 \in \{x_0\} \times \omega(x_0)$, *and assume the set of suboptimal solutions* $\omega_\varepsilon(x_0 + t_k \bar{x}_1 + t_k^2 \bar{x}_2)$ *to be weakly Lipschitz continuous, i. e., for any sequence* $t_k \downarrow 0$ *there exists a point* $y_0 \in \omega(x_0)$ *such that*

$$\rho(y_0, \omega_\varepsilon(x_0 + t_k \bar{x}_1 + t_k^2 \bar{x}_2)) \le lt_k$$

for all $k = 1, 2, \ldots$ *Then*
  (i) *for all* $\bar{x}_1 \in X$ *there exists the directional derivative*

$$\varphi'(x_0; \bar{x}_1) = \inf_{y_0 \in \omega(x_0)} \inf_{\bar{y}_1 \in \Gamma_F(z_0; \bar{x}_1)} \langle \nabla f(z_0), \bar{z}_1 \rangle \qquad (6.70)$$

$$= \inf_{y_0 \in \omega(x_0)} \sup_{\lambda \in \Lambda(z_0)} \langle \nabla_x L(z_0, \lambda), \bar{x}_1 \rangle ;$$

(ii) *for all* $\bar{x}_1, \bar{x}_2 \in X$ *there exists the second-order directional derivative in the sense of Ben-Tal and Zowe and*

$$\varphi''(x_0; \bar{x}_1, \bar{x}_2) = \inf_{(y_0, \bar{y}_1) \in \omega(x_0, \bar{x}_1)} \inf_{\bar{y}_2 \in \Gamma_F^2(z_0, \bar{z}_1; \bar{x}_2)} 2\Phi(z_0, \bar{z}_1, \bar{z}_2) \qquad (6.71)$$

$$= \inf_{(y_0,\bar{y}_1)\in\omega(x_0,\bar{x}_1)} \sup_{\lambda\in\Lambda^2(z_0,\bar{x}_1)} \{2\langle\nabla_x L(z_0,\lambda),\bar{x}_2\rangle + \langle\bar{z}_1,\nabla^2 L(z_0,\lambda)\bar{z}_1\rangle\},$$

where

$$\bar{z}_1 = (\bar{x}_1,\bar{y}_1),\ \bar{z}_2 = (\bar{x}_2,\bar{y}_2),$$
$$\omega(x_0,\bar{x}_1) = \{(y_0,\bar{y}_1)|\ y_0 \in \omega(x_0),\ \bar{y}_1 \in \Gamma_F(z_0;\bar{x}_1),$$
$$\varphi'(x_0;\bar{x}_1) = \langle\nabla f(z_0),\bar{z}_1\rangle\},$$
$$\Phi(z_0,\bar{z}_1,\bar{z}_2) = \langle\nabla f(z_0),\bar{z}_2\rangle + \tfrac{1}{2}\langle\bar{z}_1,\nabla^2 f(z_0)\bar{z}_1\rangle.$$

Proof. (i) 1. Let $D^+\varphi(x_0;\bar{x}_1)$ be attained on the sequence $t_k \downarrow 0$. We take an arbitrary $y_0 \in \omega(x_0)$, $\bar{y}_1 \in \Gamma_F(z_0;\bar{x}_1)$ and denote $\tilde{z}_k = (x_0 + t_k\bar{x}_1, y_0 + t_k\bar{y}_1)$. Let us choose k_0 such that $x_0 + t_k\bar{x}_1 \in X_0$, $t_k|\bar{y}_1| < \varepsilon_0/4$ for $k > k_0$, where $\varepsilon_0 > 2\mathrm{diam}Y_0$. Due to the Lemma 3.68 about the removal of constraints we then get

$$\varphi(x_0 + t_k\bar{x}_1) \leq f(\tilde{z}_k) + \beta d_F(\tilde{z}_k)$$

for all $k > k_0$. In view of the (R)-regularity of problem (P_x) and the continuity of h_i, $i = 1,\ldots,p$, we can find a scalar $\alpha > 0$ such that for all k sufficiently large

$$\begin{aligned}\varphi(x_0 + t_k\bar{x}_1) - \varphi(x_0) &\leq f(\tilde{z}_k) + \beta d_F(\tilde{z}_k) - f(z_0)\\ &\leq f(\tilde{z}_k) - f(z_0) + \alpha\beta\max\{0, h_i(\tilde{z}_k),\ i\in I(z_0),\ |h_i(\tilde{z}_k)|,\ i\in I_0\}.\end{aligned} \tag{6.72}$$

We denote $M = \alpha\beta$. Two cases are possible:

Case 1. There exists $\tilde{k}_0$, $\tilde{k}_0 \geq k_0$, such that $h_i(\tilde{z}_k) \leq 0$, $i \in I(z_0)$, and $h_i(\tilde{z}_k) = 0$, $i \in I_0$, for all $k \geq \tilde{k}_0$. In this case

$$\varphi(x_0 + t_k\bar{x}_1) - \varphi(x_0) \leq f(\tilde{z}_k) - f(z_0),$$

and, therefore

$$D^+\varphi(x_0;\bar{x}_1) \leq \langle\nabla f(z_0),\bar{z}_1\rangle.$$

Case 2. Without loss of generality we may assume that for each k there exist an index $i \in I(z_0)$ such that $h_i(\tilde{z}_k) > 0$ or an index $i \in I_0$ such that $|h_i(\tilde{z}_k)| > 0$. Then

$$\max\{0, h_i(\tilde{z}_k),\ i \in I(z_0),\ |h_i(\tilde{z}_k)|,\ i \in I_0\}$$
$$= \max\{h_i(\tilde{z}_k),\ i \in I(z_0),\ |h_i(\tilde{z}_k)|,\ i \in I_0\}$$
$$= \max\{t_k\langle\nabla h_i(z_0),\bar{z}_1\rangle + o(t_k),\ i \in I(z_0), |t_k\langle\nabla h_i(z_0),\bar{z}_1\rangle + o(t_k)|,\ i \in I_0\}$$
$$\leq t_k\max\{\langle\nabla h_i(z_0),\bar{z}_1\rangle,\ i \in I(z_0),\ |\langle\nabla h_i(z_0),\bar{z}_1\rangle|,\ i \in I_0\} + |\tilde{o}(t_k)|.$$

Since $\bar{z}_1 \in \Gamma_F(z_0)$, i.e.

$$\langle\nabla h_i(z_0),\bar{z}_1\rangle \leq 0,\ i \in I(z_0),\quad \langle\nabla h_i(z_0),\bar{z}_1\rangle = 0,\ i \in I_0,$$

then from (6.72) we obtain

$$\varphi(x_0 + t_k\bar{x}_1) - \varphi(x_0) \le f(\tilde{z}_k) - f(z_0) + M|\tilde{o}(t_k)|.$$

Therefore

$$D^+\varphi(x_0; \bar{x}_1) \le \langle \nabla f(z_0), \bar{z}_1 \rangle,$$

and, consequently

$$D^+\varphi(x_0; \bar{x}_1) \le \inf_{y_0 \in \omega(x_0)} \inf_{\bar{y}_1 \in \Gamma_F(z_0; \bar{x}_1)} \langle \nabla f(z_0), \bar{z}_1 \rangle. \qquad (6.73)$$

2. Let $D_+\varphi(x_0; \bar{x}_1)$ be attained on the sequence $t_k \downarrow 0$. Due to the assumption of the theorem about the weak Lipschitz continuity of the set $\omega_\varepsilon(x_0 + t_k\bar{x}_1)$ (assume $\bar{x}_2 = 0$), we can find a point $y_0 \in \omega(x_0)$ as well as a sequence $y_k \in \omega_\varepsilon(x_0 + t_k\bar{x}_1)$ such that $|y_k - y_0| \le Mt_k$ and

$$f(x_0 + t_k\bar{x}_1, y_k) \le \varphi(x_0 + t_k\bar{x}_1) + \varepsilon(t_k), \qquad (6.74)$$

where $\varepsilon(t_k)t_k^{-2} \downarrow 0$ as $t_k \downarrow 0$.

In this way, the sequence $\{t_k^{-1}(y_k - y_0)\}$ is bounded. Therefore, without loss of generality we can assume that it converges to $\bar{y}_1 \in \hat{D}_L F(z_0; \bar{x}_1) = \Gamma_F(z_0; \bar{x}_1)$ (cf. Corollary 6.41). Thus, from (6.74) we get

$$\varphi(x_0 + t_k\bar{x}_1) - \varphi(x_0) \ge f(x_0 + t_k\bar{x}_1, y_k) - f(x_0, y_0) - \varepsilon(t_k).$$

Dividing the last expression by t_k and passing to the limit, we obtain

$$D_+\varphi(x_0; \bar{x}_1) \ge \langle \nabla f(z_0), \bar{z}_1 \rangle \ge \inf_{y_0 \in \omega(x_0)} \inf_{\bar{y}_1 \in \Gamma(z_0; \bar{x}_1)} \langle \nabla f(z_0), \bar{z}_1 \rangle.$$

Comparing this result with (6.73) and applying the duality theorem, we get (6.70).

(ii) 1. Let us choose arbitrary elements $(y_0, \bar{y}_1) \in \omega(x_0, \bar{x}_1)$ and $\bar{y}_2 \in \Gamma_F^2(z_0, \bar{z}_1; \bar{x}_2)$. Due to Lemma 6.109, we get the estimation

$$\varphi(x_0 + t\bar{x}_1 + t^2\bar{x}_2) - \varphi(x_0) \le t\langle \nabla f(z_0), \bar{z}_1 \rangle + t^2 \Phi(z_0, \bar{z}_1, \bar{z}_2) + o(t^2),$$

where $\bar{z}_2 = (\bar{x}_2, \bar{y}_2)$.

Since for the chosen y_0 and $\bar{y}_1$ the equality $\langle \nabla f(z_0), \bar{z}_1 \rangle = \varphi'(x_0; \bar{x}_1)$ holds, then from the last inequality it follows that

$$\varphi(x_0 + t\bar{x}_1 + t^2\bar{x}_2) - \varphi(x_0) - t\varphi'(x_0; \bar{x}_1) \le t^2 \Phi(z_0, \bar{z}_1, \bar{z}_2) + o(t^2)$$

and, consequently

$$\frac{1}{t^2}\left[\varphi(x_0 + t\bar{x}_1 + t^2\bar{x}_2) - \varphi(x_0) - t\varphi'(x_0; \bar{x}_1)\right] \le \Phi(z_0, \bar{z}_1, \bar{z}_2) + \frac{o(t^2)}{t^2}.$$

In this way, we obtain

$$D^{2+}\varphi(x_0; \bar{x}_1, \bar{x}_2) \leq 2\Phi(z_0, \bar{z}_1, \bar{z}_2)$$

for all $(y_0, \bar{y}_1) \in \omega(x_0; \bar{x}_1)$, $\bar{y}_2 \in \Gamma^2(z_0, \bar{z}_1; \bar{x}_2)$. Therefore, in view of the (R)-regularity and due to Lemma 6.82, we have

$$D^{2+}\varphi(x_0; \bar{x}_1, \bar{x}_2) \leq \inf_{(y_0, \bar{y}_1) \in \omega(x_0, \bar{x}_1)} \min_{\bar{y}_2 \in \Gamma^2(z_0, \bar{z}_1; \bar{x}_2)} 2\Phi(z_0, \bar{z}_1, \bar{z}_2). \quad (6.75)$$

2.   Let the limit

$$D^2_+\varphi(x_0; \bar{x}_1, \bar{x}_2) = \liminf_{t\downarrow 0} \frac{2}{t^2} \left[\varphi(x_0 + t\bar{x}_1 + t^2\bar{x}_2) - \varphi(x_0) - t\varphi'(x_0; \bar{x}_1) \right]$$

be attained on the sequence $\{t_k\}$, $t_k \downarrow 0$. Due to the weak Lipschitz continuity of the set $\omega_\varepsilon(x_0 + t_k\bar{x}_1 + t_k^2\bar{x}_2)$ we can find a point $y_0 \in \omega(x_0)$ as well as a sequence $y_k \in \omega_\varepsilon(x_0 + t_k\bar{x}_1 + t_k^2\bar{x}_2)$ such that $|y_k - y_0| \leq Mt_k$, $k = 1, 2, \ldots$

Without loss of generality we can assume that $\{t_k^{-1}(y_k - y_0)\} \to \bar{y}_1$. Moreover, $\bar{y}_1 \in \hat{D}_U F(z_0; \bar{x}_1) = \Gamma_F(z_0; \bar{x}_1)$. Let us denote $x_k = x_0 + t_k\bar{x}_1 + t_k^2\bar{x}_2$, $z_k = (x_k, y_k)$. Since

$$f(x_k, y_k) \leq \varphi(x_k) + o(t_k^2),$$

then

$$\varphi(x_k) - \varphi(x_0) \geq f(x_k, y_k) - o(t_k^2) - f(x_0, y_0).$$

Due to Theorem 6.69, the function φ is Lipschitz continuous. Now, dividing the last inequality by t_k and passing to the limit, we get

$$\varphi'(x_0; \bar{x}_1) \geq \langle \nabla f(z_0), \bar{z}_1 \rangle. \quad (6.76)$$

Therefore $(y_0, \bar{y}_1) \in \omega(x_0, \bar{x}_1)$.

Applying Lemma 6.66 for an arbitrary $\lambda \in \Lambda^2(z_0; \bar{x}_1)$ and taking into account (6.76) as well as Lemma 6.82, we get

$$\varphi(x_k) - \varphi(x_0) - t_k\varphi'(x_0; \bar{x}_1) \geq L(z_k, \lambda) - L(z_0, \lambda) - t_k\langle \nabla L(z_0, \lambda), \bar{z}_1 \rangle.$$

Since $\lambda \in \Lambda^2(z_0; \bar{x}_1)$, then $\nabla_y L(z_0, \lambda) = 0$. Dividing this inequality by t_k^2 and passing to the limit, we conclude

$$D^2_+\varphi(x_0; \bar{x}_1, \bar{x}_2) \geq 2\langle \nabla_x L(z_0, \lambda), \bar{x}_2 \rangle + \langle \bar{z}_1, \nabla^2 L(z_0, \lambda)\bar{z}_1 \rangle.$$

due to the choice of $\{t_k\}$. In this way

$$D^2_+\varphi(x_0; \bar{x}_1, \bar{x}_2) \geq$$
$$\inf_{(y_0, \bar{y}_1) \in \bar{\omega}(x_0, \bar{x}_1)} \sup_{\lambda \in \Lambda^2(z_0; \bar{x}_1)} \left\{ 2\langle \nabla_x L(z_0, \lambda), \bar{x}_2 \rangle + \langle \bar{z}_1, \nabla^2 L(z_0, \lambda)\bar{z}_1 \rangle \right\}.$$

By comparing this result with (6.75) and applying Lemma 6.82, we obtain equality (6.71) stated in the theorem. ∎

Our next aim is to complement this theorem by asking the question under which assumptions the optimal solutions of problem (P_x) are directionally differentiable. First results concerning differential properties of solutions were obtained by using a second-order sufficient optimality condition and the linear independence constraint qualification together with the strict complementary slackness assumption which, in accordance with the implicit function theorem of classical analysis applied to the Kuhn-Tucker optimality conditions, guarantee differentiability of optimal solutions of perturbed problems (see Fiacco [66]).

Later, Jittorntrum ([86]) supplemented this results by proving that even without strict complementarity one may have differentiability of solutions provided that the linear independence condition and the strong second-order sufficient condition are satisfied. But it seems that Gauvin and Janin ([70]) and Shapiro ([163]) were the first who took advantage of the fact that the potential directional derivatives of the optimal solutions should solve the quadratic problem

$$(P^*(z_0, \lambda; \bar{x})) : \quad \begin{cases} \langle \bar{z}, \nabla^2 L(z_0, \lambda) \bar{z} \rangle \to \min \\ \bar{y} \in \tilde{\omega}(z_0; \bar{x}), \end{cases}$$

where $z_0 = (x_0, y_0)$, $y_0 \in \omega(x_0)$, $\lambda \in \Lambda^2(z_0; \bar{x})$, $\tilde{\omega}(z_0; \bar{x}) = \{\bar{y} \in \Gamma_F(z_0; \bar{x}) \mid \varphi'(x_0; \bar{x}) = \langle \nabla f(z_0), \bar{z} \rangle\}$. We denote its optimal value function by

$$\varphi^*(z_0, \lambda; \bar{x}) = \min \{\langle \bar{z}, \nabla^2 L(z_0, \lambda) \bar{z} \rangle \mid \bar{y} \in \tilde{\omega}(z_0; \bar{x})\}$$

and the set of optimal solutions by

$$\omega^*(z_0, \lambda; \bar{x}) = \{\bar{y} \in \tilde{\omega}(z_0; \bar{x}) \mid \langle \bar{z}, \nabla^2 L(z_0, \lambda) \bar{z} \rangle = \varphi^*(z_0, \lambda; \bar{x})\}.$$

In the following theorem we shall prove differentiability of suboptimal solutions based on some kind of second-order properties for the optimal value function. This theorem generalizes the results of Auslender and Cominetti ([9]) for the case of (R)-regular problems and without the demand that the set of optimal solutions $\omega(x)$ is a singleton.

Let us denote by $\tilde{\omega}(x_0; \bar{x})$ the points from the set $\omega(x_0)$ solving the problem

$$\min_{y_0 \in \omega(x_0)} \left[\min_{\bar{y} \in \Gamma_F(z_0; \bar{x})} \langle \nabla f(z_0), \bar{z} \rangle \right].$$

THEOREM 6.113 *Suppose the problem* (P_x) *to be* (R)-*regular at every point* $z_0 \in \{x_0\} \times \omega(x_0)$, *and assume that the optimal value function* φ *at the point* x_0 *for some* $\bar{x}$ *has the derivatives*

$$\varphi'(x_0; \bar{x}) = \min_{y_0 \in \omega(x_0)} \max_{\lambda \in \Lambda(z_0)} \langle \nabla_x L(z_0, \lambda), \bar{x} \rangle, \tag{6.77}$$

$$\varphi''(x_0; \bar{x}) = \min_{y_0 \in \omega(x_0)} \quad \min_{\bar{y} \in \tilde{\omega}(z_0; \bar{x})} \quad \max_{\lambda \in \Lambda^2(z_0; \bar{x})} \langle \bar{z}, \nabla^2 L(z_0, \lambda) \bar{z} \rangle. \qquad (6.78)$$

Then there exists $\varepsilon = o(t^2)$ such that for any $y(t) \in \omega_\varepsilon(x_0 + t\bar{x}, y_0, l) = \omega_\varepsilon(x_0 + t\bar{x}) \cap (y_0 + ltB)$ on some interval $(0, t_0)$, where $y_0 \in \tilde{\omega}(x_0; \bar{x})$, we get

$$\lim_{t \downarrow 0} \rho(t^{-1}(y(t) - y_0), \omega^*(z_0, \lambda; \bar{x})) = 0.$$

Moreover, the set of all limit points $\bar{y} = \lim_{t \downarrow 0} t^{-1}(y(t) - y_0)$ coincides with the set of optimal solutions of the problem (P^*).

Proof. We take an arbitrary $y_0 \in \tilde{\omega}(x_0; \bar{x})$. Then, due to Theorem 6.112, $\omega^*(z_0, \lambda; \bar{x}) \neq \emptyset$, and one can find $l_0 > 0$ and $t_0 > 0$ such that

$$\omega_\varepsilon(x_0 + t\bar{x}, y_0, l) \neq \emptyset$$

for every $l \geq l_0$, $t \in [0, t_0)$. Now we choose an arbitrary selection $y(t) \in \omega_\varepsilon(x_0 + t\bar{x}, y_0, l)$ for $l \geq l_0$. From the definition of $\omega_\varepsilon(x_0 + t\bar{x}, y_0, l)$ it follows that $|y(t) - y_0| \leq lt$. Therefore $y(t) \to y_0$ if $t \downarrow 0$.

Suppose the limit

$$h = \limsup_{t \downarrow 0} \rho(t^{-1}(y(t) - y_0), \omega^*(z_0, \lambda; \bar{x}))$$

to be attained on the sequence $t_k \downarrow 0$. Without loss of generality we can assume that $t_k^{-1}(y(t_k) - y_0) \to \bar{y}$, where $\bar{y} \in \hat{D}_L F(z_0; \bar{x})$. From the (R)-regularity of the mapping F the equality $\hat{D}_L F(z_0; \bar{x}) = \Gamma_F(z_0; \bar{x})$ results (see Corollary 6.41). Hence $\bar{y} \in \Gamma_F(z_0; \bar{x})$.

Repeating the proof of Theorem 6.112 and denoting $\bar{z} = (\bar{x}, \bar{y})$, we get

$$\varphi'(x_0; \bar{x}) = \langle \nabla f(z_0), \bar{z} \rangle,$$

$$D^{2+}\varphi(x_0; \bar{x}) \geq \max_{\lambda \in \Lambda^2(z_0; \bar{x})} \langle \bar{z}, \nabla^2 L(z_0, \lambda) \bar{z} \rangle,$$

and, consequently,

$$\varphi''(x_0; \bar{x}) = \max_{\lambda \in \Lambda^2(z_0; \bar{x})} \langle \bar{z}, \nabla^2 L(z_0, \lambda) \bar{z} \rangle,$$

i.e. $\bar{y} \in \omega^*(z_0, \lambda; \bar{x})$. Consequently, $h = 0$ and all limit points of the function $t^{-1}(y(t) - y_0)$ belong to $\omega^*(z_0, \lambda; \bar{x})$.

Vice versa, let $\bar{y} \in \omega^*(z_0, \lambda; \bar{x})$. Then, according to Theorem 6.112, there exists a point $\bar{y}_2 \in \Gamma_F^2(z_0, \bar{z}; 0) = \hat{D}_L^2 F(z_0, \bar{z}; 0)$, while $\varphi''(x_0; \bar{x}) = 2\Phi(z_0, \bar{z}, (0, \bar{y}_2))$. Thus we can find a quantity $o(t^2)$ such that

$$y(t) = y_0 + t\bar{y} + t^2 \bar{y}_2 + o(t^2) \in F(x_0 + t\bar{x}), \quad t \geq 0.$$

We denote $\varepsilon(t) = f(x_0 + t\bar{x}, y(t)) - \varphi(x_0 + t\bar{x})$. Then $\varepsilon(t) \geq 0$ and

$$\lim_{t\downarrow 0} \frac{1}{t^2}\varepsilon(t) = \lim_{t\downarrow 0} \frac{1}{t^2}\{[f(x_0 + t\bar{x}, y(t)) - f(x_0, y_0) - t\langle \nabla f(z_0), \bar{z} \rangle]$$

$$- [\varphi(x_0 + t\bar{x}) - \varphi(x_0) - t\varphi'(x_0, \bar{x})]\} = \Phi(z_0, \bar{z}, (0, \bar{y}_2)) - \tfrac{1}{2}\varphi''(x_0, \bar{x}) = 0,$$

i.e. $\varepsilon(t) = o(t^2)$. On the other hand, $f(x_0 + t\bar{x}, y(t)) \leq \varphi(x_0 + t\bar{x}) + \varepsilon(t)$ and, therefore $y(t) \in \omega_\varepsilon(x_0 + t\bar{x})$ for $\varepsilon = o(t^2)$.

In this way, for $\bar{y} \in \omega^*(z_0, \lambda; \bar{x})$ there exists a selection $y(t) \in \omega_\varepsilon(x_0 + t\bar{x}, y_0, l)$ such that $\bar{y} = \lim_{t\downarrow 0} t^{-1}[y(t) - y_0]$. ∎

Note that since the optimal set $\Lambda^2(z_0; \bar{x})$ depends on $\bar{x}$, then the set of optimal solution of the problem (P^*) is not necessarily continuous with respect to $\bar{x}$ even if the set $\omega^*(z_0, \lambda; \bar{x})$ is a singleton. Therefore, it can happen that although the solution $y(x)$ of the original problem (P_x) is differentiable at the point x_0, the directional derivative $y'(x_0; \bar{x})$ fails to be continuous with respect to $\bar{x}$. This then implies that $y(x)$ is not Lipschitz continuous at x_0. An example of this type is given in Shapiro [163].

EXAMPLE 6.114 *Consider the problem*

$$(P_x): \quad \begin{cases} \frac{1}{2}(y_1 - 1)^2 + \frac{1}{2}y_2^2 \to \min \\ y_1 \leq 0, \\ y_1 + y_2 x_1 + x_2 \leq 0. \end{cases}$$

Let $x_0 = (0,0)$. It is not hard to see that assumption (A1) holds. Moreover $\omega(x_0) = \{y_0\}$, where $y_0 = (0,0)$, and at the point $z_0 = (x_0, y_0)$ the regularity condition (MF) holds. Therefore, the mapping F is (R)-regular at the point z_0. The set of Lagrange multipliers is

$$\Lambda(z_0) = \{(\lambda_1, \lambda_2) \mid \lambda_1 + \lambda_2 = 1, \lambda_1 \geq 0, \lambda_2 \geq 0\}.$$

Hence $\langle \bar{y}, \nabla^2_{yy}L(z_0, \lambda)\bar{y}\rangle = \bar{y}_1^2 + \bar{y}_2^2 > 0$, i.e., condition $(SOSC_{\bar{x}})$ holds at the point z_0.

From Lemma 6.111 and Theorem 6.112 we conclude the existence of the derivatives (6.77) and (6.78) of the optimal value function. Furthermore,

$$\langle \bar{z}, \nabla^2 L(z_0, \lambda)\bar{z}\rangle = \bar{y}_1^2 + \bar{y}_2^2 + 2\lambda_2\bar{x}_1\bar{y}_2,$$

and the set $\tilde{\omega}(z_0; \bar{x})$ consists of solutions of the problem

$$\min\{ -\bar{y}_1 \mid \bar{y}_1 \leq 0, \bar{y}_1 + \bar{x}_2 \leq 0\}.$$

Dual to this problem is the maximization problem

$$\max\{\lambda_2\bar{x}_2 \mid (\lambda_1, \lambda_2) \in \Lambda(z_0)\}.$$

Let us choose the direction $\bar{x} = (1,0)$. *Then* $\Lambda^2(z_0; \bar{x}) = \Lambda(z_0)$, $\tilde{\omega}(z_0; \bar{x}) = \{(0, \bar{y}_2)\}$, $\omega^*(z_0, \lambda; \bar{x}) = \{(0,0)\}$ *and, therefore,* $y'(x_0; \bar{x}) = (0,0)$.

Now let $\bar{x} = (1, \mu)$, $\mu > 0$. *In this case* $\Lambda^2(z_0; \bar{x}) = \{(0,1)\}$, $\tilde{\omega}(z_0; \bar{x}) = \{(-\mu, \bar{y}_2)\}$, $\omega^*(z_0, \lambda; \bar{x}) = \{(-\mu, -1)\}$ *and* $y'(x_0; \bar{x}) = (-\mu, -1)$, *i. e., the directional derivative* $y'(x_0; \bar{x})$ *is not continuous with respect to* $\bar{x}$.

Bibliographical Comments

Chapter 1.

Section 1.1 Many basic results and concepts reviewed in Sections 1 and 2 are nicely presented in a recent book by Bonnans and Shapiro [33]. For more detailed information on convex analysis see [64, 75, 78, 149, 154].

Section 1.2 An overview of main results in nonsmooth analysis can also be found in [3, 13, 42, 77, 159].

Section 1.3 To get more detailed knowledge about properties of quasidifferentiable functions and programming problems associated with them see [56, 57, 60] as well as [101, 105, 178]. The latest book on the subject [62] contains the newest developments as well as generalizations of previous result in quasidifferential calculus, both in theoretical and in numerical respect.

Chapter 2.

Section 2.1 Multivalued mappings and their applications are studied in [3, 91, 94, 146]. The definition of pseudo-lipschitz continuity of mappings was introduced in [3]. Lemma 3.26 is a generalization of the corresponding result from [157]. The convex multivalued mappings are considered in detail in [149], see also [130]. Lemma 3.36 integrates the results from [42, 149]. Lemma 3.40 and some others results were obtained in [149].

Section 2.2 Information about various types of tangent cones can be found in [3, 42, 46, 76, 91, 132, 146, 149].

There were a lot of attempts to extend the conception of differentiability to multivalued mappings. The derivatives of multivalued mappings introduced by Demyanov [55] and Pshenichny [149] (called the *set of feasible* or *tangent directions*) were efficiently applied to mathematical programming problems, especially to minimax problems (see [10, 55, 122]). Note that the definition of the set of tangent directions proposed by Pshenichny coincides with the lower Dini derivative of mappings. A methodical investigation of different types of derivatives in multivalued analysis was given by Polovinkin [146], Aubin [3] as well as in [10, 121, 123, 136, 147]. Differentiability properties of the distance function have been studied in [145].

We also want to emphasize that Lemma 3.57 is a well-known result (see, e.g., [3]), following immediately from previous considerations. The result contained in Lemma 3.65 was first obtained in [146].

Section 2.3 The Lemma about the removal of constraints was proved in [116] and is very close to results from [42].

Chapter 3.

Section 3.1 Estimates of Clarke subdifferentials of marginal functions were considered in [69, 116, 117, 121, 123, 138, 155, 156, 174]. The statements of this section are based on [117] and are close to [171]. The results on metrical regularity of multivalued mappings and its connection with pseudolipschitz continuity follow the lines of [128, 132, 134, 142], see also [135].

Section 3.2 The approach applied in the book allows us to generalize some results from [55] as well as the known theorem of Pshenichny [149] about the subdifferential of the marginal function in convex programming problems. Among recent papers in this field we especially mention [45] and [49].

Chapter 4.

Section 4.1 We refer to the book [123]. Furthermore, Theorems 5.9 and 5.12 generalize some known results from [48, 136, 149].

Section 4.2 The consideration is based on [35, 123]. Theorems 5.29 and 5.30 summarize results from [122, 149, 150].

Section 4.3 The concept of strongly differentiable mappings was introduced in [14] and [173]. The properties of such mappings were further considered in [10, 123, 140], while a generalization of strong differentiability was proposed in [137]. The presentation in this section follows [131].

Chapter 5.

Sensitivity analysis, i. e. the analysis of the influence of model errors on certain characteristic quantities of the model plays an important role in mathematical modelling (optimization and control problems) as well as in economics and technology (see [3, 11, 23, 36, 44, 66, 81, 88, 113, 115, 133, 169]). The main questions of sensitivity analysis were also investigated in [13, 21, 24, 25, 26, 31, 32, 67, 82, 97, 99, 100, 164]. Particularly we want to refer to the most recent book by Bonnans and Shapiro [33], where the reader can find several results of this chapter presented with a more detailed background as well as for the case of parametrized optimization problems involving abstract constraints.

Section 5.1 The explanation follows [120]. For related work on stability properties of optimal solutions we also refer to [28, 29, 47, 98, 168].

Section 5.2 In order to obtain meaningful results on the basic questions of sensitivity analysis, the constraints of the underlying optimization problem have to satisfy so-called regularity conditions. As a regularity condition one often uses the Slater condition [52, 55, 75], the linear independence constraint qualification [69], the Mangasarian-Fromowitz regularity condition [69, 155, 156, 163] and its directional modification [9, 32, 33]. The (R)-regularity condition, which is used in this chapter, was proposed by Robinson [152] and Fedorov [65] and is very natural. The connection between (MF)- and (R)-regularity was revealed in [6, 152]. The approach applied here extends results from [36, 73]. The equivalence between the (R)-condition and Lagrange regularity (i. e. the existence of Lagrange multipliers at the optimal point) for problems convex with respect to the main variable was proved in [19]. From [65] it follows that programming problems linear in the main variable are (R)-regular. Relations between different regularity conditions can be found in [7, 122]. Pseudolipschitz continuity of (R)-regular mappings follows also from a general result of Penot [141, 142].

Section 5.3 The study of differentiability of the optimal value function is the central problem of sensitivity analysis in mathematical programming. Many results in this field are obtained hitherto, but the topic is still far from being complete. Following the historical evolution of ideas and results in this filed, we would like to mention the papers [9, 24, 55, 69, 75, 84, 123, 156] as well as the books [97, 33].

Estimates of directional derivatives of optimal value function were obtained in [123], the results on the differentiability of value function generalize the ones from [84].

The consideration in Subsection 5.3.3 follows [126], which itself generalizes results from [32, 70, 163].

Problems with vertical perturbation were considered in [155]. The connection between the stability in the sense of Clarke and the construction of an exact penalty function was first investigated in [39].

Estimates for the upper Dini derivative of the marginal function in quasidifferentiable programming problems were first developed in [102, 105], while the quasidifferential of the marginal function was studied in [103]. These results are based on quasidifferential calculus (cf. e. g. [59]), a special representation of the quasidifferential of a continual maximum function borrowed from [101] and statements concerning the existence of (modified) Lagrange multipliers [107]. The estimates generalize e. g. those from [69]. More detailed facts can be found in [102].

Section 5.4 For a review of various definitions of second-order derivatives of the optimal value function in nonlinear problems see [55, 97, 162]. In this section we study the existence of the second-order directional derivative in the sense of Ben-Tal and Zowe ([17, 162]). Lemma 6.111 and Theorem 6.112 supplement the results described in [9]. We also refer to the papers [160, 167].

First results concerning differential properties of optimal solutions were obtained in [66, 86]. Among important contributions we would also like to mention the papers [9, 22, 33, 52, 70, 163]. The explanation in this section follows the lines of [118, 129] and is closely related to [9].

References

[1] Alt W. (1991) Local Stability of Solutions to Differentiable Optimization Problems in Banach Spaces. J Optim Theory Appl 70:443-466

[2] Aubin J.P.(1984) Lipschitz Behaviour of Solutions to Convex Minimization Problems. Math Oper Res 9:7-111

[3] Aubin J.P., Ekeland I.(1984) Applied Nonlinear Analysis. Wiley & Sons, New York

[4] Aubin J.P., Frankowska H. (1990) Set-valued Analysis. Birkhäuser, Boston

[5] Auslender A. (ed) (1977) Convex Analysis and Its Applications. Proceeding of the Conference Held at Murat-le-Quaire, March 1976. Springer-Verlag, Berlin Heidelberg

[6] Auslender A.(1984) Stability in Mathematical Programming with Nondifferentiable Data. SIAM J Control Optim 22:239-254

[7] Auslender A.(1987) Regularity Theorems in Sensitivity Theory with Nonsmooth Data. Math Res 35:9-15

[8] Auslender A. (2000) Existence of Optimal Solutions and Duality Results under Weak Conditions. Math Program 88:45-59

[9] Auslender A., Cominetti R. (1990) First and Second Order Sensitivity Analysis of Nonlinear Programs under Directional Constraint Qualification Conditions. Optimization 21:351-363

[10] Auslender A., Cominetti R. (1991) A Comparative Study of Multifunction Differentiability with Applications in Mathematical Programming. Math Oper Res 10:240-258

[11] Auslender A., Coutat P. (1996) Sensitivity Analysis for Generalized Linear-quadratic Problems. J Optim Theory Appl 88:541-559

[12] Auslender A., Crouzeix J.P. (1988) Global Regularity Theorems. Math Oper Res 13:243-253

[13] Bank B., Guddat J., Klatte D., Kummer B., Tammer K. (1982) Nonlinear Parametric Optimization. Akademie Verlag, Berlin

[14] Banks H. T., Jacobs M. Q. (1970) A Differential Calculus for Multifunctions. J Math Anal Appl 29:246-272

[15] Beer K., Zenker G. (to appear) The Marginal Value in Quadratic Programming with Unbounded Solution Sets. Math Methods Oper Res

[16] Ben-Tal A., Nemirovski A. (1998) Robust Convex Optimization. Math Oper Res 23:769-805

[17] Ben-Tal A., Zowe J. (1982) Necessary and Sufficient Optimality Conditions for a Class of Nonsmooth Minimization Problems. Math. Programming 24:70-92

[18] Ben-Tal A., Zowe J. (1985) Directional Derivatives in Nonsmooth Optimization. J Optim Theory Appl 47:483-490

[19] Bereznev V.A. (1988) Interrelation Between the Lagrange Theorem and the Geometry of Feasible Sets (in Russian). Dokl AN USSR (Translated as: Doklady Mathematics) 300:1289-1291

[20] Berge C. (1997) Topological Spaces; Including a Treatment of Multi-Valued Functions, Vector Spaces and Convexity. Dover Publications, Inc., Mineola, New York

[21] Bertsekas D. (1987) Constrained Optimization and Lagrange Multiplier Methods. Academic Press, New York

[22] Bonnans J.F. (1992) Directional Derivatives of Optimal Solutions in Smooth Nonlinear Programming. J Optim Theory Appl 73:27-45

[23] Bonnans J.F. (2000) Mathematical Study of Very High Voltage Power Networks III. The optimal AC power flow problem. Comput Optim Appl 16:83-101

[24] Bonnans J.F., Cominetti R. (1996) Perturbed Optimization in Banach Space I: a General Theory Based on a Weak Directional Constraint Qualification. SIAM J Control Optim 34:1151-1171

[25] Bonnans J.F., Cominetti R. (1996) Perturbed Optimization in Banach Space II: a Theory Based on a Strong Directional Constraint Qualification. SIAM J Control Optim 34:1172-1189

[26] Bonnans J.F., Cominetti R., Shapiro A. (1998) Sensitivity Analysis of Optimization Problems under Second Order Regular Conditions. Math Oper Res 23:806-831

[27] Bonnans J.F., Cominetti R., Shapiro A. (1999) Second Order Optimality Conditions Based on Parabolic Second Order Tangent Sets. SIAM J Optim 9:466-492

[28] Bonnans J.F., Ioffe A.D. (1995) Quadratic Growth and Stability in Convex Programming Problems with Multiple Solutions. J Convex Anal. 2 (Special issue dedicated to R.T.Rockafellar), 41-57

[29] Bonnans J.F., Ioffe A. (1995) Second-order Sufficiency and Quadratic Growth for Nonisolated Minima. Math Oper Res 20:801-817

[30] Bonnans J.F., Ioffe A.D., Shapiro A. (1992) Développement de Solutions Exactes et Approchées en Programmation non Linéaire. Comptes Rendus Hebdomadaires des Seances d'Academie des Sciences, Paris, Serie I 315:119-123

[31] Bonnans J.F., Shapiro A. (1992) Sensitivity Analysis of Parametrized Programs under Cone Constraints. SIAM J Control Optim 30:1409-1422

[32] Bonnans J.F., Shapiro A. (1996) Optimization Problems with Perturbations: a Guided Tour. Unité de recherche INRIA Rocquencourt, France N 2872

[33] Bonnans J.F., Shapiro A. (2000) Perturbation Analysis of Optimization Problems. Springer-Verlag, New York

[34] Bonnans J.F., Sulem A. (1995) Pseudopower Expansion of Solutions of Generalized Equations and Constrained Optimization Problems. Math Program 70:123-148

[35] Borisenko O.F., Minchenko L.I. (1992) Directional Derivatives of the Maximum Function (in Russian). Cybernet Systems Anal 28:309-312

[36] Borwein J.M. (1986) Stability and Regular Points of Inequality Systems. J Optim Theory Appl 48:9-52

[37] Borwein J.M., Lewis A.S. (2000) Convex Analysis and Nonlinear Optimization: Theory and Examples. Springer-Verlag, New York

[38] Borwein J.M., Zhuang D.M. (1988) Verifiable Necessary and Sufficient Conditions for Openness and Regularity of Set-Valued and Single-Valued Maps. J Math Anal Appl 134:441-459

[39] Burke J. (1991) Calmness and Exact Penalization. SIAM J Control Optim 29:968-998

[40] Burke J., Ferris M. (1993) Weak Sharp Minima in Mathematical Programming. SIAM J Control Optim 31:1340-1359

[41] Caballero R., Ruiz F., Steuer R.F. (eds) (1997) Advances in Multiple Objective and Goal Programming. Springer-Verlag, Berlin

[42] Clarke F.H. (1983) Optimization and Nonsmooth Analysis. Wiley, New York

[43] Clarke F.H., Demyanov V.F., Gianessi F.(eds) (1989) Nonsmooth Optimization and Related Topics. Plenum Press, New York

[44] Clarke F.H., Ledyaev Y.S., Stern R.J., Wolenski P.R. (1998) Nonsmooth Analysis and Control Theory. Springer-Verlag, Berlin

[45] Combari C., Laghdir M., Thibault L. (1999) On Subdifferential Calculus for Convex Functions Defined on Locally Convex Spaces. Ann Sci Math Québec 23:23-36

[46] Cominetti R. (1990) Metric Regularity, Tangent Sets and Second Order Optimality Conditions. Appl Math Optim 21:265-287

[47] Cornet B., Laroque G. (1987) Lipschitz Properties of Solutions in Mathematical Programming. J Optim Theory Appl 53:407-427

[48] Correa R., Jofre A. (1989) Tangentially Continuous Directional Derivatives in Nonsmooth Analysis. J Optimiz Theory Appl 61:1-21

[49] Correa R., Jofre A., Thibault L. (1995) Subdifferential Characterization of Convexity. In: Recent Advances in Nonsmooth Optimization (Du D.-Z., ed.), World Sci Publishing, Singapore

[50] Craven B.D., Janin R. (1993) Regularity Properties of the Optimal Value Function in Nonlinear Programming. Optimization 28:1-7

[51] Danskin J.M. (1967) The Theory of Max Min. Springer-Verlag, Berlin

[52] Dempe S. (1993) Directional Differentiability of Optimal Solutions under Slater's condition. Math Program 59:49-69

[53] Dempe S., Pallaschke D. (1997) Quasidifferentiability of Optimal Solutions in Parametric Nonlinear Optimization. Optimization 40:1-24

[54] Dempe S., Schmidt H. (1995) On an Algorithm Solving Two-Level Programming Problems with Nonunique Lower Level Solutions. Comput Optim Appl 6:227-249

[55] Demyanov V.F. (1974) Minimax: Directional Differentiability (in Russian). Leningrad University Press, Leningrad

[56] Demyanov V.F., Dixon L.C.W. (eds) (1986) Quasidifferential Calculus. Math Program Study 29:1-19

[57] Demyanov V.F., Pallaschke D.(eds) (1987) Nondifferentiable Optimization - Methods and Applications. Springer-Verlag, Berlin

[58] Demyanov V.F., Rubinov A.M. (1980) On Quasidifferentiable Functionals (in Russian). Dokl Akad Nauk SSSR (Translated as: Doklady Mathematics) 250:21-25

[59] Demyanov V.F., Rubinov A.M. (1986) Quasidifferential Calculus. Optimization Software. Springer-Verlag, New York

[60] Demyanov V.F., Rubinov A. M. (1990) Nonsmooth Analysis and Quasidifferentiable Calculus. Nauka, Moscow

[61] Demyanov V.F., Rubinov A.M. (1995) Constructive Nonsmooth Analysis. Verlag Peter Lang, Frankfurt am Main

[62] Demyanov V.F., Rubinov A.M. (eds) (2000) Quasidifferentiability and Related Topics. Ser.: Nonconvex Optimization and Its Applications, Vol. 43. Kluwer Academic Publishers, Dordrecht

[63] Demyanov V.F., Vasilev L.V. (1986) Nondifferentiable Optimization. Optimization Software, New York [Translated from the Russian]

[64] Ekeland I., Temam R. (1976) Convex Analysis and Variational Problems. North-Holland, Amsterdam

[65] Fedorov V.V. (1979) Numerical Methods of Max-Min Problems (in Russian). Nauka, Moscow

[66] Fiacco A.V. (1983) Introduction to Sensitivity and Stability Analysis in Nonlinear Programming. Academic Press, New York

[67] Fiacco A.V., Yshizuka Yo. (1991) Sensitivity and Stability Analysis for Nonlinear Programming. Ann Oper Res 27:215-235

[68] Floudas Ch.A. (1995) Nonlinear and Mixed-Integer Optimization. Oxford University Press, New York

[69] Gauvin J., Dubeau F. (1982) Differential Properties of the Marginal Function in Mathematical Programming. Math Program Study 19: 101-119

[70] Gauvin J., Janin R. (1988) Directional Behaviour of the Optimal Solution in Nonlinear Mathematical Programming. Math Oper Res 13:629-649

[71] Gauvin J., Janin R. (1989) Directional Lipschitzian Optimal Solutions and Directional Derivative for Optimal Value Function in Nonlinear Mathematical Programming. Analyse non-linéaire, Gauthiers-Villars, Paris, 305-324

[72] Gauvin J., Janin R. (1990) Directional Derivative of the Value Function in Parametric Optimization. Ann Oper Res 27:237-252

[73] Gauvin J., Tolle J.W. (1977) Differential Stability in Nonlinear Programming. SIAM J Control Optim 15: 294-311

[74] Gollan B. (1984) On the Marginal Function in Nonlinear Programming. Math Oper Res 9:208-221

[75] Gol'shtein E.G. (1972) Theory of Convex Programming. Transactions of Mathematical Monographs 36, American Mathematical Society, Providence

[76] Gorohovik V.V. (1990) Convex and Nonsmooth Problems of Vector Optimization (in Russian). Nauka i Technika, Minsk

[77] Guddat J., Jongen H., Nožička F., Still G., Twilt F. (eds) (1997) Parametric Optimization and Related Topics IV. Verlag Peter Lang, Frankfurt am Mein

[78] Hiriart-Urruty J.-B., Lemarechal C. (1993) Convex Analysis and Minimization Algorithms. Parts I and II. Springer-Verlag, Berlin Heidelberg

[79] Hoffman A. (1952) On Approximate Solutions of Systems of Linear Inequalities. J Research Nat Bur Standards 49:263-265

[80] Horst R., Tuy H. (1993) Global Optimization. 2nd rev. ed. Springer-Verlag, Berlin Heidelberg

[81] Insua D.R. (1990) Sensitivity Analysis in Multiobjective Decision Making. Springer-Verlag, Berlin Heidelberg

[82] Ioffe A.D. (1994) On Sensitivity Analysis of Nonlinear Programs in Banach Spaces: the Approach via Composite Unconstrained Optimization. SIAM J Optim 4:1-43

[83] Ioffe A.D., Tikhomirov V.M. (1979) Theory of Extremal Problems. North-Holland, Amsterdam [Translated from the Russian]

[84] Janin R. (1984) Directional Derivative of the Marginal Function in Nonlinear Programming. Math Program Study 21:110-126

[85] Janin R., Mado J.C., Naraganinsamy J. (1991) Second Order Multipliers and Marginal Function in Nonlinear Programs. Optimization 22:163-176

[86] Jittorntrum K. (1984) Solution Point Differentiability Without Strict Complementarity in Nonlinear Programming. Math Program Study 21:127-138

[87] Jongen H.Th., Klatte D., Tammer K. (1990) Implicit Functions and Sensitivity of Stationary Points. Math Program 19:123-138

[88] Jongen H.Th., Weber G.W. (1991) On Parametric Nonlinear Programming. Ann Oper Res 27:253-283

[89] Jourani A., Thibault L. (1993) Approximation and Metric Regularity in Mathematical Programming in Banach Space. Math Oper Res 18:390-401

[90] Jourani A., Thibault L. (1996) Extentions of Subdifferential Calculus Rules in Banach Spaces. Canada J Math 48:834-848

[91] Kirilyuk V.S. (1991) About One Tangent Cone, Properties of Multivalued Mappings and Marginal Functions (in Russian). Cybernetics and System Analysis 6:89-96

[92] Klatte D., Tammer K. (1990) Strong Stability of Stationary Solutions and Karush-Kuhn-Tucker Points in Nonlinear Optimization. Ann Oper Res 27:285-308

[93] Kuntz, L., Scholtes, S. (1993) Constraint qualifications in quasidifferentiable optimization. Math. Program 60:339-347

[94] Kuntz L., Scholtes S. (1994) Structural Analysis of Nonsmooth Mappings, Inverse Functions and Metric Projections. J Math Anal Appl 188:346-386

[95] Kuratovsky K. (1966) Topology (in Russian). Nauka, Moscow

[96] Laurent P. (1972) Approximation et Optimisation. Hermann, Paris

[97] Levitin E.S. (1994) Perturbation Theory in Mathematical Programming and its Applications. Wiley, Chichester

[98] Levy A.B., Poliquin R.A., Rockafellar R.T. (2000) Stability of Locally Optimal Solutions. SIAM J Optim 10:580-604

[99] Levy A.B., Rockafellar R.T. (1994) Sensitivity Analysis of Solutions to Generalized Equations. Trans Amer Math Soc 345:661-671

[100] Liu J. (1995) Sensitivity Analysis in Nonlinear Programs and Variational Inequalities via Continuous Selections. SIAM J Control Optim 34

[101] Luderer B. (1986) On the Quasidifferential of a Continual Maximum Function. Optimization 17:447-452

[102] Luderer B. (1987) Primale Dekomposition quasidifferenzierbarer Optimierungsaufgaben. Habilitation, Technical University of Chemnitz, Germany

[103] Luderer B. (1989) The Quasidifferential of an Optimal Value Function in Nonsmooth Programming. Optimization 20:597-613

[104] Luderer B. (1989) Quasidifferenzierbare Optimierungsprobleme und Primale Dekomposition. Mitteilungen der Math. Gesellschaft der DDR, 3-4, 19-32

[105] Luderer B. (1991) Directional Derivative Estimates for the Optimal Value Function of a Quasidifferentiable Programming Problem. Math Program, Ser. A 51:333-348

[106] Luderer B. (1992) Does the Special Choice of Quasidifferentials Influence Necessary Minimum Conditions?, Proc. 6th French-German Conf. Optimization (Lambrecht, Germany, 1991), Lecture Notes in Econom. and Math. Systems 382, Springer Verlag, Berlin, 256-266

[107] Luderer B., Eppler K. (1987) The Lagrange Principle and Quasidifferential Calculus. Wiss. Zeitschrift TU Karl-Marx-Stadt 29:187-192

[108] Luderer B., Eppler K. (2001) Some Remarks on Sufficient Conditions for Nonsmooth Functions. Optimization (to appear)

[109] Luderer B., Rösiger R. (1990) On Shapiro's Results in Quasidifferential Calculus. Math Program 46:403-407

[110] Luderer B., Rösiger R., Würker U. (1991) On Necessary Minimum Conditions in Quasidifferential Calculus: Independence of the Specific Choice of Quasidifferentials. Optimization 22:643-660

[111] Luderer B., Weigelt J. (1994) A Generalized Steepest Descent Method for Continuously Subdifferentiable Functions. Optimization 30:119-135

[112] Luenberger D. (1989) Linear and Nonlinear Programming. Addison-Wesley Longman, Reading, Massachusetts

[113] Malanowsky K. (1987) Stability of Solutions to Convex Problems of Optimization. Springer-Verlag, Berlin

[114] Malanowsky K. (1992) Second Order Conditions and Constraint Qualifications in Stability and Sensitivity Analysis of Solutions to Optimization Problems in Hilbert Spaces. Appl Math Optim 25:51-79

[115] Mangasarian O.L., Shiau T.M. (1987) Lipschitzian Continuity of Solutions of Linear Inequalities: Programs and Complementarity Problems. SIAM J Control Optim 25:583-595

[116] Minchenko L.I. (1988) Estimates for Subdifferentials of a Maximum Function (in Russian). Vesti Akad Navuk BSSR Ser Fiz-Mat Navuk 2:25-28

[117] Minchenko L.I. (1990) Subdifferentials of Marginal Functions of Multivalued Mappings (in Russian). Kibernetika (Kiev) 1:116-118

[118] Minchenko L.I. (1990) On Directional Stability with Respect to Parameters in Mathematical Programming Problems (in Russian). Dokl Akad Nauk Belarusi 34:978-981

[119] Minchenko L.I.(1991) About Directional Differentiability of Marginal Functions in Mathematical Programming Problems (in Russian). Cybernet Systems Anal 6:70-77

[120] Minchenko L.I., Bondarenko S.V. (1999) Sensitivity Analysis of Parametrical Programming Problems (in Russian). Dokl Nats Akad Nauk Belarusi 43: 34-38

[121] Minchenko L.I., Bondarevsky V.G. (1986) On Properties of Differentiable Multivalued Mappings (in Russian). Cybernetics 2:77-79

[122] Minchenko L.I., Borisenko O.F. (1983) On Directional Differentiability of the Maximum Function (in Russian). Comput Math Math Phys 23:567-575

[123] Minchenko L.I., Borisenko O.F. (1992) Differential Properties of Marginal Functions and Their Applications to Optimization Problems (in Russian). Nauka i Technika, Minsk

[124] Minchenko L.I., Borisenko O.F., Gritsay S.P. (1993) Multivalued Analysis and Perturbed Problems of Nonlinear Programming (in Russian). Nauka i Technika, Minsk

[125] Minchenko L.I., Gordienia A.N. (to appear) On Optimization Problems with Weak Sharp Minimizers. Oper Res

[126] Minchenko L.I., Sakolchik P.P. (1996) Hoelder Behaviour of Optimal Solutions and Directional Differentiability of Marginal Functions in Nonlinear Programming. J Optim Theory Appl 90:559-584

[127] Minchenko L.I., Satsura T.V. (1997) Calculation of Directional Derivatives in Max-Min Problems. Comput Math Math Phys 37:16-20

[128] Minchenko L.I., Satsura T.V. (1998) Pseudohoelder Continuity and Metrical Regularity of Multivalued Mappings. Dokl Nats Akad Nauk Belarusi 42:30-35

[129] Minchenko L.I., Satsura T.V. (1999) On the Parametric Differentiability of Optimal Solutions in Nonlinear Programming. Nonlinear Phenom Complex Systems, an Interdisciplinary Journal 2:60-63

[130] Minchenko L.I., Tesluk V.N. (1995) On Controllability of Convex Processes with Delay. J Optim Theory Appl 86:191-197

[131] Minchenko L.I., Volosevich A.A. (2000) Strongly Differentiable Multifunctions and Directional Differentiability of Marginal Functions. In [62]

[132] Mordukhovich B.S. (1988) Approximation Methods in Optimization and Control Problems (in Russian). Nauka, Moscow

[133] Mordukhovich B.S. (1991) Sensitivity Analysis in Nonsmooth Optimization. In: Theoretical Aspects of Industrial Design, SIAM, New York

[134] Mordukhovich B.S. (1993) Complete Characterization of Openness, Metric Regularity and Lipschitzian Properties of Multifunctions. Trans Amer Math Soc 340:1-36

[135] Mordukhovich B.S., Yongheng Shao. (1997) Stability of Set-valued Mappings in Infinite Dimensions: Point Criteria and Applications. SIAM J Control Optim 35:285-314

[136] Nikolsky M.S. (1988) The Contingent Directional Derivative in Nonsmooth Analysis (in Russian). Vestnik MGU. Ser 15 3:50-53

[137] Nurminski E.A. (1987) On the Differentiability of Set-valued Mappings. Kibernetika (Kiev) 4:111-113

[138] Outrata J.V. (1990) On Generalized Gradients in Optimization Problems with Set-valued Constraints. Math Oper Res 15:626-639

[139] Pallaschke D., Urbanski R. (1994) Reduction of Quasidifferentials and Minimal Representations. Math Program 66:161-180

[140] Pecherskaya N.A. (1986) Quasidifferentiable Mappings and the Differentiability of Maximum Functions. Math Program Study 29:145-159

[141] Penot J.P. (1982) On Regularity Conditions in Mathematical Programming. Math Program Study 19:167-199

[142] Penot J.P. (1989) Metric Regularity, Openness and Lipschitzian Behaviour of Multifunctions. Nonlinear Anal 13:629-645

[143] Penot J.P. (1998) Second-order Conditions for Optimization Problems with Constraints. SIAM J Control Optim 37:303-318

[144] Phelps R. (1993) Convex Functions, Monotone Operators and Differentiability. Springer-Verlag, Berlin Heidelberg

[145] Poliquin A., Rockafellar R.T., Thibault L. (2000) Local Differentiability of Distance Functions. Trans Amer Math Soc 352: 5231-5249.

[146] Polovinkin E.S. (1983) The Theory of Multivalued Mappings (in Russian). Nauka, Moscow

[147] Polovinkin E.S., Smirnov G. V. (1986) Differentiation of Multivalued Mappings and Properties of Solutions to Differential Inclusions. Dokl Akad Nauk SSSR 288:296-301

[148] Pshenichny B.N. (1971) Necessary Conditions for Extremum Problems. Dekker, New York [Translated from the Russian]

[149] Pshenichny B.N. (1980) Convex Analysis and Extremal Problems (in Russian). Nauka, Moscow

[150] Pshenichny B.N., Kirilyuk V.S. (1985) Differentiability of a Minimum Function with Connected Constraints (in Russian). Cybernetics 1:123-125

[151] Ralph D., Dempe S. (1995) Directional Derivatives of the Solution of a Parametric Nonlinear Program. Math Program 70:159-172

[152] Robinson S.M. (1976) Regularity and Stability for Convex Multivalued Functions. Math Oper Res 1:130-143

[153] Robinson S.M. (1982) Generalized Equations and Their Solutions, part 2. Math Program Study 12:200-221

[154] Rockafellar R.T. (1970) Convex Analysis. Princeton University Press, Princeton

[155] Rockafellar R.T. (1982) Lagrange Multipliers and Subderivatives of Optimal Value Functions in Nonlinear Programming. Math Program Study 17:28-66

[156] Rockafellar R.T. (1984) Directional differentiability of the optimal value function in a nonlinear programming problem. Math Program Study 21:213-226

[157] Rockafellar R.T. (1985) Lipschitzian Properties of Multifunctions. Nonlinear Anal 9:867-885

[158] Rockafellar R.T. (1989) Second-order Optimality Conditions in Nonlinear Programming Obtained by Way of Epi-derivatives. Math Oper Res 14:462-484

[159] Rockafellar R.T.(1990) Nonsmooth Analysis and Parametric Optimization. Springer, Berlin

[160] Rockafellar R.T., Tyrrell R. (2000) Second-order Convex Analysis. J Nonlinear Convex Anal 1:1-16

[161] Rubinov A.M. (1987) Approximation of Multivalued Mappings and Differentiability of Marginal Functions (in Russian). Dokl AN USSR (Translated as: Doklady Mathematics) 292:269-272

[162] Seeger A. (1988) Second-Order Directional Derivatives in Parametric Optimization Problems. Math Oper Res 13:124-139

[163] Shapiro A. (1988) Sensitivity Analysis of Nonlinear Programs and Differentiability Properties of Metric Projections. SIAM J Control Optim 26:628-645

[164] Shapiro A. (1988) Perturbation Theory of Nonlinear Programs When the Set of Optimal Solutions is Not a Singleton. Appl Math Optim 18:215-229

[165] Shapiro A. (1990) On Concepts of Directional Differentiability. J Optim Theory Appl 66:477-487

[166] Shapiro A. (1992) Perturbation Analysis of Optimization Problems in Banach Spaces. Numer Funct Anal Optim 13:97-116

[167] Shapiro A. (1994) Second Order Derivatives of Extremal-Value Functions and Optimality Conditions for Semi-Infinite Programs. Math Oper Res 10:207-219

[168] Shapiro A. (1994) On Lipschitzian Stability of Optimal Solutions of Parametrized Semi-Infinite Programs. Math Oper Res 19:743-752

[169] Shapiro A. (1994) Sensitivity Analysis of Parametrized Programs via Generalized Equations. SIAM J Control Optim 32:553-571

[170] Tamiz M. (ed) (1996) Multi-Objective Programming and Goal Programming. Springer-Verlag, Berlin

[171] Thibault L. (1991) On Subdifferentials of Optimal Value Functions. SIAM J Control Optim 29:1019-1036

[172] Tikhomirov V.M. (1986) Fundamental Principles of the Theory of Extremal Problems. Wiley, Chichester [Translated from the German]

[173] Tyurin Y. N. (1965) A Simplified Model of Production Planning (in Russian). Econom Math Meth 1:391-410

[174] Treiman J.S. (1986) Clarke's Gradient and Epsilon-Subgradient in Banach Spaces. Trans Amer Math Soc 294:65-78

[175] Ward D.E. (1991) A Constraint Qualification in Quasidifferentiable Programming. Optimization 22:661-668

[176] Ward D.E. (1994) Characterizations of Strict Local Minima and Necessary Conditions for Weak Sharp Minima. J Optim Theory Appl 80:551-571

[177] Xia Z.-Q. (1988) Some results on quasidifferentiable functions. Working Paper. Dalian University of Technology, Department of Applied Mathematics, Dalian (China)

[178] Xia Z.-Q. (1990) On quasi-differential kernels. Quaderni del Dipartimento di Matematica, Statistica, Informatica e Applicazioni 13, Istituto Universitario di Bergamo, Bergamo (Italy)

[179] Yin H., Xu C. (1999) Generalized K-T conditions and penalty functions for quasidifferentiable programming. Appl. Math. J. Chinese Univ. Ser. B 14:85-89

Index

Nonconvex Optimization and Its Applications

Nonconvex Optimization and Its Applications

Nonconvex Optimization and Its Applications

Nonconvex Optimization and Its Applications

65. M. Tawarmalani and N.V. Sahinidis: *Convexification and Global Optimization in Continuous and Mixed-Integer Nonlinear Programming.* Theory, Algorithms, Software, and Applications. 2002 ISBN 1-4020-1031-1
66. B. Luderer, L. Minchenko and T. Satsura: *Multivalued Analysis and Nonlinear Programming Problems with Perturbations.* 2002 ISBN 1-4020-1059-1

KLUWER ACADEMIC PUBLISHERS – DORDRECHT / BOSTON / LONDON